NOW YOU CAN CHOOSE THE PERFECT FIT FOR ALL YOUR CURRICULUM NEEDS.

The new Prentice Hall Science program consists of 19 hardcover books, each of which covers a particular area of science. All of the sciences are represented in the program so you can choose the perfect fit to *your* particular curriculum needs.

The flexibility of this program will allow you to teach those topics you want to teach, and to teach them *in-depth*. Virtually any approach to science—general, integrated, coordinated, thematic, etc.—is possible with Prentice Hall Science.

Above all, the program is designed to make your teaching experience easier and more fun.

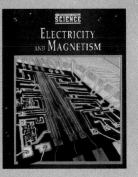

ELECTRICITY AND MAGNETISM
Ch. 1. Electric Charges and Currents
Ch. 2. Magnetism
Ch. 3. Electromagnetism
Ch. 4. Electronics and Computers

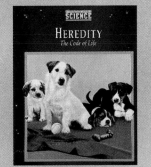

HEREDITY: THE CODE OF LIFE
Ch. 1. What is Genetics?
Ch. 2. How Chromosomes Work
Ch. 3. Human Genetics
Ch. 4. Applied Genetics

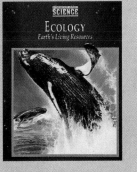

ECOLOGY: EARTH'S LIVING RESOURCES
Ch. 1. Interactions Among Living Things
Ch. 2. Cycles in Nature
Ch. 3. Exploring Earth's Biomes
Ch. 4. Wildlife Conservation

PARADE OF LIFE: MONERANS, PROTISTS, FUNGI, AND PLANTS
Ch. 1. Classification of Living Things
Ch. 2. Viruses and Monerans
Ch. 3. Protists
Ch. 4. Fungi
Ch. 5. Plants Without Seeds
Ch. 6. Plants With Seeds

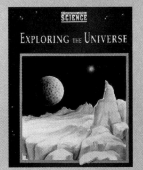

EXPLORING THE UNIVERSE
Ch. 1. Stars and Galaxies
Ch. 2. The Solar System
Ch. 3. Earth and Its Moon

EVOLUTION: CHANGE OVER TIME
Ch. 1. Earth's History in Fossils
Ch. 2. Changes in Living Things Over Time
Ch. 3. The Path to Modern Humans

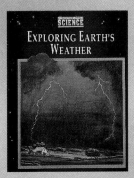

EXPLORING EARTH'S WEATHER
Ch. 1. What Is Weather?
Ch. 2. What Is Climate?
Ch. 3. Climate in the United States

THE NATURE OF SCIENCE
Ch. 1. What is Science?
Ch. 2. Measurement and the Sciences
Ch. 3. Tools and the Sciences

ECOLOGY: EARTH'S NATURAL RESOURCES

Ch. 1. Energy Resources
Ch. 2. Earth's Nonliving Resources
Ch. 3. Pollution
Ch. 4. Conserving Earth's Resources

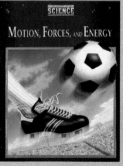

MOTION, FORCES, AND ENERGY

Ch. 1. What Is Motion?
Ch. 2. The Nature of Forces
Ch. 3. Forces in Fluids
Ch. 4. Work, Power, and Simple Machines
Ch. 5. Energy: Forms and Changes

PARADE OF LIFE: ANIMALS

Ch. 1. Sponges, Cnidarians, Worms, and Mollusks
Ch. 2. Arthropods and Echinoderms
Ch. 3. Fish and Amphibians
Ch. 4. Reptiles and Birds
Ch. 5. Mammals

CELLS: BUILDING BLOCKS OF LIFE

Ch. 1. The Nature of LIfe
Ch. 2. Cell Structure and Function
Ch. 3. Cell Processes
Ch. 4. Cell Energy

DYNAMIC EARTH

Ch. 1. Movement of the Earth's Crust
Ch. 2. Earthquakes and Volcanoes
Ch. 3. Plate Tectonics
Ch. 4. Rocks and Minerals
Ch. 5. Weathering and Soil Formation
Ch. 6. Erosion and Deposition

MATTER: BUILDING BLOCK OF THE UNIVERSE

Ch. 1. General Properties of Matter
Ch. 2. Physical and Chemical Changes
Ch. 3. Mixtures, Elements, and Compounds
Ch. 4. Atoms: Building Blocks of Matter
Ch. 5. Classification of Elements: The Periodic Table

CHEMISTRY OF MATTER

Ch. 1. Atoms and Bonding
Ch. 2. Chemical Reactions
Ch. 3. Families of Chemical Compounds
Ch. 4. Chemical Technology
Ch. 5. Radioactive Elements

HUMAN BIOLOGY AND HEALTH

Ch. 1. The Human Body
Ch. 2. Skeletal and Muscular Systems
Ch. 3. Digestive System
Ch. 4. Circulatory System
Ch. 5. Respiratory and Excretory Systems
Ch. 6. Nervous and Endocrine Systems
Ch. 7. Reproduction and Development
Ch. 8. Immune System
Ch. 9. Alcohol, Tobacco, and Drugs

EXPLORING PLANET EARTH

Ch. 1. Earth's Atmosphere
Ch. 2. Earth's Oceans
Ch. 3. Earth's Fresh Water
Ch. 4. Earth's Landmasses
Ch. 5. Earth's Interior

HEAT ENERGY

Ch. 1. What Is Heat?
Ch. 2. Uses of Heat

SOUND AND LIGHT

Ch. 1. Characteristics of Waves
Ch. 2. Sound and Its Uses
Ch. 3. Light and the Electromagnetic Spectrum
Ch. 4. Light and Its Uses

A COMPLETELY INTEGRATED LEARNING SYSTEM...

The Prentice Hall Science program is an *integrated* learning system with a variety of print materials and multimedia components. All are designed to meet the needs of diverse learning styles and your technology needs.

THE STUDENT BOOK

Each book is a model of **excellent writing and dynamic visuals**—designed to be exciting and motivating to the student *and* the teacher, with relevant examples integrated throughout, and more opportunities for many different activities which apply to everyday life.

Problem-solving activities emphasize the thinking process, so problems may be more open-ended.

"Discovery Activities" throughout the book foster active learning.

Different sciences, and other disciplines, are integrated throughout the text and reinforced in the "Connections" features (the connections between computers and viruses is one example).

TEACHER'S RESOURCE PACKAGE

In addition to the student book, the complete teaching package contains:

ANNOTATED TEACHER'S EDITION

Designed to provide **"teacher-friendly"** support regardless of instructional approach:

■ **Help is readily available** if you choose to teach thematically, to integrate the sciences, and/or to integrate the sciences with other curriculum areas.

■ **Activity-based learning** is easy to implement through the use of Discovery Strategies, Activity Suggestions, and Teacher Demonstrations.

■ **Integration of all components** is part of the teaching strategies.

■ For instant accessibility, all of the teaching suggestions are wrapped around the student pages to which they refer.

ACTIVITY BOOK

Includes a **discovery activity for each chapter**, plus other activities including problem-solving and cooperative-learning activities.

THE REVIEW AND REINFORCEMENT GUIDE

Addresses **students' different learning styles** in a clear and comprehensive format:

■ Highly visual for visual learners.

TEACHER'S RESOURCE PACKAGE

FOR THE PERFECT FIT TO YOUR TEACHING NEEDS.

■ Can be used in conjunction with the program's audiotapes for auditory and language learners.

■ More than a study guide, it's a guide to comprehension, with activities, key concepts, and vocabulary.

ENGLISH AND SPANISH AUDIOTAPES
Correlate with the Review and Reinforcement Guide to aid auditory learners.

LABORATORY MANUAL ANNOTATED TEACHER'S EDITION
Offers **at least one additional hands-on opportunity per chapter** with

answers and teaching suggestions on lab preparation and safety.

TEST BOOK
Contains **traditional and up-to-the-minute strategies for student assessment.** Choose from performance-based tests in addition to traditional chapter tests and computer test bank questions.

STUDENT LABORATORY MANUAL
Each of the 19 books also comes with its own Student Lab Manual.

ALSO INCLUDED IN THE INTEGRATED LEARNING SYSTEM:

■ Teacher's Desk Reference

■ English Guide for Language Learners

■ Spanish Guide for Language Learners

■ Product Testing Activities

■ Transparencies

■ Computer Test Bank (IBM, Apple, or MAC)

■ VHS Videos

■ Videodiscs

■ Interactive Videodiscs (Level III)

■ Interactive Videodiscs/ CD ROM

■ Courseware

All components are integrated in the teaching strategies in the Annotated Teacher's Edition, where they directly relate to the science content.

THE PRENTICE HALL SCIENCE
INTEGRATED LEARNING SYSTEM

The following components are integrated in the teaching strategies for
PARADE OF LIFE: MONERANS, PROTISTS, FUNGI, AND PLANTS.

- **Spanish Audiotape English Audiotape**
- **Activity Book**
- **Review and Reinforcement Guide**
- **Test Book**—including Performance-Based Tests
- **Laboratory Manual, Annotated Teacher's Edition**

- **Product-Testing Activities:**
  Testing Yogurt
  Testing Jeans
- **Laboratory Manual**
- **English Guide for Language Learners**
- **Spanish Guide for Language Learners**
- **Transparencies:**
  The Bacteriophage Virus
  Protists
  Mushroom Development
  Bread Mold

- **Interactive Videodiscs:**
  On Dry Land: The Desert Biome
  ScienceVision: EcoVision
- **Interactive Videodiscs/ CD ROM:**
  Paul ParkRanger and the Mystery of the Disappearing Ducks
  Virtual BioPark
  Amazonia

INTEGRATING OTHER SCIENCES

Many of the other 18 Prentice Hall Science books can be integrated into **PARADE OF LIFE: MONERANS, PROTISTS, FUNGI, AND PLANTS.** The books you will find suggested most often in the Annotated Teacher's Edition are EVOLUTION: CHANGE OVER TIME; HUMAN BIOLOGY AND HEALTH; CELLS: BUILDING BLOCKS OF LIFE; ECOLOGY: EARTH'S LIVING RESOURCES; EXPLORING PLANET EARTH; DYNAMIC EARTH; MATTER: BUILDING BLOCK OF THE UNIVERSE; HEAT ENERGY; EXPLORING THE UNIVERSE; CHEMISTRY OF MATTER; and ECOLOGY: EARTH'S NATURAL RESOURCES.

INTEGRATING THEMES

Many themes can be integrated into **PARADE OF LIFE: MONERANS, PROTISTS, FUNGI, AND PLANTS.** Following are the ones most commonly suggested in the Annotated Teacher's Edition:
SCALE AND STRUCTURE, EVOLUTION, UNITY AND DIVERSITY, and PATTERNS OF CHANGE.

For more detailed information on teaching thematically and integrating the sciences, see the Teacher's Desk Reference and teaching strategies throughout the Annotated Teacher's Edition.

For more information, call 1-800-848-9500 or write:

P R E N T I C E   H A L L

Simon & Schuster Education Group
113 Sylvan Avenue Route 9W
Englewood Cliffs, New Jersey 07632
Simon & Schuster A Paramount Communications Company

Annotated Teacher's Edition

Prentice Hall Science

Parade of Life
Monerans, Protists, Fungi, and Plants

Anthea Maton
Former NSTA National Coordinator
Project Scope, Sequence,
    Coordination
Washington, DC

Jean Hopkins
Science Instructor and Department
    Chairperson
John H. Wood Middle School
San Antonio, Texas

Susan Johnson
Professor of Biology
Ball State University
Muncie, Indiana

David LaHart
Senior Instructor
Florida Solar Energy Center
Cape Canaveral, Florida

Charles William McLaughlin
Science Instructor and Department
    Chairperson
Central High School
St. Joseph, Missouri

Maryanna Quon Warner
Science Instructor
Del Dios Middle School
Escondido, California

Jill D. Wright
Professor of Science Education
Director of International Field
    Programs
University of Pittsburgh
Pittsburgh, Pennsylvania

Prentice Hall
A Division of Simon & Schuster
Englewood Cliffs, New Jersey

ISBN 0-13-225608-8

  2 3 4 5 6 7 8 9 10   97 96 95 94 93

Contents of Annotated Teacher's Edition

To the Teacher   T–3

About the Teacher's Desk Reference   T–3

Integrating the Sciences   T–4

Thematic Overview   T–4

Thematic Matrices   T–5

Comprehensive List of Laboratory Materials   T–11

To the Teacher

Welcome to the *Prentice Hall Science* program. *Prentice Hall Science* has been designed as a complete program for use with middle school or junior high school science students. The program covers all relevant areas of science and has been developed with the flexibility to meet virtually all your curriculum needs. In addition, the program has been designed to better enable you—the classroom teacher—to integrate various disciplines of science into your daily lessons, as well as to enhance the thematic teaching of science.

The *Prentice Hall Science* program consists of nineteen books, each of which covers a particular topic area. The nineteen books in the *Prentice Hall Science* program are

The Nature of Science
Parade of Life: Monerans, Protists, Fungi, and Plants
Parade of Life: Animals
Cells: Building Blocks of Life
Heredity: The Code of Life
Evolution: Change Over Time

Ecology: Earth's Living Resources
Human Biology and Health
Exploring Planet Earth
Dynamic Earth
Exploring Earth's Weather
Ecology: Earth's Natural Resources
Exploring the Universe
Matter: Building Block of the Universe
Chemistry of Matter
Electricity and Magnetism
Heat Energy
Sound and Light
Motion, Forces, and Energy

Each of the student editions listed above also comes with a complete set of teaching materials and student ancillary materials. Furthermore, videos, interactive videos and science courseware are available for the *Prentice Hall Science* program. This combination of student texts and ancillaries, teacher materials, and multimedia products makes up your complete *Prentice Hall Science* Learning System.

About the Teacher's Desk Reference

The *Teacher's Desk Reference* provides you, the teacher, with an insight into the workings of the *Prentice Hall Science* program. The *Teacher's Desk Reference* accomplishes this task by including all the standard information you need to know about *Prentice Hall Science*.

The *Teacher's Desk Reference* presents an overview of the program, including a full description of each ancillary available in the program. It gives a brief summary of each of the student textbooks available in the *Prentice Hall Science* Learning System. The *Teacher's Desk Reference* also demonstrates how the seven science themes incorporated into *Prentice Hall Science* are woven throughout the entire program.

In addition, the *Teacher's Desk Reference* presents a detailed discussion of the features of the Student Edition and the features of the Annotated Teacher's Edition, as well as an overview section that summarizes issues in science education and offers a message about teaching special students. Selected instructional essays in the *Teacher's Desk Reference* include English as a Second Language (ESL), Multicultural Teaching, Cooperative-Learning Strategies, and Integrated Science Teaching, in addition to other relevant topics. Further, a discussion of the Multimedia components that are part of *Prentice Hall Science*, as well as how they can be integrated with the textbooks, is included in the *Teacher's Desk Reference*.

The *Teacher's Desk Reference* also contains in blackline master form a booklet on Teaching Graphing Skills, which may be reproduced for student use.

Integrating the Sciences

The *Prentice Hall Science* Learning System has been designed to allow you to teach science from an integrated point of view. Great care has been taken to integrate other science disciplines, where appropriate, into the chapter content and visuals. In addition, the integration of other disciplines such as social studies and literature has been incorporated into each textbook.

On the reduced student pages throughout your Annotated Teacher's Edition you will find numbers within blue bullets beside selected passages and visuals. An Annotation Key in the wraparound margins indicates the particular branch of science or other discipline that has been integrated into the student text. In addition, where appropriate, the name of the textbook and the chapter number in which the particular topic is discussed in greater detail is provided. This enables you to further integrate a particular science topic by using the complete *Prentice Hall Science* Learning System.

Thematic Overview

When teaching any science topic, you may want to focus your lessons around the underlying themes that pertain to all areas of science. These underlying themes are the framework from which all science can be constructed and taught. The seven underlying themes incorporated into *Prentice Hall Science* are

Energy
Evolution
Patterns of Change
Scale and Structure
Systems and Interactions
Unity and Diversity
Stability

A detailed discussion of each of these themes and how they are incorporated into the *Prentice Hall Science* program are included in your *Teacher's Desk Reference*. In addition, the *Teacher's Desk Reference* includes thematic matrices for the *Prentice Hall Science* program.

A thematic matrix for each chapter in this textbook follows. Each thematic matrix is designed with the list of themes along the left-hand column and in the right-hand column a big idea, or overarching concept statement, as to how that particular theme is taught in the chapter.

The primary themes in this textbook are Evolution, Patterns of Change, Scale and Structure, and Unity and Diversity. Primary themes throughout *Prentice Hall Science* are denoted by an asterisk.

CHAPTER 1

Classification of Living Things

ENERGY	• Organisms may be classified according to how they obtain energy.
***EVOLUTION**	• Evolutionary relationships are the basis for the modern classification system.
***PATTERNS OF CHANGE**	• In any classification system, the number of kinds of organisms decreases as we move from the largest grouping (for example, kingdom) to the smallest grouping (for example, species).
***SCALE AND STRUCTURE**	• Classification groups form a hierarchy in which the largest groups are the most general and the smallest are the most specific. The smallest classification group (species) refers to just one organism.
SYSTEMS AND INTERACTIONS	• All heterotrophs ultimately rely on autotrophs for food.
***UNITY AND DIVERSITY**	• The different organisms in the same classification group share certain important characteristics.
STABILITY	• Binomial nomenclature gives every organism a unique name that can be used and understood all over the world.

CHAPTER 2

Viruses and Monerans

ENERGY	• Bacteria meet their energy needs in a variety of ways. Some bacteria are autotrophs; others are heterotrophs.
***EVOLUTION**	• Bacteria were the first forms of life on Earth.
***PATTERNS OF CHANGE**	• Bacteria perform many of the chemical changes involved in nutrient cycles.
***SCALE AND STRUCTURE**	• Viruses are noncellular; they consist of a core of hereditary material surrounded by a protein coat. • Bacteria are unicellular organisms that lack a nucleus.
SYSTEMS AND INTERACTIONS	• Viruses and bacteria cause a number of diseases. • Bacteria interact with their environment in a variety of ways. Some bacteria are producers. Others are parasites. Still others are decomposers.
***UNITY AND DIVERSITY**	• Some bacteria are helpful; others are harmful.
STABILITY	• Bacteria help to maintain the supply of nutrients in the environment.

CHAPTER 3

Protists

ENERGY	• Protists meet their energy needs in a variety of ways. Some protists are autotrophs that use light energy to make food from simple raw materials. Others are heterotrophs.
*EVOLUTION	• The first protists may have been the result of a symbiosis among several types of bacteria. • Ancient protists are the ancestors of modern protists and multicellular organisms.
*PATTERNS OF CHANGE	• Sporozoans have complex life cycles that involve more than one host. • Slime molds undergo many transformations during their life cycles.
*SCALE AND STRUCTURE	• Protists are unicellular organisms that possess a nucleus.
SYSTEMS AND INTERACTIONS	• Plantlike protists produce most of the Earth's supply of oxygen. • Protists interact with other organisms in many ways. Some live inside a host and help it. Others are parasites. Still others are a source of food for larger organisms.
*UNITY AND DIVERSITY	• Although protists are quite diverse in appearance and habits, they are similar in basic structure.
STABILITY	• Plantlike protists help to maintain the supply of oxygen in the atmosphere.

CHAPTER 4

Fungi

ENERGY	• Fungi are heterotrophs that obtain energy by absorbing food. Some fungi obtain food from living organisms; others are decomposers.
*EVOLUTION	• Fungi, like other organisms, are formally classified in a way that best shows the evolutionary relationships among the members of the group.
*PATTERNS OF CHANGE	• Multicellular fungi undergo many transformations during their life cycles.
*SCALE AND STRUCTURE	• Most fungi are made up of hyphae.
SYSTEMS AND INTERACTIONS	• Some fungi are decomposers, which help to recycle materials from dead organisms. • Fungi interact with other organisms in many ways. Some are involved in helpful symbiotic relationships. Others are parasites. A few are predators. Still others are a source of food for other organisms.
*UNITY AND DIVERSITY	• Although fungi vary greatly in appearance, they are similar in the way they obtain their food, in basic structure, and in the way they reproduce.
STABILITY	• Some fungi break down dead organisms. The broken-down materials can then be used by other organisms. This keeps materials from being "locked up" in the bodies of dead organisms.

CHAPTER 5

Plants Without Seeds

ENERGY	• Algae are autotrophs; they meet their energy needs by capturing light energy and converting it into chemical energy.
*EVOLUTION	• Land plants evolved from green algae. • Land plants evolved in ways that made them better suited to meet the challenges of life on land. • Vascular plants are better adapted for a terrestrial existence than are algae or mosses and their relatives.
*PATTERNS OF CHANGE	• Like other plants, green algae use the carbon dioxide that animals exhale as a waste product; and they produce oxygen, which animals can breathe.
*SCALE AND STRUCTURE	• Green algae may be organized on unicellular, colonial, or multicellular levels. • Vascular plants contain a system of tubes used to transport materials throughout the body of the plant. • Vascular plants have true roots, stems, and leaves.
SYSTEMS AND INTERACTIONS	• Plants without seeds are used by humans in many different ways. • Some green algae are engaged in symbioses with heterotrophs.
*UNITY AND DIVERSITY	• Plants without seeds make up a diverse assemblage that includes unicellular, multicellular, vascular, and nonvascular forms.
STABILITY	• Like other plants, green algae use carbon dioxide, a waste product from animals, and produce oxygen, a waste product that is used by animals. • Land plants have evolved in response to the challenges of their environment.

CHAPTER 6

Plants With Seeds

ENERGY	• Light energy is used to change water and carbon dioxide into glucose and oxygen during the process of photosynthesis.
*EVOLUTION	• Gymnosperms were the first group of seed plants to evolve. • Angiosperms evolved about 100 million years ago. • Seed plants have evolved in ways that make them better adapted to life on land than are other kinds of plants.
*PATTERNS OF CHANGE	• Plants grow toward or away from certain stimuli. • After fertilization, ovules develop into seeds, and ovaries develop into fruit.
*SCALE AND STRUCTURE	• The internal structure of roots, stems, leaves, and flowers reflects and enhances their function. • Vascular tissue is composed of tubes of xylem and phloem. • Seed plants have true roots, stems, and leaves. • Seeds consist of a seed coat, stored food, and an embryo.
SYSTEMS AND INTERACTIONS	• Many angiosperms rely on animals for pollination and seed dispersal. • Photosynthesis produces oxygen, which animals and most other living things need.
*UNITY AND DIVERSITY	• Seed plants are divided into two main groups, based on the presence or absence of an ovary. • The same plant structures are adapted for a variety of different conditions and special functions, and thus look quite dissimilar in different plants. • Plants can be grouped according to how long they live.
STABILITY	• Photosynthesis helps to maintain the balance of carbon dioxide and oxygen in the environment.

Comprehensive List of Laboratory Materials

Item	Quantities per Group	Chapter
Apple	1 piece	4
Beaker, large glass	1	3
Bowl, small glass	1	3
Bread	1 slice	4
Brown alga plant	1	5
Cheese	1 slice	4
Clay, modeling	small piece	6
Container, large covered	1	4
Corn seeds	4	6
Coverslip	3	5
Dissecting needle	1	3
Fern plant	1	5
Filter paper containing slime mold	1	3
Hand lens	1	5
Magnifying glass	1	3, 4
Medicine dropper	1	3, 5
Metric ruler	1	5
Microscope	1	5
Moss plant	1	5
Oatmeal flakes	small box	3
Paper towel	several sheets	3, 4, 6
Pencil, glass-marking	1	2, 6
Petri dish		
with sterile nutrient agar	1	6
Scissors	1 pair	5, 6
Slide, microscope	3	5
Soap	1 bar	2
Tape, masking	1 small roll	6

PARADE OF LIFE
Monerans, Protists, Fungi, and Plants

Anthea Maton
Former NSTA National Coordinator
Project Scope, Sequence, Coordination
Washington, DC

Jean Hopkins
Science Instructor and Department Chairperson
John H. Wood Middle School
San Antonio, Texas

Susan Johnson
Professor of Biology
Ball State University
Muncie, Indiana

David LaHart
Senior Instructor
Florida Solar Energy Center
Cape Canaveral, Florida

Charles William McLaughlin
Science Instructor and Department Chairperson
Central High School
St. Joseph, Missouri

Maryanna Quon Warner
Science Instructor
Del Dios Middle School
Escondido, California

Jill D. Wright
Professor of Science Education
Director of International Field Programs
University of Pittsburgh
Pittsburgh, Pennsylvania

Prentice Hall
Englewood Cliffs, New Jersey
Needham, Massachusetts

Prentice Hall Science

Parade of Life: Monerans, Protists, Fungi, and Plants

Student Text and Annotated Teacher's Edition
Laboratory Manual
Teacher's Resource Package
Teacher's Desk Reference
Computer Test Bank
Teaching Transparencies
Product Testing Activities
Computer Courseware
Video and Interactive Video

The illustration on the cover, rendered by Keith Kasnot, provides a glimpse of a few of the many organisms in a forest ecosystem.

Credits begin on page 193.

SECOND EDITION

© 1994, 1993 by Prentice-Hall, Inc., Englewood Cliffs, New Jersey 07632.

ISBN 0-13-225590-1

2 3 4 5 6 7 8 9 10   97 96 95 94 93

Prentice Hall
A Division of Simon & Schuster
Englewood Cliffs, New Jersey 07632

STAFF CREDITS

Editorial:	Harry Bakalian, Pamela E. Hirschfeld, Maureen Grassi, Robert P. Letendre, Elisa Mui Eiger, Lorraine Smith-Phelan, Christine A. Caputo
Design:	AnnMarie Roselli, Carmela Pereira, Susan Walrath, Leslie Osher, Art Soares
Production:	Suse F. Bell, Joan McCulley, Elizabeth Torjussen, Christina Burghard
Photo Research:	Libby Forsyth, Emily Rose, Martha Conway
Publishing Technology:	Andrew Grey Bommarito, Deborah Jones, Monduane Harris, Michael Colucci, Gregory Myers, Cleasta Wilburn
Marketing:	Andrew Socha, Victoria Willows
Pre-Press Production:	Laura Sanderson, Kathryn Dix, Denise Herckenrath
Manufacturing:	Rhett Conklin, Gertrude Szyferblatt

Consultants

Kathy French	National Science Consultant
Jeannie Dennard	National Science Consultant
Brenda Underwood	National Science Consultant
Janelle Conarton	National Science Consultant

CONTENTS

PARADE OF LIFE: MONERANS, PROTISTS, FUNGI, AND PLANTS

CHAPTER 1   **Classification of Living Things** 10

1–1   History of Classification 12
1–2   Classification Today 19
1–3   The Five Kingdoms 25

CHAPTER 2   **Viruses and Monerans** 34

2–1   Viruses 36
2–2   Monerans 43

CHAPTER 3   **Protists** 58

3–1   Characteristics of Protists 60
3–2   Animallike Protists 63
3–3   Plantlike Protists 73
3–4   Funguslike Protists 77

CHAPTER 4   **Fungi** 84

4–1   Characteristics of Fungi 86
4–2   Forms of Fungi 90
4–3   How Fungi Affect Other Organisms 94

CHAPTER 5   **Plants Without Seeds** 104

5–1   Plants Appear: Multicellular Algae 106
5–2   Plants Move Onto Land: Mosses,
        Liverworts, and Hornworts 114
5–3   Vascular Plants Develop: Ferns 119

CHAPTER 6   **Plants With Seeds** 128

6–1   Structure of Seed Plants 130
6–2   Reproduction in Seed Plants 143
6–3   Gymnosperms and Angiosperms 148
6–4   Patterns of Growth 153

SCIENCE GAZETTE

Colleen Cavanaugh Explores the Underwater
  World of Tube Worms 160
Pests or Pesticides: Which Will It Be?.............162
The Corn Is as High as a Satellite's Eye 165

Activity Bank/Reference Section

For Further Reading   168
Activity Bank   169
Appendix A: The Metric System   185
Appendix B: Laboratory Safety: Rules and Symbols   186
Appendix C: Science Safety Rules   187
Glossary   189
Index   191

Features

Laboratory Investigations
Whose Shoe Is That?   30
Examining Bacteria   54
Examining a Slime Mold   80
Growing Mold   100
Comparing Algae, Mosses, and Ferns   124
Gravitropism   156

Activity: Discovering
Classifying Living Things   25
Bacteria for Breakfast   44
Food Spoilage   48
Making Models of Multicellular Organisms   88
Fit to Be Dyed   131
Plant-Part Party   140
Seed Germination   146
Pick a Plant   155

Activity: Doing
Classification of Rocks   13
All in the Family   17
Protist Models   62
Capturing Food   67
Making Spore Prints   93
That's About the Size of It   108
Dead Ringer   137

Activity: Calculating
Bacterial Growth   51
How Big Is Big?   68
Divide and Conquer   71
It Starts to Add Up   138

Activity: Thinking
Moss-Grown Expressions   117

Activity: Writing
A Secret Code   21
Harmful Microorganisms   37
Benefiting From Bacteria   53
Human Fungal Diseases   96
Taking a "Lichen" to It   97

Activity: Reading
The Universe in a Drop of Water   75
A Secret Invasion   94
The Secret of the Red Fern   121
The Birds and the Bees   151
A World of Fun With Plants   152

Problem Solving
Classifying the Dragons of Planet Nitram   24
A Hot Time for Yeast   92
They Went Thataway!   147

Connections
What's in a Name?   29
Computer Viruses: An Electronic Epidemic   42
Revenge of the Protist   72
Murderous Mushrooms   99
I Scream, You Scream,
  We All Scream for Ice Cream   123
Plant Power for Power Plants   142

Careers
Zoo Keeper   15
Bacteriologist   52
Laboratory Technician   69
Mushroom Grower   98

CONCEPT MAPPING

Throughout your study of science, you will learn a variety of terms, facts, figures, and concepts. Each new topic you encounter will provide its own collection of words and ideas—which, at times, you may think seem endless. But each of the ideas within a particular topic is related in some way to the others. No concept in science is isolated. Thus it will help you to understand the topic if you see the whole picture; that is, the interconnectedness of all the individual terms and ideas. This is a much more effective and satisfying way of learning than memorizing separate facts.

Actually, this should be a rather familiar process for you. Although you may not think about it in this way, you analyze many of the elements in your daily life by looking for relationships or connections. For example, when you look at a collection of flowers, you may divide them into groups: roses, carnations, and daisies. You may then associate colors with these flowers: red, pink, and white. The general topic is flowers. The subtopic is types of flowers. And the colors are specific terms that describe flowers. A topic makes more sense and is more easily understood if you understand how it is broken down into individual ideas and how these ideas are related to one another and to the entire topic.

It is often helpful to organize information visually so that you can see how it all fits together. One technique for describing related ideas is called a **concept map**. In a concept map, an idea is represented by a word or phrase enclosed in a box. There are several ideas in any concept map. A connection between two ideas is made with a line. A word or two that describes the connection is written on or near the line. The general topic is located at the top of the map. That topic is then broken down into subtopics, or more specific ideas, by branching lines. The most specific topics are located at the bottom of the map.

To construct a concept map, first identify the important ideas or key terms in the chapter or section. Do not try to include too much information. Use your judgment as to what is really important. Write the general topic at the top of your map. Let's use an example to help illustrate this process. Suppose you decide that the key terms in a section you are reading are School, Living Things, Language Arts, Subtraction, Grammar, Mathematics, Experiments, Papers, Science, Addition, Novels. The general topic is School. Write and enclose this word in a box at the top of your map.

SCHOOL

Now choose the subtopics—Language Arts, Science, Mathematics. Figure out how they are related to the topic. Add these words to your map. Continue this procedure until you have included all the important ideas and terms. Then use lines to make the appropriate connections between ideas and terms. Don't forget to write a word or two on or near the connecting line to describe the nature of the connection.

Do not be concerned if you have to redraw your map (perhaps several times!) before you show all the important connections clearly. If, for example, you write papers for Science as well as for Language Arts, you may want to place these two subjects next to each other so that the lines do not overlap.

One more thing you should know about concept mapping: Concepts can be correctly mapped in many different ways. In fact, it is unlikely that any two people will draw identical concept maps for a complex topic. Thus there is no one correct concept map for any topic! Even

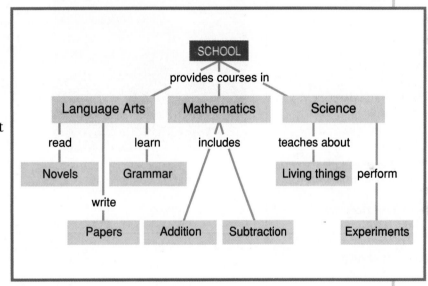

though your concept map may not match those of your classmates, it will be correct as long as it shows the most important concepts and the clear relationships among them. Your concept map will also be correct if it has meaning to you and if it helps you understand the material you are reading. A concept map should be so clear that if some of the terms are erased, the missing terms could easily be filled in by following the logic of the concept map.

Parade of Life: Monerans, Protists, Fungi, and Plants

TEXT OVERVIEW

In this textbook students will learn about the ways that living things are classified by scientists. First, they will meet the simplest living things, viruses and monerans, and learn about their structure and reproduction. Next, they will examine different kinds of protists. Then students will learn about fungi, including mushrooms, yeasts, and molds. Next, they are introduced to plants without seeds: algae, mosses and their relatives, and ferns. They will find out about the adaptations plants made to live on land. Finally, students will discover seed plants and learn about their structures, reproductive processes, and growth patterns.

TEXT OBJECTIVES

1. Explain the use of classification systems, and list the seven major classification groups.
2. Identify parts of a virus and describe viral reproduction.
3. Identify parts of a moneran and explain how it obtains energy.
4. List and describe three kinds of protists.
5. Describe the characteristics of several types of fungi.
6. Analyze the adaptations necessary for plants to live on land.
7. Compare ferns to mosses and algae.
8. Describe the process of photosynthesis.
9. Discuss the patterns of growth in seed plants and the factors that affect growth.

PARADE
OF LIFE

Monerans, Protists, Fungi, and Plants

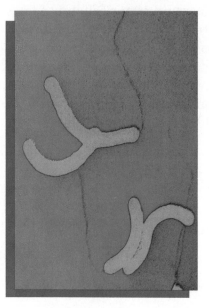

The first marchers to appear in the parade of life were monerans, more commonly known as bacteria.

Think about the last time you saw a parade. You may have seen marching bands, beautiful floats, enormous balloons, showers of ticker tape, or waving flags.

At this moment, you are in the middle of the biggest, oldest, most spectacular parade on Earth—the parade of life. This parade began its journey through time more than 3.5 billion years ago, and its marchers are all the living things on Earth—past, present, and future. In this book, you will begin your exploration of the parade of life—an adventure that will take you all over the world.

Two billion years after bacteria started the parade of life, the next group of marchers appeared. This group consists of living things known as protists. Protists, such as wheel-shaped diatoms, are made of a single, complex cell.

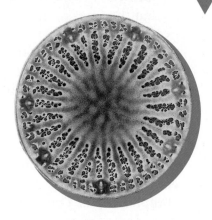

INTRODUCING PARADE OF LIFE

USING THE TEXTBOOK

Begin your introduction of the textbook by having students examine the title of the textbook and the textbook-opening photographs and captions. Before students read the textbook introduction, ask them the following questions.
• **Are any of the names listed under the main title familiar to you? What do you know about them?** (Accept all answers.)

• **What do you think is meant by "parade of life"?** (Accept all answers.)
• **You have seen mushrooms and flowers, but you probably haven't seen monerans and protists. Why not?** (Bacteria and other one-celled living things cannot be seen by the human eye.)

Have students read the textbook introduction on pages B8–9.
• **What is the major difference between the monerans and protists on the one hand and the fungi and flowering plants on the other?** (Monerans and protists are

CHAPTERS

1 Classification of Living Things
2 Viruses and Monerans
3 Protists
4 Fungi
5 Plants Without Seeds
6 Plants With Seeds

Living things that are composed of many cells are relatively recent additions to the parade of life. Many-celled living things include fungi, such as mushrooms, as well as plants.

Flowering plants are newcomers to the parade of life. Many flowers are actually clusters of flowers. Each yellow bump in the center and each "petal" of a painted daisy is a single tiny flower.

Discovery Activity

Taking a Closer Look

1. Obtain a sample of pond water. What does the pond water look like? Can you see living things in the water?

2. Using a medicine dropper, place a drop of the pond water in the center of a glass microscope slide. Cover the drop with a coverslip.

3. Examine the drop of water with a microscope under low and high powers. What do you observe?
   - How do you think the inventor of the microscope felt when he first looked at pond water with his invention?
   - What do your observations tell you about some of the living things that march alongside you in the parade of life?
   - Why are microscopes necessary tools for studying the parade of life?

B ■ 9

CHAPTER DESCRIPTIONS

1 Classification of Living Things In Chapter 1 taxonomic classification is introduced, followed by a brief history of classification. Next, binomial nomenclature is explained, and the link between evolutionary theory and classification is explored. The seven major levels of taxonomic grouping are discussed, and finally some general characteristics of the biological kingdoms are analyzed.

2 Viruses and Monerans Chapter 2 begins with an analysis of the structure of viruses. Then the events in the viral reproduction cycle are explained. Next, parts of monerans are described, and monerans are contrasted according to how they make their food. Finally, helpful and harmful effects of monerans are discussed.

3 Protists Chapter 3 deals with the characteristics of protists. They are classified according to their animallike, plantlike, and funguslike attributes.

4 Fungi In Chapter 4 the characteristics of fungi are explained; mushrooms, yeasts, and molds are described; and lichens are analyzed.

5 Plants Without Seeds Chapter 5 focuses on a group of plants that range from algae to mosses and their relatives to ferns. The adaptations necessary for plants to live on land are discussed, followed by an introduction to the concept of vascular and nonvascular plants.

6 Plants With Seeds In Chapter 6 the nature and structure of seed plants are analyzed, including gymnosperms and angiosperms. The discussion focuses on the processes of plant reproduction, plant life cycles, and factors that affect plant growth.

---

one-celled living things; most fungi and flowering plants are many-celled.)
• **How can this difference help to explain these living things' places in the "parade of life"?** (Life evolved from one-celled to many-celled organisms. As one-celled living things, monerans and protists would have been early marchers in the parade; as many-celled living things, fungi and flowering plants would have joined the parade later.)

DISCOVERY ACTIVITY
Taking a Closer Look

Begin your introduction to the textbook by having students perform the Discovery Activity. You may wish to divide the class into as many small groups as there are available microscopes. Each group could prepare a slide and then take turns examining it and recording their findings. Make sure that each student records his or her own observations and answers the questions in the activity. Afterward, en-

courage students to share their thoughts about the activity with their classmates.

Once students have seen the abundant life in a drop of water, they will have a new appreciation not only for the wide variety of life on Earth, but also for the power of the microscope that can reveal that life to us. Point out that they will learn about some of these living things in the chapters that follow.

Chapter 1 CLASSIFICATION OF LIVING THINGS

SECTION	HANDS-ON ACTIVITIES
1–1 History of Classification pages B12–B18 Multicultural Opportunity 1–1, p. B12 ESL Strategy 1–1, p. B12	**Student Edition** ACTIVITY (Doing): Classification of Rocks, p. B13 ACTIVITY (Doing): All in the Family, p. B17 **Activity Book** CHAPTER DISCOVERY: Organizing a Junk Drawer, p. B9 ACTIVITY BANK: Sorting It Out, p. B12 **Teacher Edition** Diversity of Life, p. B10d
1–2 Classification Today pages B19–B25 Multicultural Opportunity 1–2, p. B19 ESL Strategy 1–2, p. B19	**Student Edition** ACTIVITY (Discovering): Classifying Living Things, p. B25 LABORATORY INVESTIGATION: Whose Shoe Is That? p. B30 **Laboratory Manual** Developing a Classification System for Seeds, p. B7
1–3 The Five Kingdoms pages B25–B29 Multicultural Opportunity 1–3, p. B25 ESL Strategy 1–3, p. B25	**Student Edition** ACTIVITY BANK: A Key to the Puzzle, p. B170 **Laboratory Manual** Identifying Invertebrates Using Classification Keys, p. B11 **Teacher Edition** Classification of Organisms, p. B10d
Chapter Review pages B30–B33	

OUTSIDE TEACHER RESOURCES

Books

Abbott, Lois A., et al. *Taxonomic Analysis in Biology: Computers, Models and Databases,* Columbia University Press.

Clark, M.E. *Contemporary Biology: Concepts and Implications,* W.B. Saunders.

Dixon B. *What Is Science For,* Collins.

Samuel, E. *Order in Life,* Prentice Hall.

Audiovisuals

Carolus Linnaeus, film, EBE

Classifying Living Things, video, EBE

Classifying Plants and Animals, film or video, Coronet

Courseware

Classifying Mammals, Prentice Hall

OTHER ACTIVITIES	MEDIA AND TECHNOLOGY
Activity Book ACTIVITY: Early Taxonomists, p. B15 ACTIVITY: A Trip to the Natural History Museum, p. B21 **Review and Reinforcement Guide** Section 1–1, p. B5	**Interactive Videodisc/CD ROM** Amazonia **English/Spanish Audiotapes** Section 1–1
Student Edition ACTIVITY (Writing): A Secret Code, p. B21 **Activity Book** ACTIVITY: Making a Classification System, p. B19 ACTIVITY: Classifying the States, p. B23 ACTIVITY: Fun With Fictitious Animals, p. B25 ACTIVITY: Going Around in Circles, p. B31 ACTIVITY: A Diversity of Dragons, p. B35 **Review and Reinforcement Guide** Section 1–2, p. B9	**Interactive Videodisc/CD ROM** Virtual BioPark **English/Spanish Audiotapes** Section 1–2
Activity Book ACTIVITY: How Are Organisms Classified? p. B13 ACTIVITY: Classifying Organisms, p. B17 ACTIVITY: Classifying Creatures, p. B29 ACTIVITY: Making a Pictograph, p. B33 **Review and Reinforcement Guide** Section 1–3, p. B11	**English/Spanish Audiotapes** Section 1–3
Test Book Chapter Test, p. B9 Performance-Based Tests, p. B129	**Test Book** Computer Test Bank Test, p. B15

*All materials in the Chapter Planning Guide Grid are available as part of the Prentice Hall Science Learning System.

CHAPTER OVERVIEW

To date, scientists have identified millions of various species of organisms on Earth. Such great diversity among living things called for a system of naming and ordering the organisms in a logical manner. Early classification systems were developed and used by Aristotle in the fourth century BC. In the eighteenth century, a more specific system was devised by a Swedish botanist named Carolus Linnaeus.

Linnaeus's system of naming organisms is called binomial nomenclature. In this system each organism is assigned a two-part scientific name that identifies its genus and species.

Linnaeus grouped organisms according to similar body structures. The groups that make up Linnaeus's classification system are kingdom, phylum, class, order, family, genus, and species.

Taxonomy, the science of naming and grouping living things, illustrates evolutionary relationships among organisms.

Linnaeus's two-kingdom classification system has evolved to a five-kingdom classification system widely used today.

1–1 HISTORY OF CLASSIFICATION
THEMATIC FOCUS

The purpose of this section is to show the diversity of life and the need for a classification system of some type throughout history. There has always been a need for classification systems. Thousands of years ago, classifying living things according to observable characteristics helped people survive.

The science of classification is a branch of biology, or life science, known as taxonomy. Taxonomy has a long history, during which many classification systems were developed, changed to fit new facts and theories, and even rejected and replaced with better systems. Early classification systems were relatively simple compared to the complexity and specificity of today's systems. But these systems were useful in their time and served as a foundation on which modern classification systems are built.

The themes that can be focused on in this section are unity and diversity and stability.

*Unity and diversity: Early classification systems had many different organisms within the same group. Even though these organisms were different, they shared certain important characteristics. In these early classification systems, for example, even though dogs and cats are quite different, they were grouped together because both are animals that can walk.

Stability: A classification system developed in the eighteenth century named living things with a system known as binomial nomenclature. This system assigns every organism a unique name that can be used and understood by scientists from all over the world. Binomial nomenclature allowed a more specific, universal sharing of knowledge of living things.

PERFORMANCE OBJECTIVES 1–1
1. Describe Aristotle's early classification system of living things.
2. Explain how binomial nomenclature is used to classify living things.

SCIENCE TERMS 1–1
binomial nomenclature p. B17
genus p. B17
species p. B17

1–2 CLASSIFICATION TODAY
THEMATIC FOCUS

The purpose of this section is to explain how taxonomy provides information regarding the evolutionary relationships of organisms. These evolutionary relationships are the basis for the modern system of biological classification. Modern taxonomists try to classify living things in such a way that each classification group contains organisms that evolved from the same ancestor. The invention of the microscope has enabled the body of knowledge from which taxonomists work to grow.

This section also explains the twofold function of today's classification system: It gives each organism a unique name, and it groups organisms according to basic characteristics that reflect their evolutionary relationships. The major groups into which all living things are classified include kingdom, phylum, class, order, family, genus, and species.

The themes that can be focused on in this section are evolution, patterns of change, and scale and structure.

*Evolution: The basis for the modern classification system is the evolutionary relationships of organisms. Living things are classified in such a way that each classification group contains organisms that evolved from the same ancestor.

*Patterns of change: In any classification system, the number of different organisms decreases as we move from the largest grouping to the smallest grouping. In modern classification, the kingdom grouping contains the greatest number of organisms, and the species grouping the least.

*Scale and structure: Classification groups form a hierarchy in which the largest groups are the most general and the smallest are the most specific. In modern classification, the kingdom group is most general and the species group is most specific.

PERFORMANCE OBJECTIVES 1–2
1. State the two major functions of modern classification systems.
2. List in correct sequence the seven major classification groups.

1–3 THE FIVE KINGDOMS

THEMATIC FOCUS

The purpose of this system is to identify the five-kingdom system of classification. The discovery of new life forms since the time of Linnaeus led to the development of this system of classification. The five-kingdom system of classification is generally accepted by scientists, but not unanimously accepted. It consists of individual kingdoms named monerans, protists, fungi, plants, and animals. Monerans are unicellular organisms that do not contain a nucleus. Like other organisms, monerans can be grouped into two categories, depending on how they obtain energy: Organisms that obtain energy by making their own food are called autotrophs; organisms that cannot make their own food are called heterotrophs. Protists are unicellular organisms that have a nucleus. A number of protists are capable of movement and are autotrophs or heterotrophs. Fungi are generally multicellular organisms with a unique cell wall. They are also heterotrophs. Plants are generally multicellular and are autotrophic. Animals are multicellular, heterotrophic organisms with specialized tissues.

The themes that can be focused on in this section are energy and systems and interactions.

Energy: Organisms can be classified according to how they obtain energy. Organisms must either make food or obtain food. Organisms that obtain energy by making their own food are called autotrophs. The prefix *auto-* means "self," and the root word *-troph* means "food." Organisms that cannot make their own food are called heterotrophs. The prefix *hetero-* means "other."

Systems and interactions: Most plants are examples of autotrophs, and most animals are examples of heterotrophs. Ultimately, all heterotrophs rely on autotrophs for food.

PERFORMANCE OBJECTIVES 1–3

1. Distinguish between multicellular and unicellular organisms.

2. Distinguish between organisms that are autotrophs and organisms that are heterotrophs.

3. List characteristics of each kingdom in the five-kingdom classification system.

SCIENCE TERMS 1–3

autotroph p. B26
heterotroph p. B26

Discovery *Learning*

TEACHER DEMONSTRATIONS MODELING

Diversity of Life

Have all class members stand. Read each of the following physical characteristics aloud: over five feet tall, brown eyes, female, left-handed. Only those students possessing each trait should remain standing.

The result will be an extremely small group of students left standing. (If no students are standing at this point, modify the criteria to such a degree that one or several students will be left standing.)

• **Are the students still standing or the students sitting more alike?** (Answers will vary.)
• **Explain your reasoning.** (Explanations will vary.)

Lead students to understand that great differences exist among living things. Grouping organisms according to similar traits makes it easier to understand the diversity of life.

You may choose to repeat the activity having individual students develop a scheme whereby they will be the last person standing.

Classification of Organisms

Collect some easily obtained specimens and/or pictures of organisms from each of the five kingdoms. Place the specimens or photographs at several stations around the classroom, labeling those that might be unfamiliar to students. Allow ample time for students to move from station to station to observe the displays. As they do so, have them collect this information about each:
1. Does it appear to be multicellular or unicellular?
2. If it appears unicellular, does it have a nucleus?
3. Can it move about?
4. Can it manufacture its own food?

Share with students the correct answers to the questions for each station. Mention that they have completed some basic taxonomy skills, or classification skills. In this chapter they will learn more about classification and classification skills.

CHAPTER 1
Classification of Living Things

INTEGRATING SCIENCE

This life science chapter provides you with numerous opportunities to integrate other areas of science, as well as other disciplines, into your curriculum. Blue numbered annotations on the student page and integration notes on the teacher wraparound pages alert you to areas of possible integration.

In this chapter you can integrate physical science and chemistry (p. 13), earth science and geology (pp. 13, 14), language arts (pp. 13, 17, 21, 29), life science and marine biology (p. 14), music (p. 15), and life science and evolution (pp. 19, 20, 27).

SCIENCE, TECHNOLOGY, AND SOCIETY/COOPERATIVE LEARNING

In Greek mythology, many creatures existed that were hybrids of real-life animals. For example, Pegasus was a horse with wings, a griffin had the body of a lion and the head and wings of an eagle, and a centaur was part man and part horse. Although these creatures were mythical, today's scientists have the technological know-how to produce composite organisms. Techniques such as embryo splicing and microsurgical combination of embryos have enabled scientists to produce unusual organisms.

As early as 1982, a goat-sheep "chimera" was born at Cambridge University. The animal, produced from spliced embryos, had the body of a goat, the legs of a sheep, and various combinations of body parts. The goal of this type of reproductive research is not to produce animals from mythology but to improve farm production, to help solve economic problems on a global scale, and to preserve endangered species.

Many species of organisms are in danger of extinction. Supporters of reproductive technologies hope that the reproductive rates of endangered animals in captivity can be increased. Reproductive technologies are also being used to enable childless couples to increase their likelihood for having children of their own.

INTRODUCING CHAPTER 1

DISCOVERY LEARNING

▶ *Activity Book*

Begin teaching the chapter by using the Chapter 1 Discovery Activity from the *Activity Book*. Using this activity, students will develop a classification scheme for the contents of a junk drawer.

USING THE TEXTBOOK

Have students observe the photograph on page B10 and read the caption. Call attention to the fish's name—coelacanth. Point out that because of its fearsome appearance, sometimes the coelacanth is mistaken for a shark. Coelacanths, however, are not closely related to sharks.
• **What else is unusual about this fish?** (Accept all answers, but lead students to observe its fleshy fins.)

Classification of Living Things

Guide for Reading

After you read the following sections, you will be able to

1–1 History of Classification
- Give examples of the ways classification is used in science and in everyday life.
- Explain how binomial nomenclature is used to name living things.

1–2 Classification Today
- Relate biological classification to evolution.
- List the seven major classification groups.

1–3 The Five Kingdoms
- Describe some general characteristics of each of the five kingdoms.

Mary Courtnay-Latimer had never seen anything like the creature the crew of a fishing boat had captured near the mouth of the Chalumna River in South Africa. The monstrous fish stretched more than 1.5 meters, or about six times the length of this page. Large spiny steel-blue scales covered its oily body. A powerful jaw hung down from a frightening face. Most peculiar of all, its fins were attached to what appeared to be stubby legs!

After searching through many books but finding no description of the strange fish, Latimer remembered someone she thought might be able to solve the riddle. That someone was Dr. James L. B. Smith, a well-known fish expert at a nearby university.

The scientist was shocked. He later wrote, "I would hardly have been more surprised if I met a dinosaur on the street." Smith was looking at an animal thought to have become extinct more than 60 million years ago. Yet, in a flash, he had been able to identify the fish as a coelacanth (SEE-luh-kanth). Smith gave the coelacanth its scientific name: *Latimeria chalumnae*. To do so, he used biological classification—a special system that helps scientists identify and name organisms. And it is this special system that you will learn about in the pages that follow.

Journal *Activity*

You and Your World You probably know someone who often puts people into categories such as "jocks" or "brains." In your journal, discuss the practice of putting people into categories.

◀ *The coelacanth is the only surviving member of the ancient group of fishes from which modern four-footed land animals are thought to have evolved.*

B ■ 11

1–1 History of Classification

Point out that many scientific terms, including those used to classify organisms, have Latin or Greek roots. Have students use the following combining forms to create the names of various organisms, then relate the definition of the name. For example, the word *arthropod* is made of which two combining forms?

ESL STRATEGY 1-1

Make sure students understand that the word *form,* used in the textbook to describe Linnaeus's method of grouping plants and animals, refers to structure.

Have students write several short sentences identifying Linnaeus. Check for accuracy; then have students state the order in which names are given in his naming system. Next, introduce the word *surname,* (meaning "family name") and reinforce the word by explaining that Linnaeus's order is similar to using one's own surname first. Ask if this is the custom in any of the students' native countries; then ask students to think of examples when they have used their surname first.

Activity Bank

Sorting It Out, Activity Book, p. B181. This activity can be used for ESL and/or Cooperative Learning.

Guide for Reading

Focus on these questions as you read.

▶ How and why do people use classification systems?

▶ How is binomial nomenclature used to name living things?

1-1 History of Classification

Thousands of years ago, as people made observations about the world around them, they began to recognize that there were different groups of living things. There were animals and there were plants. Of the animals, some had claws and sharp teeth and roamed the land. Others had feathers and beaks and flew in the air. Still others had scales and fins and swam in the water.

Plants too showed a wide range of differences. Not only did they vary in shape, size, and color, but some were good to eat whereas others were poisonous. People soon learned that poisonous plants were best avoided. In a similar way, people learned that some animals, such as those with sharp teeth, were dangerous. Others, such as those with feathers, were relatively harmless.

What these early people were doing is something you often do in your daily life. They were giving order to the world around them by putting things in groups or categories based on certain characteristics. In other words, they were developing simple systems of classification. **Classification is the grouping of things according to similar characteristics.**

Stop and think for a moment about the ways in which you classify things every day. Perhaps you group your clothing by season—lightweight, cool items for summer and heavy, warm items for winter.

12 ■ B

TEACHING STRATEGY 1-1

FOCUS/MOTIVATION

Collect and display pictures of these animals: dog, wolf, fox, lion, tiger, house cat. Ask the class to compare these animals.

• **In what ways are all the organisms similar?** (Students will likely suggest that all are flesh-eating animals, or carnivores.)

• **How might these animals be placed into two different groups?** (Lead students to suggest that the dog, wolf, and fox are put into one group, and the tiger, lion, and cat into another group.)

• **Why do you think the animals are grouped in this way?** (Lead the class to respond that similar characteristics were used to group the animals. Dogs are more similar to wolves and foxes than they are to lions, tigers, and cats, and so on.)

Point out that classification systems based on structural similarities have been used since early times to group and name plants and animals.

• **Are there other ways in which we could** have grouped these animals? (Yes.)

• **What other criteria could be used to group these animals in different ways?** (Students might suggest that the dog and the house cat can be grouped together because they are common, domesticated pets.)

Explain that any system for classifying life is arbitrary. Classification was invented by humans as a convenience to help them keep track of living things. Different classification schemes are the result of different criteria being selected when grouping organisms.

Figure 1–1 *When humans first came to North America tens of thousands of years ago, they found a number of unusual animals. How did classifying these animals help humans survive?*

Or perhaps by type—pants, sweaters, skirts, jackets. What are some other things that you sort in a meaningful way? 2

Classification is important to all fields of science, not just to the subject of biology you are now studying. For example, geologists classify rocks, soils, and fossils. Meteorologists (people who study the weather) classify clouds, winds, and types of storms. And all of the 109 known chemical elements are classified into a system that helps chemists understand 1 their behavior.

Classification is important in subjects other than science. In English, parts of speech are categorized 3 as nouns, verbs, adjectives, and adverbs, to name a few. In mathematics, you work with odd numbers and even numbers, circles, rectangles, and triangles. In history, you group people and events according to time periods or geographic locations. You know that music can be rock-and-roll, rhythm and blues, country and western, or classical.

These are but a few examples of the important role that classification plays in all phases of life. For the people living thousands of years ago, classifying living things according to observable characteristics often helped them to survive. For you, classifying objects probably makes life easier and more meaningful. For scientists, classification systems provide a

ACTIVITY
DOING

Classification of Rocks

1. Obtain ten different rocks.
2. Examine the rocks. Notice how they are similar and how they are different. 2
3. Decide which characteristics of the rocks are most important. Use these characteristics to create a classification system for the rocks.
4. Notice that in this case there is no single correct way to classify your rocks. However, geologists do classify rocks in a particular way. Some characteristics that are important to geologists are the ways rocks are formed, the kind of chemicals that make up the rocks, and the shape of the crystals in the rocks.

B ■ 13

BACKGROUND INFORMATION
MODERN TAXONOMY

Classification systems evolve, or change, during the course of time. It seems that as the guiding philosophy of a classification system changes, so too does the classification of certain specimens. Ancient Greek biologists classified dolphins as fish. Modern biologists classify dolphins as mammals, because they have lungs and mammary glands.

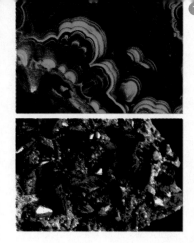

Figure 1–2 *Geologists classify minerals according to the chemical compounds that make them up. Malachite (top) and azurite (bottom) are both forms of copper carbonate.*

means of learning more about life on Earth and of discovering the special relationships that exist between different kinds of living things.

But no matter who is doing the classifying or what is being classified, a classification system is always based on observable characteristics a group of things share. Good classification systems are meaningful, easily understood, and readily communicated among people.

Biological classification systems name and organize living things in a logical, meaningful way. To date, scientists have identified more than 2.5 million different types of living things—and their job is not even close to being finished! Some biologists estimate that there may be at least another 7 million different kinds of organisms living in tropical rain forests and in the depths of the Earth's oceans. In order to bring some order to this great diversity of living things, biologists have developed systems of classification.

The science of classification is a branch of biology known as taxonomy (taks-AH-nuh-mee). Scientists

Figure 1–3 *The dizzying variety of corals, fishes, and algae in a coral reef represents only a tiny fraction of the Earth's living things. Why is it necessary for biologists to classify living things?* ❶

1–1 (continued)

CONTENT DEVELOPMENT

Write the word *taxonomy* on the chalkboard. Beside the word, write its root words, *taxis* and *-nomy*. Explain that the word *taxonomy* is derived from the Greek terms *taxis*, meaning arrangement, and *nomos*, meaning law.

• **Why do you think the science of classification is called taxonomy?** (The science of taxonomy provides an orderly set of rules, or laws, for arranging organisms into groups.)

Mention that a scientist who specializes in taxonomy is called a taxonomist.

Ask students to answer true or false to the following statements:

• **Taxonomy establishes a standard system for naming newly discovered organisms.** (True.)

• **Taxonomy helps to solve biological riddles of the past.** (True.)

● ● ● ● **Integration** ● ● ● ●

Use Figure 1–2 to integrate concepts of geology into your life science lesson.

Use Figure 1–3 to integrate concepts of marine biology into your life science lesson.

♪ Media and Technology

Students can explore the concept and process of biological classification by using

the Interactive Videodisc called Amazonia. Students will examine how organisms from the tropical rain forest are classified and named. In subsequent lessons and chapters, students can continue their exploration and discover more about the evolution, behavior, and interactions of specific organisms in Amazonia. They can also learn more about the ecological interactions and cycles of this fascinating ecosystem.

GUIDED PRACTICE

Skills Development
Skill: Identifying relationships

Divide the class into groups of four to six students per group. Write the following list of food products on the chalkboard:

margarine	English muffins
flour	raisins
rye bread	grapes
honey	cheddar cheese
waffles	chicken legs
fish	crackers

Figure 1–4 *The instruments in a marching band can be classified as woodwinds, percussion, and brass. How are some other everyday things classified?*

who work in this field are called taxonomists. Taxonomy has a long history, during which many classification systems were developed, changed to fit new facts and theories, and even rejected and replaced with better systems. This fact should not surprise you if you remember that science is an ongoing process that is often marked by change. This ability to change when new knowledge becomes available is one of science's greatest strengths.

The First Classification Systems

In the fourth century BC, the Greek philosopher Aristotle proposed a system to classify living things. He divided organisms into two groups: plants and animals. He also placed animals into three groups according to the way they moved. One group included all animals that flew, another group included all those that swam, and a third group included all those that walked.

Although this system was useful, it had some problems. Can you see why Aristotle's system for classifying animals is not perfect? It ignores the ways in which animals are similar and different in form. According to Aristotle, both a bird and a bat would be placed in the same group of flying animals. Yet in some basic ways, birds and bats are quite different. For example, birds are covered with feathers, whereas bats are covered with hair.

CAREERS

Zoo Keeper

Today is a big day at the zoo. The arrival of a male and a female giraffe is expected from Africa. Once tests show they are healthy, the **zoo keeper** will help to move them to their new home.

The zoo keeper will also be responsible for giving food and water to the giraffes. Other responsibilities include keeping the animals' area clean and observing and recording their behavior. The zoo keeper usually is the first to notice any medical problems.

Of the many people working at the zoo, zoo keepers have the most direct contact with the animals. For further career information, write to the American Association of Zoo Keepers National Headquarters, 635 Gage Boulevard, Topeka, KS 66606.

ANNOTATION KEY

Answers

❶ Answers might include it would be difficult to share research if living things were not classified. (Making generalizations)

❷ Answers will vary. (Identifying patterns)

Integration

❶ Earth Science: Geology. See *Dynamic Earth*, Chapter 4.

❷ Life Science: Marine Biology. See *Exploring Planet Earth*, Chapter 2.

❸ Music

cereal	yogurt
lettuce	potatoes
muffins	milk
macaroni	oranges

Ask groups to classify these items into the sections where they would be found in a supermarket. Then have groups discuss their results.

CONTENT DEVELOPMENT

Remind students that in the fourth century, Aristotle proposed a system to classify life. This system placed animals into three groups: animals that flew, animals that swam, and animals that walked.

• **Using Aristotle's classification system, name an animal from each group.** (Answers might include bird, fish, and horse.)

• **Do you think this classification system is a good system? Why or why not?** (Lead students to suggest that this system was not good in the sense that a particular animal could be classified into more than one group.)

• **Name an animal that can be classified into two of these groups.** (Responses might include an otter because it can swim and walk.)

• **Can you think of an animal that can fly, walk, and swim?** (Answers might include duck.)

• **How do you think Aristotle dealt with these apparent inconsistencies in his system of classification?** (Answers will vary.)

● ● ● ● **Integration** ● ● ● ●

Use Figure 1–2 to integrate music into your life science lesson.

SCIENTIFIC NAMES

Most of the names used within the classification of living things are from Latin and are quite similar to the English names we use. Point out the following similarities:

English	Latin
animal	Animalia
chordate	Chordata
mammal	Mammalia
carnivore	Carnivora
felid	Felidae
panther	Panthera

Figure 1–5 *Although the false vampire bat can fly like a bird, it belongs to the same class as the mouse it is about to eat (left). The egret belongs to a separate class of animals (right). How are the wings and other characteristics of the bat and egret different?* ❶

Although the system devised by Aristotle would not satisfy today's taxonomists, it was one of the first attempts to develop a scientific and orderly system of classification. In fact, Aristotle's classification system was used for almost 2000 years. And until the middle of the twentieth century, scientists continued to use Aristotle's system of classifying living things as either plants or animals.

By the seventeenth century, biologists had started to classify organisms according to similarities in form and structure. They examined the organism's internal anatomy as well as its outward appearance. This helped them to place animals and other organisms into groups that were more meaningful than those Aristotle had created.

The classification system we use today is based on the work of the eighteenth-century Swedish scientist Carolus Linnaeus. Linnaeus built upon the work of previous scientists to develop his new system of classification. Like Aristotle, Linnaeus identified all living things as either plants or animals. Like the seventeenth-century biologists, he grouped plants and animals according to similarities in form. And like almost all other previous taxonomists, he used a system that consisted of groups within larger groups within still larger groups. Linnaeus spent the major part of his life using his classification system to describe all known plants and animals.

Linnaeus also developed a simple system for naming organisms. His system is such a logical and easy way of naming organisms that it is still used today.

1–1 (continued)

REINFORCEMENT/RETEACHING

▶ *Activity Book*

Students who need help on the concept of classification systems should complete the chapter activity Early Taxonomists. In this activity students will explore the contributions to classification systems made by early taxonomists.

GUIDED PRACTICE

Skills Development

Skill: Making comparisons

Divide the class into groups of three to five students per group. Each group should create a list of 25 different animals, then classify those animals using Aristotle's classification system.

• **How many animals that you classified fit into more than one classification group?** (Answers will vary.)

• **How could Aristotle's system of classification be improved?** (Lead students to suggest that it should contain more specific groups of living things.)

CONTENT DEVELOPMENT

In the eighteenth century, Carolus Linnaeus developed a new system of species classification. Explain that Linnaeus proposed that animals of the same kind should be placed in the same species and that similar species should be placed together to form a genus. Similar genera should form an order of animals, and similar orders should be placed together to form a class. Point out that Linnaeus devised a simple naming system that he called binomial nomenclature. Explain that *binomial* means "name consisting of two terms" and that *nomenclature* means "a system of terms." Therefore, *binomial nomenclature* means "a system of names consisting of two terms."

Naming Living Things

Before Linnaeus developed his naming system, plants and animals had been identified by a series of Latin words. These words, sometimes numbering as many as 12 for one organism, described the physical features or appearance of the organism. To make things even more confusing, the names of plants and animals were rarely the same from book to book or from place to place.

BINOMIAL NOMENCLATURE The naming system devised by Linnaeus is called **binomial nomenclature** (bigh-NOH-mee-uhl NOH-muhn-klay-cher). In this ① system, each organism is given two names—which is exactly what the term binomial (consisting of two names) nomenclature (system of naming) means. The two names are a **genus** (plural: genera) name and a **species** name.

To help you understand this method of naming, think of the genus name as your family name. The species name could then be thought of as your first name. Your family name and your first name are the two names that identify you—the two names that most people know you by. They represent the most specific way of identifying you by name.

NAMING ORGANISMS TODAY Each kind of organism is given its very own two-part scientific name. An organism's scientific name is made up of its genus name and species name. The genus name is capitalized, but the species name begins with a small letter. Both names are printed in italics, which will help you recognize scientific names as you do your reading. Here is an example: The genus and species name for a wolf is *Canis lupus*. These two names identify this organism—which, by the way, has become very rare in the wild. Although most scientific names are in Latin, some are in Greek. Now think about the scientific name of the honeybee, *Apis mellifera*. What is the honeybee's genus name? Its species name? ②

Each organism has only one scientific name. And no two organisms can have the same scientific name. To understand why this is important, consider the following story. In North and South America, a certain large cat is called a mountain lion by some people, a cougar by others, and a puma by still others.

Figure 1–6 *The honeybee was once called* Apis pubescens, thorace subgriseo, abdomine fusco, pedibus posticis glabris utrinque margine ciliatus. *This means "fuzzy bee, light gray middle, brown body, smooth hind legs that have a small bag edged with tiny hairs." Linnaeus named the honeybee* Apis mellifera, *which means "honey-bearing bee."*

ACTIVITY DOING

All in the Family

Use reference books to learn about the various families in the plant and animal kingdoms. Choose the family that you find most interesting. Make a collage of the family using pictures from old magazines and newspapers. Present the collage to your class.

ACTIVITY DOING

ALL IN THE FAMILY

Skills: Research, classifying

Materials: old newspapers and magazines

This activity will help students grasp the modern classification system by allowing them to select an organism and determine its basic classification. By creating collages of the family they choose, students will be able to compare the similarities and differences in a particular family of living things.

• **What is your name?** (Answers will vary.)
• **Do you have a first and a last name?** (Yes.)
• **Does everyone you know have a first and a last name?** (Yes.)

Stress that Linnaeus gave each kind of plant and animal two names—a first and a last name. Remind students that the first name is the species name or the individual's name in the family; the last name is called a genus (or family) name, just as students' last names tell what family they belong to.

• • • • **Integration** • • • •

Use the introduction of binomial nomenclature to integrate language arts concepts into your science lesson.

ENRICHMENT

You may wish to point out that for organisms to be considered members of a single species, they must be capable of interbreeding, and their young must be fertile. Although horses and donkeys can interbreed to produce mules, the parent animals are considered two different species because mules are not fertile. On the other hand, organisms may be quite dissimilar in appearance and yet be members of the same species. For example, it is possible for a boxer and a collie to produce fertile offspring. Even though boxers and collies look quite different, they are classified as different breeds, or varieties, within the same species.

FACTS AND FIGURES
TAXONOMISTS

Taxonomists catalog about 11 million new specimens each year.

1–1 (continued)

INDEPENDENT PRACTICE

▶ *Activity Book*
Students who need practice on the concept of classification should complete the chapter activity A Trip to the Natural History Museum. In this activity students will study and classify 15 different museum organisms.

INDEPENDENT PRACTICE

Section Review 1–1
1. Answers will vary. Students might respond that their album, cassette-tape, or compact-disc collection is classified into a particular order.
2. Taxonomy is the science of classification.
3. Binomial nomenclature is a system of assigning each organism two names—a genus name and a species name.
4. Answers may include that early classification systems ignored similarities and differences in structure and form.

Figure 1–7 *The wolf and the puma are known by many different common names. However, each organism has its own unique scientific name. What is the scientific name for the wolf? The puma?* 1

18 ■ B

If these people were to talk to one another about this animal, they might get rather confused, thinking they were talking about three different animals. But scientists cannot afford to have such confusion. So scientists throughout the world know this large cat by only one name, *Felis concolor.* This name easily identifies the cat to all scientists, no matter where they live or what language they speak.

Keep in mind that it is not necessary to memorize the scientific names of different organisms. Even biologists know only a few names by heart, and most of these names are of organisms they have studied for years. What is important for you to know is that each organism has a scientific name that is used and understood all over the world, and that this name is related to the way the organism is classified.

1–1 Section Review

1. Describe some ways in which you use classification in everyday life.
2. What is taxonomy?
3. What is binomial nomenclature? How is it used?

Critical Thinking—*Evaluating Systems*
4. Discuss three problems with the classification and naming systems that existed before Linnaeus.

REINFORCEMENT/RETEACHING

Monitor students' responses to the Section Review questions. If students appear to have difficulty with any of the questions, review the appropriate material in the section.

CLOSURE

▶ *Review and Reinforcement Guide*
At this point have students complete Section 1–1 in the *Review and Reinforcement Guide.*

TEACHING STRATEGY 1–2

FOCUS/MOTIVATION

Show students a picture of a cat. Write "Katze," "chat," "kot," and "gato" on the chalkboard.
• **What do you think these words mean?** (Lead students to respond that the words are words for "cat" in other languages.)

Point out that it would be difficult to understand a conversation between two people if one person was talking about

1-2 Classification Today

In the 200 years since Linnaeus developed his classification and naming systems, knowledge of the living world has grown enormously. And as the understanding of organisms improved, it became necessary to adjust the system of biological classification. Two things in particular have had a large effect on biological classification. One of these is Charles Darwin's theory of evolution. The other is advances in technology that have enabled scientists to take a better look at organisms.

Evolution and Classification

As you can see in Figure 1–8, wolves and lions both developed from a meat-eating animal that existed about 60 million years ago. And lions and house cats both developed from a catlike animal that lived about 15 million years ago. During the long history of life on Earth, organisms have changed, or evolved. You can think of evolution as the process in

Guide for Reading

Focus on this question as you read.

▶ *What are the classification groups from largest to smallest?*

Figure 1–8 *Because they evolved from a shared ancestor, lions, cats, wolves, and the catlike animal belong to the same classification group (order Carnivora). Cats and lions also belong to a smaller classification group (family Felidae), which contains all the descendants of the catlike animal. Why aren't wolves placed in the family Felidae?* ②

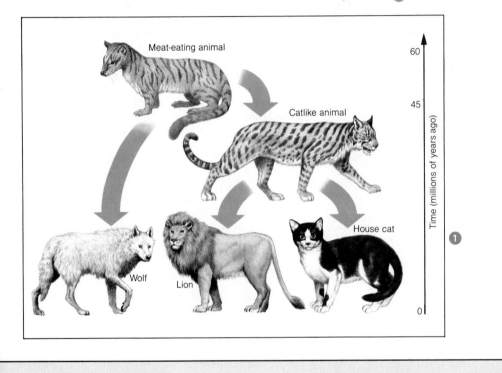

Meat-eating animal

Catlike animal

House cat

Wolf

Lion

Time (millions of years ago)

60

45

0

①

1-2 Classification Today

MULTICULTURAL OPPORTUNITY 1-2

Creating their own family tree will help students understand the nature of classification. Students, especially those from cultures in which the extended family is important, will better understand the nature of hierarchy in the Linnaean classification system.

ESL STRATEGY 1-2

After students have completed the Find Out by Writing feature in this section, have them complete the names of classification groups, given the first letter of each group: s_____, p_____, c_____, g_____, k_____, o_____, f_____. Then have students rearrange the groups into correct order beginning with the largest.

BACKGROUND INFORMATION

THE DARWIN-WALLACE THEORY OF EVOLUTION

The theory of evolution by natural selection is properly called the Darwin-Wallace theory. Although Charles Darwin had formulated the theory by 1838, he did not publish his ideas. Twenty years later, Alfred Wallace independently arrived at the same ideas and unknowingly "scooped" Darwin. Because both scientists felt that the other had a strong claim to the theory, they wanted to share the credit. Together, they presented their papers at a meeting in London in 1858.

---

their "Katze" and the other person was talking about their "gato." Each person would not know what the other person was talking about because each person was speaking a different language. Stress that a system of classification allows scientists to speak the same language.
• **Which language do you think each of these words for "cat" is from?** (*Katze*—German; *chat*—French; *kot*—Russian; *gato*—Spanish.)

CONTENT DEVELOPMENT

Explain to students that our modern system of classification is based on the evolutionary relationships of organisms. Modern taxonomists try to classify living things in such a way that each classification group contains organisms that evolved from the same ancestor.

Make sure students understand that the modern system of classification does two things: First, it groups organisms according to their basic characteristics; second, it gives a unique name to each organism.

● ● ● ● **Integration** ● ● ● ●

Use Figure 1–8 and its caption to integrate the concept of evolution into your lesson.

INDEPENDENT PRACTICE

▶ *Activity Book*

Students who need practice on the concept of classification should complete the chapter activity Fun With Fictitious Animals. In this activity students will use a taxonomic key to answer classification questions.

LINNAEAN SYSTEM OF CLASSIFICATION

Linnaeus published a system for classifying plants in 1753 and one for classifying animals in 1758. Initially, his grouping system consisted of three levels: kingdom, genus, and species. The phylum, class, family, and order levels were added later. Jean Baptiste Lamarck, who is best known for his pre-Darwinian theory of evolution, was largely responsible for expanding and refining the Linnaean system.

Figure 1–9 *As knowledge of the evolutionary relationships among animals improved, the lesser panda (bottom) and the giant panda (top center) were classified and reclassified. The data now available support a classification scheme in which the giant panda belongs to the same family as other bears, such as the grizzly (top left). The lesser panda is placed in the same family as raccoons (top right).*

20 ■ B

which new kinds of organisms develop from previously existing kinds of organisms.

Evolutionary relationships, such as those between the ancient catlike animal and house cats or between wolves and lions, are extremely important to modern taxonomists. Evolutionary relationships are the basis for the modern system of biological classification. Modern taxonomists try to classify living things in such a way that each classification group contains organisms that evolved from the same ancestor.

The knowledge of evolution has changed the nature of biological classification groups. It has also changed the job of taxonomists. Because they did not know about evolution, taxonomists of the past felt free to classify organisms in any manner that made sense to them. They did not classify prehistoric organisms because they did not know about them. And they could choose any characteristics that they thought were important as their basis for classification.

Present-day taxonomists, on the other hand, classify organisms in a way that shows evolutionary relationships. They must consider organisms that existed in the distant past as well as those that exist in the present. And they classify organisms using characteristics that are proven to be good indicators of evolutionary relationships.

Technology and Classification

Today, 200 years after Linnaeus completed his work, scientists consider many factors when classifying organisms. Of course, they still examine the

---

1–2 (continued)

GUIDED PRACTICE

Skills Development

Skill: Identifying relationships

Have students examine Figure 1–9.
• **List one characteristic that all these animals share.** (Students might suggest that each animal is covered by hair, or fur.)

Have students read the caption for Figure 1–9.
• **How many different families do these four animals represent?** (Two.)
• **The giant panda and the grizzly bear belong to the same family. Name a char-**acteristic that these two animals share. (Answers will vary.)
• **Name a characteristic that makes the giant panda and the grizzly bear different from one another.** (Answers will vary.)
• **Why do you think the giant panda and the grizzly bear belong to the same family?** (Because of their evolutionary relationship.)
• **The lesser panda and the raccoon belong to the same family. Name a characteristic that these two animals share.** (Answers will vary.)

• **Name a characteristic that makes the lesser panda and the raccoon different from one another.** (Answers will vary.)
• **Why do you think the lesser panda and the raccoon belong to the same family?** (Because of their evolutionary relationship.)

CONTENT DEVELOPMENT

Emphasize that taxonomists today consider several factors other than structural similarities when classifying organisms. A few factors that might be mentioned are these.

large internal and external structures, but they also rely on other observations. The invention of the microscope has allowed scientists to examine tiny structures hidden within the cells of an organism. It has also allowed them to examine organisms at their earliest stages of development. And special chemical tests have been developed that enable scientists to analyze the chemical building blocks of all living things. All of these techniques are important tools that help scientists group and name organisms.

Classification Groups

At first glance, the modern classification system may seem complicated to you. However, it is really quite simple, especially when you keep in mind its purpose. The system of classification used today does two jobs. First, it gives each organism a unique name that scientists all over the world can use and understand. Second, it groups organisms according to basic characteristics that reflect their evolutionary relationships.

All living things are classified into seven major groups: kingdom, phylum, class, order, family, genus, and species. The largest and most general group is the kingdom. For example, all animals belong to the animal kingdom. The second largest group is the phylum (FIGH-luhm; plural: phyla). A phylum includes a large number of very different organisms. However, these organisms share some important characteristics. A species is the smallest and most specific group in the classification system. Members of the same species share many characteristics and

A Secret Code

You can remember the correct classification sequence from the largest to the smallest group by remembering this sentence: Kings play cards on fat green stools. The first letter of each word is the same as the first letter of each classification group.

Write two additional sentences that can help you to remember the classification sequence.

Figure 1–10 *The male (right) and female (left) grand eclectus parrots look so different that they were once thought to be separate species.*

B ■ 21

1. Fossil evidence: The fossil record often reveals that certain organisms are related because of common ancestors.
2. Early development: Scientists believe that similarities in the early embryos of organisms indicate relationships.
3. Chemical makeup of an organism: Studies have revealed that certain proteins have a similar chemical makeup in similar species. By studying the chemical structure of these proteins, scientists can get an idea of how closely related various organisms may be.

4. Genetic makeup of an organism: Relationships among species can be found by comparing the structure of DNA—the hereditary material of their cells. The more closely related the species are, the more similar their DNA will be.

● ● ● ● **Integration** ● ● ● ●

Use the discussion about the basis of our system of classification to integrate the concept of evolution into your lesson.

HOMOLOGOUS VERSUS ANALOGOUS

When classifying organisms, taxonomists are careful to distinguish between homologous structures and analogous structures. Homologous structures have a similar structure and development pattern. The function of homologous structures however, may be different. The wing of a bird and the human arm are homologous structures due to their similar structure and pattern of development. Analogous structures appear similar and perform similar functions: however, their structure and developmental pattern are quite different. The wing of a bird and the wing of a butterfly are analogous structures.

1–2 (continued)

INDEPENDENT PRACTICE

▶ *Activity Book*

Students who need practice on the concept of classification systems should complete the chapter activity Making a Classification System. In this activity students will develop a classification system for 50 different objects in their classroom.

Media and Technology

Have students find out about the evolution and taxonomy of the animals featured in the Interactive Videodisc called Virtual BioPark. In subsequent lessons and chapters, students can explore other aspects of the biology and ecology of the cheetah, pigeon, giant orb weaving spider, prairie dog, and rattlesnake. Challenge students to discover how these organisms interact with monerans, protists, fungi, and plants.

CONTENT DEVELOPMENT

Have students examine Figure 1–11 and obtain a college dictionary or its equivalent.

Ask students to look up the word *animal* in their dictionary.

• **Which of the organisms in the first** row of pictures, or kingdom level, belong to the animal (or Animalia) kingdom? (All of them.)

Remind students that the lion is the guide and that they are to find similarities and differences between the lion and the other pictured animals at each level of classification.

Ask students to look up the word *chordate* in their dictionary.

• **Which animals pictured in the phylum level fit the description of a chordate?** (All of them.)

• **Which pictured animals were removed at the phylum level because they do not fit the description of a chordate?** (Starfish and insect.)

Ask sudents to look up the word *mammal* in their dictionary.

• **Which pictured animals in the mammal (or Mammalia) level fit the description of a mammal?** (All of them.)

• **What was removed from the mammal level that is in the phylum level?** (Tunicate and lizard.)

CLASSIFICATION OF THE LION

Kingdom Animalia	
Phylum Chordata	
Class Mammalia	
Order Carnivora	
Family Felidae	
Genus *Panthera*	
Species *leo*	

Figure 1–11 *This chart shows several other organisms that are in the same classification groups as the lion. To what class do lions belong?* ❶

are similar to one another in appearance and behavior. In addition, members of the same species can interbreed and produce offspring. These offspring can in turn produce offspring of their own.

Ideally, the largest classification groups represent the earliest ancestors and the most ancient branches of life's family tree. And the smallest classification groups contain organisms that evolved from shared ancestors that lived in the relatively recent past. But because people do not know everything there is to know about evolution, these categories are not perfect in real life. As scientists learn more and more about evolutionary relationships and about the history of life on Earth, the classification system is changed. Sometimes the changes are tiny. Other times, the changes are quite large. And once in a while, taxonomists choose to keep an old group because it is particularly useful and logical, even if it does not perfectly reflect evolutionary history.

Biologists often think of these classification groups as forming a tree in which the trunk represents the kingdom, the main branches represent the phyla within the kingdom, and the tiny twigs at the tips of the branches represent species. You can also

Figure 1–12 *The relationships among classification groups can be represented as a tree. This classification tree shows the major groups of animals. What animals are in the same phylum as the centipede and the red beetle?* ②

Species Genus Family Order Class Phylum Kingdom

HISTORICAL NOTE
CAROLUS LINNAEUS

Linnaeus did not intentionally establish the system of binomial nomenclature. He actually preferred to use the longer descriptive name for plants in his works; nevertheless, he did write a shorter two-name word for each organism in the margin of his notes. Other researchers found this shorthand so useful that they adopted it.

its larynx and cannot purr continuously like those of the *Felis* genus. Cats in the *Panthera* genus all roar.)
• **What must be removed from the genus level to satisfy this description of roaring?** (House cats.)

Point out that leo is the species name for a lion of the *Panthera* genus.

GUIDED PRACTICE

Skills Development

Skills: Making comparisons, identifying patterns

At this point have students complete the in-text Chapter 1 Laboratory Investigation: Whose Shoe Is That? In this investigation students will develop a classification system for shoes.

ENRICHMENT

▶ *Activity Book*

Students will be challenged by the Chapter 1 activity in the *Activity Book* called Going Around in Circles. In this activity students will explore the use of a Venn diagram classification scheme.

Ask students to look up the word *carnivore* in their dictionary.
• **Which pictured animals in the level of mammals were eliminated because they were not flesh-eating?** (Monkey, elephant, and whale.)

Ask students to look up the word *felid* in their dictionary.
• **Which pictured animals from the order level were eliminated because they were not cats?** (Skunk, walrus, and wolf.)
• **Are all the animals in the family level cats?** (Yes.)

• **Are all the cats the same type of cat?** (No.)
• **What is different about the cats?** (Accept all logical answers.)

Ask students to look up the word *panther* in their dictionary.

Explain to students that the cat genus of *Felis* is a purring cat because of the structure of its larynx.
• **Why do you think the lion is placed in the genus *Panthera* instead of the genus *Felis*?** (Point out that a cat placed in the *Panthera* genus has a different structure to

PROBLEM SOLVING

CLASSIFYING THE DRAGONS OF PLANET NITRAM

This feature enables students to reinforce and extend their classification skills.
1.–2. Answers will vary. See A Diversity of Dragons answers 1 and 2 on p. B42 in the *Activity Book* for possible answers.
3. These data indicate that sanjorges have juvenile coloring and features. Students may revise their classification systems so that smaugs, drakos, and sanjorges are in the same species. Alternatively, they may assume that juveniles of the smaug species and the drako species look alike.

ACTIVITY
DISCOVERING
CLASSIFYING LIVING THINGS

Discovery Learning

Skills: Making comparisons, classifying

Materials: notebook, pen or pencil

This activity will reinforce the ability of each student to classify organisms by their basic characteristics. Using field guides when necessary, students should have little trouble identifying various organisms. Students will develop a better understanding of the nature of classification by comparing and grouping the organisms according to their similarities and differences.

PROBLEM ??? Solving

Classifying the Dragons of Planet Nitram

Imagine that you are a famous space explorer and biologist. You have recently arrived on the planet Nitram. Your job is to study the Nitramian animals and develop a system for classifying them.

You decide to begin your work by classifying Nitram's "dragons." You think that this name is not very scientific. However, you must admit that the animals look a lot like the monsters of Earth legends.

You ask the computer to give you a brief summary of all the information it has on Nitramian dragons. The computer produces the following printout.

THE DRAGONS OF NITRAM

Drako: About 5 to 6 meters tall. Four legs. Two batlike wings. Able to fly. Body covered in scales. Lives in mountains. Feeds on large animals.

Quetzalcoatl: About 2 meters long. Two legs. Two birdlike wings. Able to fly. Green and red feathers. Long feathers of many colors on top of head and around neck. Green scales on snakelike tail. Lives in tropical jungles. Eats fruit and small animals.

Sanjorge: About 2 to 5 meters tall. Four legs. Two small batlike wings. Not able to fly. Body has reddish-brown and greenish-brown scales. Lives in mountains. Probably eats animals.

Smaug: Largest known dragon. About 6 to 9 meters tall. Four legs. Two batlike wings. Able to fly. Body red, belly orange. Seems to be covered in scales. Extremely aggressive. Lives in mountains. Feeds on large animals, including humans.

Tailoong: About 4 meters long. Long, thin, snakelike body covered with gold scales. Four legs. Lionlike mane of long, colorful feathers. Lives in forests near lakes. Feeds on flowers and water plants.

Wyvern: About 2 to 3 meters long. Two legs. Two birdlike wings. Able to fly. Head, neck, and tail covered with red, yellow, and brown scales. Long red feathers around base of neck. Wings and body covered with brown feathers. Lives in mountains. Feeds on small animals. May also feed on the remains of dead animals.

Classifying Animals

1. Develop a classification system for the Nitramian dragons. Explain how you devised your classification scheme.

2. Like the animals of Earth, dragons are the result of millions of years of evolution. Which types of dragons seem to be most closely related? Explain.

3. When a sanjorge gets to be about 5 meters tall, its wings begin to grow rapidly. Within a year or two, it can fly. At the same time, its color changes to either red or green. How do these new data affect your classification system?

1–2 (continued)

ENRICHMENT

▶ *Activity Book*

Students will be challenged by the Chapter 1 activity in the *Activity Book* called A Diversity of Dragons. In this activity students will revise an existing classification system to accommodate new organisms and data.

CONTENT DEVELOPMENT

Remind students that our modern system of classification gives each organism a unique name and that it groups organisms according to basic characteristics that reflect an organism's evolutionary relationships.
• **What are the seven major groups into which all living things are classified?** (Kingdom, phylum, class, order, family, genus, and species.)
• **Which is the largest, and most general, of these groups?** (Kingdom.)

• **Which is the smallest, and most specific, of these groups?** (Species.)

GUIDED PRACTICE

▶ *Laboratory Manual*

Skills Development

Skill: Making observations

At this point you may want to have students complete the Chapter 1 Laboratory Investigation in the *Laboratory Manual* called Developing a Classification System for Seeds. In this investigation students will

think of the classification groups as boxes within boxes. For example, you might open a huge kingdom box to discover several large phyla boxes, each of which contains a number of still smaller class boxes. Each time you opened a box, you would find one or more smaller boxes. If you opened the smallest boxes (the species boxes), you would find a number of individuals all of the same type.

By knowing which branches to climb, you can eventually arrive at one particular twig. (Provided, of course, that the branches are strong enough to bear your weight!) If you know which boxes to open in a set of boxes within boxes, you will sooner or later find the one tiny box you are looking for. Similarly, if you go through each level of classification groups, you will finally arrive at one species; that is, one specific kind of organism.

1–2 Section Review

1. List the classification groups from largest to smallest.
2. What is evolution? How does evolution affect the way organisms are classified?

Critical Thinking—*Making Inferences*
3. Explain why knowing the classification of an unfamiliar organism can tell you a lot about that organism.

1–3 The Five Kingdoms

The discoveries of new living things and the changing ideas about the most effective ways of classifying life forms have resulted in the five-kingdom classification system we use now. **Today, the most generally accepted classification system contains five kingdoms: monerans, protists, fungi, plants, and animals.**

As is often the case in science, not all scientists agree on this classification system. And this is an important idea for you to keep in mind. More research may someday show that different systems make more

---

ACTIVITY
DISCOVERING

Classifying Living Things

1. Go for a long walk outside. Take along a pencil and a small notebook.
2. Write down the names of all the living things you see on your walk. (You should notice at least 15 different kinds of organisms.)
■ Develop a classification system for the organisms on your list.

Guide for Reading

Focus on this question as you read.
▶ What are the five kingdoms of living things?

Activity Bank

A Key to the Puzzle, p.170

1–3 The Five Kingdoms

In order to help students better understand classification, have them list their ten favorite foods and then classify them according to national or regional origin. Students should also list two food dishes from each of the following countries: Mexico, Japan, China, Germany, Italy.

Suggest that students think of dishes that represent their own ethnic background as well. Have volunteers explain how some of these dishes are prepared.

ESL STRATEGY 1–3

Remind students of the secret code they worked with in a previous Find Out by Writing feature. Then have them work in small groups to complete the following activities.
1. Create their own code for recalling the five-kingdom classification sequence.
2. Make a chart that lists the five kingdoms in order, identifying each as unicellular or multicellular, describing each as autotroph and/or heterotroph, and giving one additional characteristic of each.

Students can exchange codes and charts to check for accuracy. Then ask groups to use their charts as outlines for writing a description of each kingdom.

CLOSURE

▶ *Review and Reinforcement Guide*
Students may now complete Section 1–2 in the *Review and Reinforcement Guide.*

TEACHING STRATEGY 1–3

FOCUS/MOTIVATION

Divide the class into groups of two to four students per group. Explain that the classification system of living things used today contains five kingdoms and that two of these kingdoms are the plant and animal kingdoms. Have each group assign a note taker, then allow five minutes for each group to name as many plants as they can. Repeat the activity, having each group name as many animals as they can. You may then wish to have groups exchange lists to determine the accuracy of the lists.

---

use the observable characteristics of different seeds to classify them in an orderly way.

INDEPENDENT PRACTICE
Section Review 1–2
1. Kingdom, phylum, class, order, family, genus, species.
2. Evolution is the process by which new kinds of organisms develop from previously existing kinds of organisms; evolutionary relationships are the basis for the modern system of biological classification.

3. Knowing the classification of an unfamiliar organism helps to explain where that organism fits into the scheme of living things and helps to explain its relationship to other living things.

REINFORCEMENT/RETEACHING

Review students' responses to the Section Review questions. Reteach any material that is still unclear, based on students' responses.

OTHER CLASSIFICATION SYSTEMS

The use of classification systems is not performed exclusively by life scientists. Earth scientists, for example, classify rocks into three main groups: sedimentary, igneous, and metamorphic. Students may enjoy listing other classification systems used by scientists.

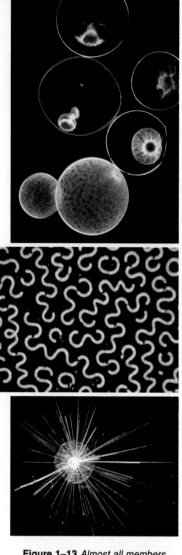

Figure 1–13 *Almost all members of the kingdoms Monera and Protista are microscopic. The blue-green bacterium* Anabaena, *a moneran, forms squiggly chains of cells. The glow-in-the-dark spheres and foamy starburst are types of protists. What is the main difference between monerans and protists?* ②

26 ■ B

sense and better represent how living things evolved. But for now, this five-kingdom system is a useful tool for studying living things—and that's exactly what taxonomy is all about.

MONERANS Bacteria are placed in the kingdom Monera. Monerans are unicellular organisms, or organisms that consist of only one cell. A moneran's cell does not have its hereditary material enclosed in a nucleus, the structure that in other cells houses this important material. In addition to a nucleus, moneran cells lack many other structures found in other cells. Because of their unique characteristics, monerans are considered to be very distantly related to the other kingdoms.

Like other organisms, monerans can be placed into two categories based on how they obtain energy. Organisms that obtain energy by making their own food are called **autotrophs.** Such a name makes a lot of sense since the prefix *auto-* means self and the root word *-troph* means food. Organisms that cannot make their own food are called **heterotrophs.** The prefix *hetero-* means other. Can you explain why heterotroph is a good name for such organisms? Heterotrophs may eat autotrophs in order to obtain food or they may eat other heterotrophs. But all heterotrophs ultimately rely on autotrophs for food. ①

Scientists have evidence that monerans were the earliest life forms on Earth. They first appeared about 3.5 billion years ago.

PROTISTS The kingdom Protista includes most of the unicellular (one-celled) organisms that have a nucleus. The nucleus controls the functions of the cell and also contains the cell's hereditary material. In addition, the cell of a protist has special structures that perform specific functions for the cell.

A number of protists are capable of animallike movement but also have some distinctly plantlike characteristics. Specifically, they are green in color and can use the energy of light to make their own food from simple substances. As you can imagine, such organisms are difficult, if not impossible, to classify using a two-kingdom classification system. They are neither plants nor animals. Or perhaps they are both plants and animals! These puzzling

---

1–3 (continued)

CONTENT DEVELOPMENT

Explain that when Linnaeus developed an early system of classification, he used only two kingdoms—the plant kingdom and the animal kingdom.
• **Name several characteristics that make a plant different from an animal.** (Accept all logical responses. Students might respond that animals have the ability to move, whereas plants do not; plants have the ability to make food, whereas animals do not.)
• **Linnaeus classified living things into only two kingdoms—plant and animal.**

Why do you think this is not a very good classification system? (Lead students to suggest that there are many organisms that do not fit into either kingdom.)
• **How many kingdoms are included in today's classification system?** (Five.)
• **Name the kingdoms used in today's classification system.** (Plants, animals, protists, monerans, and fungi.)
• **Why do you think all scientists do not agree that there should be a five-kingdom system of classification?** (Answers will vary. Lead students to recognize that the num-

ber of different kingdoms that best represent how living things evolved is arguable.)
• **Do you think this type of disagreement within the scientific community is "healthy"?** (Lead students to understand that generally speaking, argument or disagreement within scientific circles is productive.)

● ● ● ● **Integration** ● ● ● ●

Use the discussion of ancient types of protists to integrate the concept of evolution into your lesson.

organisms are one of the reasons scientists finally decided to abandon the two-kingdom system.

Protists were the first kind of cells that contain a nucleus. Ancient types of protists that lived millions and millions of years ago are probably the ancestors ➊ of fungi, plants, animals, and modern protists.

FUNGI As you might expect, the world's wide variety of fungi make up the kingdom Fungi. Most fungi are multicellular organisms, or organisms that consist of many cells. Although you may not realize it, you may be quite familiar with fungi. Mushrooms are fungi. So are the molds that sometimes grow on leftover foods that have remained too long in the refrigerator. And the mildews that may appear as small black spots in damp basements and bathrooms are also fungi.

Figure 1–14 *The mushroom (left), shelf fungi (top right), and starfish stinkhorn (bottom right) all belong to the kingdom Fungi. The brownish slime on which the flies are feasting contains the stinkhorn's spores. If the spores end up in a favorable place after being deposited in the flies' droppings, they will grow into new stinkhorns.*

INTEGRATION
HEALTH AND MEDICINE

Mold is a form of fungi, one of the five kingdoms of living things. Penicillin is an antibiotic substance extracted from the cultures of certain molds. Have interested students research the relationship between molds and medicine and report their findings to the class.

R. H. WHITTAKER

The five-kingdom system of classification was suggested by R. H. Whittaker in 1969. A major difference in this classification scheme lies in grouping fungi as a separate kingdom. Because fungi lack photosynthetic pigments and obtain their nutrients by absorbing them from the environment, Whittaker believed they should be placed in a separate kingdom.

ECOLOGY NOTE

DECOMPOSERS

Decomposers exist in various sizes and forms. Have students research the impact of decomposers on the environment. Then, knowing that there are five kingdoms of plants and animals, students should attempt to find an example of a decomposer for each of the five kingdoms.

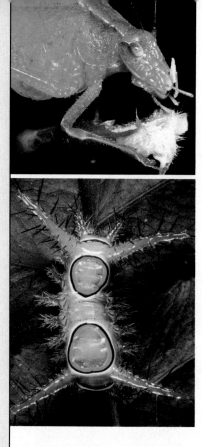

Figure 1–15 *The haired saddleback caterpillar and the pink katydid are both animals that feed on plants. How are animals similar to plants? How are they different?* ❶

For many years, fungi were classified as plants. However, they are quite different from plants in some basic ways. Their cells fit together in a different way. Their cell wall, a tough protective layer that surrounds the cell, is made of a different substance than the cell wall of plants. And most importantly, plants are able to use the energy of light to make their own food from simple substances. Fungi cannot. Like animals, fungi must obtain their food and energy from another source. Are fungi autotrophs or heterotrophs? ❷

PLANTS Plants make up the kingdom Plantae. Most members of the plant kingdom are multicellular (many-celled) autotrophs. You are probably quite familiar with members of this kingdom as it includes all the plants you have come to know by now— flowering plants, mosses, ferns, and certain algae, to name a few.

ANIMALS Animals are multicellular organisms that comprise the kingdom Animalia. Like other multicellular organisms such as plants and certain fungi, animals have specialized tissues, and most have organs and organ systems. Unlike plants, animals are heterotrophs.

1–3 Section Review

1. Name the five kingdoms of living things.
2. List three important characteristics for each of the five kingdoms.
3. How is the way an autotroph gets food different from the way a heterotroph gets food?

Connection—*Classifying Organisms*
4. Suppose that creatures from a distant planet are multicellular heterotrophs whose cells lack cell walls. Which kingdom of the Earth's organisms do these creatures most closely resemble?

28 ■ B

1–3 (continued)

GUIDED PRACTICE

Skills Development

Skill: Making classifications

Divide the class into groups of four to six students per group. Have each group create a list of 20 organisms they have seen firsthand. (Students should recall organisms they have seen while visiting a zoo, park, beach, the mountains, and so on.) After groups have completed their lists, have them classify each entry according to its kingdom. Groups can check the accuracy of their classifications with a taxonomic guide.

REINFORCEMENT/RETEACHING

▶ *Activity Book*

Students who need practice on the concept of classification of living things

should complete the chapter activity How Are Organisms Classified? In this activity students will complete a classification scheme of living things.

CONTENT DEVELOPMENT

Point out that protists include most unicellular organisms that have a well-defined nucleus.

Monerans, like protists, are unicellular, but they do not have a nucleus. All the Earth's bacteria are found in the kingdom Monera.

Fungi share many characteristics with plants. Fungi, however, do not contain the green substance chlorophyll; so, unlike plants, fungi cannot make their own food.

REINFORCEMENT/RETEACHING

▶ *Activity Book*

Students who need practice on the concept of classification kingdoms should complete the chapter activity Classifying Organisms. In this activity students will determine which kingdom various organisms belong to.

CONNECTIONS

What's in a Name? ❶

You may be familiar with the German *fairy tale* "Rumpelstiltskin." In this story, a magical dwarf saves a queen's life by helping her spin straw into gold. To pay the dwarf for his help, the queen must give him her baby—or guess the dwarf's secret name. In the end, she learns the dwarf's name. And everyone (except the dwarf) lives happily ever after.

Tales like this one reflect the ancient idea that names are extremely important. People once believed that knowing something's true name gave you magical power over it.

Knowing the true, or scientific, names of organisms is not going to give you magical power over them. But knowing what the scientific names mean can help give you a different kind of power—the power to figure out unfamiliar words.

As you look over the following examples of scientific names, remember that it is not important to memorize names. And don't let the strangeness of the names scare you. In time, you will become more familiar with scientific names and learn to see—and use—the patterns in them. You will discover that knowing the secrets of scientific names can be informative and even fun.

▲ *Naja melanoleuca* is a black-and-white-lipped cobra, a venomous snake. In India and Southeast Asia, a *naja*, or *naga*, is an imaginary beast that usually takes the form of a giant snake with a human head. A *naga* can also take the form of a human or a snake. (*melano-* means black; *leuca-* means white.)

▲ *Narcissus poeticus* is a sweet-smelling flower. According to a Greek myth, Narcissus was a handsome young man who fell in love with his own reflection. Eventually, the gods transformed Narcissus into a flower. In the spring, narcissus flowers may sometimes be found near a pond or stream. The long stems of the flowers may cause them to lean over the water so that it looks like they are admiring their own reflection. The story of Narcissus has added words other than the name of a flower to the English language. A person who is completely in love with himself (or herself) is called a narcissist. Linnaeus's name for the flower means "narcissus of the poets."

B ■ 29

ANNOTATION KEY

Answers

❶ Plants are multicellular autotrophs. Animals are multicellular heterotrophs with specialized tissues. (Relating facts)
❷ Heterotrophs. (Applying definitions)

Integration

❶ Language Arts

CONNECTIONS
WHAT'S IN A NAME?

This activity helps students understand some of the patterns inherent in many scientific names. Point out that without practice, scientific names can prove difficult to pronounce and understand. But with practice, these names become much easier to pronounce and understand. Familiarity of the patterns inherent in many scientific names can make completely unfamiliar names easier to understand. Interested students might want to find other examples of scientific names and present these examples to the class.

If you are teaching thematically, you may want to use the Connections feature to reinforce the themes of evolution, patterns of change, scale and structure, and unity and diversity.

Integration: Use the Connections feature to integrate language arts into your science lesson.

INDEPENDENT PRACTICE

Section Review 1–3

1. Monerans, protists, fungi, plants, animals.
2. Monerans: unicellular, no nucleus, autotroph or heterotroph; protists: unicellular, nucleus, some capable of movement, autotroph or heterotroph; fungi: multicellular, specialized cell wall, heterotroph; plants: multicellular, autotroph, not capable of movement; animals: multicellular, specialized tissue, heterotroph.
3. Autotrophs make their own food. Heterotrophs eat heterotrophs and/or autotrophs for food.
4. Fungi.

REINFORCEMENT/RETEACHING

Monitor students' responses to the Section Review questions. If students appear to have difficulty with any of the questions, review the appropriate material in the section.

CLOSURE

▶ *Review and Reinforcement Guide*

At this point have students complete Section 1–3 in the *Review and Reinforcement Guide.*

Laboratory Investigation

WHOSE SHOE IS THAT?

BEFORE THE LAB

1. Divide the class into groups of six students, if possible, depending on class size.
2. If desks or tables in your classroom are too small, have the groups perform the activity on the floor.

PRE-LAB DISCUSSION

Have students read the complete laboratory procedure.

• **What is the purpose of this laboratory activity?** (To develop a system for classifying shoes.)

• **Name some characteristics that can be used to divide and subdivide shoes into groups.** (Answers will vary, but lead students to discuss appropriate characteristics. A brief list of appropriate characteristics might be placed on the chalkboard.)

• **How will we know when the activity is finished?** (At the end of the activity, each person's shoe will be separate from all the others.)

Laboratory Investigation

Whose Shoe Is That?

Problem

How can a group of objects be classified?

Materials *(per group of six)*

students' shoes
pencil and paper

Procedure

1. At your teacher's direction, remove your right shoe and place it on a work table.
2. As a group, think of a characteristic that will divide all six shoes into two kingdoms. For example, you may first divide the shoes by the characteristic of color into the brown shoe kingdom and the nonbrown shoe kingdom.
3. Place the shoes into two separate piles based on the characteristic your group has selected.
4. Working only with those shoes in one kingdom, divide that kingdom into two groups based on a new characteristic. The brown shoe kingdom, for example, may be divided into shoes with rubber soles and shoes without rubber soles.
5. Further divide these groups into subgroups. For example, the shoes in the rubber-soled group may be separated into a shoelace group and a nonshoelace group.

6. Continue to divide the shoes by choosing new characteristics until you have only one shoe left in each group. Identify the person who owns this shoe.
7. Repeat this process working with the non-brown shoes.
8. Draw a diagram similar to the one shown to represent your classification system.

Observations

1. How many groups are there in your classification system?
2. Was there more than one way to divide the shoes into groups? How did you decide which classification groups to use?

Analysis and Conclusions

1. Was your shoe classification system accurate? Why or why not?
2. If brown and nonbrown shoe groups represent kingdoms, what do each of the other groups in your diagram represent?
3. Compare your classification system to the classification system used by most scientists today.
4. **On Your Own** Follow a similar procedure to classify all the objects in a closet or a drawer in your home.

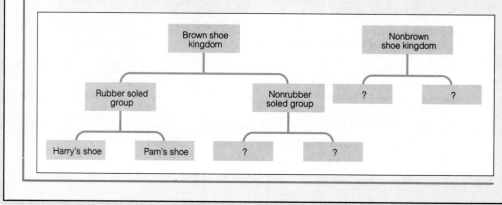

Study Guide

Summarizing Key Concepts

1–1 History of Classification

▲ Classification is the grouping of things according to similar characteristics.

▲Classification is important in all fields of science, in subjects other than science, and in everyday life.

▲ Good classification systems are meaningful, easily understood, and readily communicated among people.

▲ Classification systems organize and name living things in a logical, meaningful way.

▲ The science of biological classification is called taxonomy.

▲ Aristotle invented one of the first systems of biological classification. This system was used for about 2000 years.

▲ The classification and naming systems used today are based on the work of Carolus Linnaeus.

▲ In binomial nomenclature, which was invented by Linnaeus, each kind of organism is given a unique two-part name. This scientific name is used and understood all over the world. The scientific name is also related to the way the organism is classified.

1–2 Classification Today

▲ As our understanding of living things improves, it becomes necessary to revise our system of biological classification.

▲ Evolutionary relationships are the basis for the modern system of biological classification.

▲ Advances in technology have increased our knowledge of living things and thus have had an effect on how organisms are classified.

▲ The classification groups from largest to smallest are: kingdom, phylum, class, order, family, genus, and species. If you go through each level of classification groups, you finally arrive at one specific species.

▲ Each of an organism's classification groups tells you something about its characteristics.

1–3 The Five Kingdoms

▲ Today, the most generally accepted classification system contains five kingdoms: monerans, protists, fungi, plants, and animals.

▲ Autotrophs can make food from simple raw materials. Heterotrophs cannot make their own food.

Reviewing Key Terms

Define each term in a complete sentence.

1–1 History of Classification
binomial nomenclature
genus
species

1–3 The Five Kingdoms
autotroph
heterotroph

Part 1

Assign students to different groups and have them repeat the activity. At the conclusion of the activity, have students compare the second classification system they developed with the first. Establish the idea that even though all groups were working on the same task, many classification systems can result if different criteria are used when grouping and re-grouping the shoes.

Part 2

Have students create a branching chart to classify approximately six organisms or things with which they are familiar. For example, have them try and classify (a) pets, (b) band or orchestra instruments, and (c) different kinds of balls.

---

cific groups. For example, **kingdom is divided into five smaller groups: monerans, protists, fungi, plants, and animals. How is the classification system you developed similar to these smaller, more specific groups?** (Answers will vary—comparing, classifying.)

OBSERVATIONS

1. Answers will vary, depending on individual teams.

2. Yes; answers will vary.

ANALYSIS AND CONCLUSIONS

1. Answers will vary. In the event of inaccurate classification systems, lead teams in a discussion of how the systems could have been made more accurate.

2. Smaller, more specific groups within those kingdoms.

3. Comparisons will vary. Lead students to observe that good classification systems narrow, or become more specific, as they develop.

4. Classifications will vary.

Chapter Review

ALTERNATIVE ASSESSMENT

The *Prentice Hall Science* program includes a variety of testing components and methodologies. Aside from the Chapter Review questions, you may opt to use the Chapter Test or the Computer Test Bank Test in your *Test Book* for assessment of important facts and concepts. In addition, Performance-Based Tests are included in your *Test Book*. These Performance-Based Tests are designed to test science process skills, rather than factual content recall. Since they are not content dependent, Performance-Based Tests can be distributed after students complete a chapter or after they complete the entire textbook.

CONTENT REVIEW

Multiple Choice

1. d
2. b
3. d
4. c
5. c
6. a
7. d
8. d

True or False

1. F, species
2. F, kingdom, phylum, class, order, family, genus, species
3. F, Unicellular
4. F, monerans
5. F, autotrophs
6. F, Heterotrophs
7. T
8. F, genus

Concept Mapping

Row 1: Science
Row 2: Taxonomy

CONCEPT MASTERY

1. To organize all living things into one classification scheme and to be able to communicate to different scientists by using a universal language; also, to help determine where new organisms that are discovered fit into the scheme of living things and to better understand the relationship among living things.
2. Similarities in structure and form.
3. Both systems use binomial nomenclature. Scientists today, however, classify organisms by more than structure and form. Scientists today also group organisms into five kingdoms.
4. Plants: mostly multicellular and autotrophic; animals: all multicellular and heterotrophic; protists: almost all unicellular with well-defined nuclei; monerans: all unicellular without nucleus; fungi: similar to plants but do not contain chlorophyll, so they are not autotrophic. Examples: animal (lion), plant (tulip), protist (paramecium), monera (bacteria), fungi (mushroom).
5. An autotroph can make its own food, usually by using energy from the sun in a process called photosynthesis. A heterotroph cannot make its own food and must eat plants or heterotrophs. An example of a heterotroph is a whale, and an example of an autotroph is a maple tree.
6. It caused organisms to be classified so that the biological classification "tree" reflected the evolutionary relationships among organisms as accurately as possible.
7. The modern classification system assigns each organism a unique name, and it groups organisms according to basic

---

Content Review

Multiple Choice

Choose the letter of the answer that best completes each statement.

1. Which of the following is not a characteristic of a good classification system?
   a. It shows relationships among objects.
   b. It is meaningful.
   c. It is readily communicated among people.
   d. It creates confusion.
2. The branch of biology that deals with naming and classifying organisms is called
   a. exobiology.   c. phylum.
   b. taxonomy.   d. binomial nomenclature.
3. The largest classification group is the
   a. species.   c. phylum.
   b. order.   d. kingdom.
4. A genus can be divided into
   a. phyla.   c. species.
   b. orders.   d. families.
5. Carolus Linnaeus classified plants and animals according to similarities in
   a. color.   c. structure.
   b. habits.   d. size.
6. Which organisms have cells that do not contain a nucleus?
   a. monerans   c. plants
   b. fungi   d. protists
7. Which of the following statements about animals in the same species is false?
   a. They evolved from a shared ancestor.
   b. They share certain characteristics.
   c. They can interbreed and produce offspring.
   d. They have identical characteristics.
8. Which organism is an autotroph?
   a. frog   c. lion
   b. mushroom   d. maple tree

True or False

If the statement is true, write "true." If it is false, change the underlined word or words to make the statement true.

1. In Linnaeus's classification system, the smallest group was the <u>genus</u>.
2. The classification groups from largest to smallest are: <u>kingdom, phylum, class, family, order, species, genus</u>.
3. <u>Multicellular</u> organisms are composed of only one cell.
4. In a five-kingdom classification system, bacteria are classified as <u>plants</u>.
5. Green plants are <u>heterotrophs</u>.
6. <u>Autotrophs</u> are organisms that cannot make their own food.
7. The science of classification is called <u>taxonomy</u>.
8. The first word in a scientific name is the <u>species</u>.

Concept Mapping

Complete the following concept map for Section 1–1. Refer to pages B6–B7 to construct a concept map for the entire chapter.

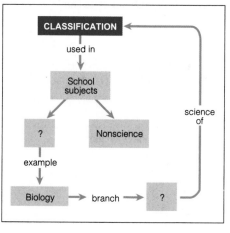

Concept Mastery

Discuss each of the following in a brief paragraph.

1. Why do scientists classify organisms?
2. What type of characteristics did Linnaeus use to develop his classification system of living things?
3. How is the classification system used by scientists today different from the classification system developed by Linnaeus? How is it similar?
4. Describe each of the kingdoms used in the five-kingdom classification system.

Give an example of an organism in each kingdom.

5. How do an autotroph and a heterotroph differ? Give an example of each.
6. How did the knowledge of evolution affect the way organisms are classified?
7. What are the two major jobs of the modern classification system? Explain why these jobs are important.

Critical Thinking and Problem Solving

Use the skills you have developed in this chapter to answer each of the following.

1. **Identifying relationships** Which two of the following three unicellular organisms are most closely related: *Entamoeba histolytica, Escherichia coli, Entamoeba coli?* Explain your answer.
2. **Developing a model** Design a classification system for objects that might be found in your closet. Then draw a diagram that illustrates your classification system.
3. **Making comparisons** In what ways were the classification and naming systems developed by Linnaeus an improvement over previous systems?
4. **Applying concepts** Why is it that scientists do not classify animals by what they eat or where they live?
5. **Relating cause and effect** Explain why advances in technology may change the way organisms are classified.
6. **Applying concepts** Suppose you discovered a new single-celled organism. This organism has a nucleus and a long taillike structure that it uses to move itself through the water in which it lives. It also has a large cup-shaped structure that is filled with a green substance. This green substance is involved in making food from simple substances. In what kingdom

would you place this organism? What are your reasons?

7. **Using the writing process** Some experts estimate that there are more unknown organisms in the tropical rain forests than there are known organisms in the world. These rain forests may be destroyed before the organisms in them can be studied and classified. It is possible that some of these organisms may be helpful in medicine, farming, and industry. Write a script for a television news program protesting the destruction of rain forests. Offer reasons why rain forests should be protected.

Chapter 2 VIRUSES AND MONERANS

SECTION	HANDS-ON ACTIVITIES
2–1 Viruses pages B36–B42 Multicultural Opportunity 2–1, p. B36 ESL Strategy 2–1, p. B36	**Student Edition** ACTIVITY BANK: A Germ That Infects Germs, p. B171 **Activity Book** CHAPTER DISCOVERY: Observing Bacteria, p. B47 **Teacher Edition** Exploring Antibiotics, p. B34d
2–2 Monerans pages B43–B53 Multicultural Opportunity 2–2, p. B43 ESL Strategy 2–2, p. B43	**Student Edition** ACTIVITY (Discovering): Bacteria for Breakfast, p. B44 ACTIVITY (Discovering): Food Spoilage, p. B48 LABORATORY INVESTIGATION: Examining Bacteria, p. B54 ACTIVITY BANK: Yuck! What Are Those Bacteria Doing in My Yogurt? p. B172 **Laboratory Manual** Identifying Bacteria and How They Grow, p. B17 Bacteria That Dine on Vegetables, p. B23 **Product Testing Activity** Testing Yogurts **Teacher Edition** Observing the Growth of Bacteria, p. 34d
Chapter Review pages B54–B57	

OUTSIDE TEACHER RESOURCES
Books
Brock, T.D., D.W. Smith, and M.T. Madigan. *Biology of Microorganisms*, 4th ed., Prentice-Hall.
Hyde, M.D., and E.H. Forsyth. *AIDS: What Does It Mean to You*, Walker.
Jacobs, Francine. *Breakthrough: The True Story of Penicillin*, Dodd.
Nourse, Alan E. *Viruses*, Watts.
Scott, Andrew. *Pirates of the Cell: The Story of Viruses From Molecule to Microbe*, Blackwell.
Singleton, Paul, and Diana Sainsbury. *Introduction to Bacteria*. Wiley.

Audiovisuals
Bacteria, 2nd ed., film or video, EBE
Bacteria, film or video, National Geographic
Bacteria, Fungi, and Viruses, filmstrip with cassette, National Geographic
Microorganisms That Cause Disease, film or video, Coronet

Simple Organisms: Bacteria, rev., film or video, Coronet
Viruses: The Mysterious Enemy, two filmstrips with cassettes, Carolina Biological Supply Co.
Viruses: What They Are and How They Work, film or video, EBE

OTHER ACTIVITIES	MEDIA AND TECHNOLOGY
Student Edition ACTIVITY (Writing): Harmful Microorganisms, p. B37 **Activity Book** ACTIVITY: Discovery of Viruses, p. B51 **Review and Reinforcement Guide** Section 2–1, p. B17	**Transparency Binder** The Bacteriophage Virus **English/Spanish Audiotapes** Section 2–1
Student Edition ACTIVITY (Calculating): Bacterial Growth, p. B44 ACTIVITY (Writing): Benefiting From Bacteria, p. B53 **Activity Book** ACTIVITY: How Small is Small? p. B53 ACTIVITY: Pictorial Riddles, p. B55 ACTIVITY: Viruses and Bacteria, p. B57 ACTIVITY: Cheesemaking, p. B59 ACTIVITY: Get the Number of That Bacterium! p. B61 ACTIVITY: How Large Is Large? p. B63 **Review and Reinforcement Guide** Section 2–2, p. B21	**Courseware** Agents of Infection (Supplemental) **English/Spanish Audiotapes** Section 2–2
Test Book Chapter Test, p. B29 Performance-Based Tests, p. B129	**Test Book** Computer Test Bank Test, p. B35

*All materials in the Chapter Planning Guide Grid are available as part of the Prentice Hall Science Learning System.

CHAPTER OVERVIEW

Viruses are parasitic noncellular particles that invade living cells. When a virus attaches itself to and invades a host cell, it injects its genetic material into the cell. Using the genetic machinery and supplies in the host cell, the genetic material of the virus causes the host cell to manufacture more viruses. After some time, the host cell bursts, and the newly manufactured viruses are freed to invade other cells. Viruses cause infections in living things. Because a virus requires a host cell, experts disagree as to whether viruses are alive.

Monerans, or bacteria, are unicellular (one-celled) organisms.

Bacteria are the oldest forms of life on Earth. They have many different shapes and colors, roles in the environment, and ways of obtaining energy. Some bacteria are autotrophs, which are able to make their own food through processes such as photosynthesis. Other bacteria are heterotrophs.

Many people tend to think of bacteria as organisms that are harmful to themselves and to the environment. This is true only in a few cases. Many bacteria are actually beneficial to people and the environment.

2–1 VIRUSES
THEMATIC FOCUS

The purpose of this section is to introduce students to the structure, function, and life cycles of viruses. Viruses are tiny particles that invade living cells. Viruses are not cells because they cannot perform the life functions of living cells. Because of this, experts disagree as to whether viruses should be called living or nonliving things.

A virus requires a host organism, or living cell, to reproduce. Viruses are considered parasites because they not only require a host cell, they also harm it.

A virus has two basic parts: a core of hereditary material and an outer coat of protein. A virus attaches itself to a host cell to reproduce. After injecting its hereditary material into the host cell, the protein coat is left behind, and the injected genetic material begins to use the supplies in the host cell to manufacture new viruses. After a period of time, the host cell bursts open, releasing these new viruses, which look for new hosts so that the process can begin again.

In human beings, viruses can cause annoying diseases such as colds, as well as fever blisters and warts. They can also cause more serious, even fatal, diseases such as AIDS, measles, influenza, hepatitis, smallpox, polio, encephalitis, mumps, and herpes.

The themes that can be focused on in this section are scale and structure and systems and interactions.

*Scale and structure: Viruses are noncellular and consist of a core of hereditary material surrounded by a protein coat. The protein coat protects the virus and allows it to attach itself to another living cell, or host. The genetic material allows the virus to reproduce after it is injected into a host.

Systems and interactions: Viruses cause a number of diseases. Some of the diseases caused by viruses are more annoying than serious, such as colds. But others, such as AIDS and polio, for example, are extremely serious and can lead to permanent disability or death.

PERFORMANCE OBJECTIVES 2–1

1. List the life functions that viruses are both capable and incapable of performing.
2. Describe the sequence of events in the reproduction of a bacteriophage.
3. List a number of virus-caused conditions or diseases.

SCIENCE TERMS 2–1
virus p. B37
host p. B38
parasite p. B38

2–2 MONERANS
THEMATIC FOCUS

The purpose of this section is to introduce monerans, or bacteria. Thousands of times larger than viruses, and much more complex, bacteria are difficult to find with a light microscope except at its highest magnification.

Bacteria are the oldest life forms and among the most numerous organisms on Earth. Bacteria generally appear in three basic shapes, although their colors are more varied.

Almost all bacteria have a cell wall that serves to surround, support, protect, and shape the cell. Lining the inside of the cell wall is the cell membrane, which serves to control specific substances from leaving or entering the cell. Within the cell membrane is the cytoplasm, or hereditary material of the cell.

Most bacteria are not able to move on their own, although a few kinds possess a structure called a flagella, or whiplike tail, which propels them about in a liquid environment.

Although bacteria are simple organisms, they carry on the same life processes as more advanced organisms do. Most bacteria are heterotrophs that feed on dead things, although some are autotrophs that use light energy or chemical energy to produce food.

Bacteria are both harmful and beneficial to humans and the environment. We could not live without these tiny unicellular organisms.

The themes that can be focused on in this section are evolution, unity and diversity, patterns of change, energy, and stability.

*Evolution: Bacteria were the first life forms on Earth. They are also among the Earth's most numerous organisms. Bacteria can be found in three basic shapes, in many different colors, and in water, air, soil, and the bodies of other organisms.

*Unity and diversity: Some bacteria are harmful to humans and the environment, whereas others are beneficial. Regardless of their contribution, the presence of bacteria allows life as we know it to continue and flourish on Earth. Without bacteria, life processes would grind to a halt.

*Patterns of change: Bacteria perform many of the chemical changes involved in the nutrient cycle. Many plants and animals die each day on the Earth, yet the Earth does not appear cluttered with these remains. Bacteria help to break down, or decompose, dead things into simpler substances for use by other living things.

Energy: Bacteria meet their energy needs in a variety of ways. Some bacteria are autotrophs—producing food for energy from raw materials. Other bacteria are heterotrophs—producing energy from the food they eat. Though bacteria are simple cells, they perform the same life functions as more complex, sophisticated cells do.

Stability: Bacteria help to maintain the supply of nutrients in the environment. The nutrients supplied by bacteria are used directly or indirectly by every living organism on Earth.

PERFORMANCE OBJECTIVES 2–2

1. Identify the major structures in a bacterium cell.
2. Describe the ways in which bacteria obtain energy.
3. Describe the method by which bacteria reproduce.
4. List several ways in which bacteria can be harmful and several ways in which they can be beneficial.

SCIENCE TERMS 2–2
decomposer p. B46
symbiosis p. B48
antibiotic p. B52

Discovery Learning

TEACHER DEMONSTRATIONS MODELING

Exploring Cold Remedies

From your medicine cabinet, collect all the remedies you may have for a cold and then cut out any number of advertisements for the same remedies from magazines. Mount these in an appealing way at the front of the room. Allow students time to examine the products and advertisements.

• What are all these products? (Lead students to respond that they are all cold remedies.)
• Which product does the best job of curing a cold? (Expect any number of responses based on customer loyalty and advertising effectiveness.)
• How does a product for colds "cure" a cold? (The products claim to reduce fever and aches and pains, as well as relieve congestion and general stuffiness.)
• When your cold symptoms start to disappear, is this a sign that you are cured? (Students may tend to suggest yes; however, the correct answer is no.)

Point out that these medications provide relief from symptoms, not a cure. A cure would have to disable the cold virus, and none of the products can disable or eliminate the virus.

Have students note the various ingredients in several products and try to relate them to the activity by noticing how they all go about treating symptoms. In their own words, have students explain the difference between relief and a cure.

Mention that we basically "survive" a cold until we develop antibodies. Remedies simply try to make us more comfortable.

Observing the Growth of Bacteria

1. Obtain four disposable plastic petri dishes that contain nutrient agar.
2. Using a source of nonpathogenic bacteria, such as yogurt or buttermilk, dip a sterile cotton swab into the bacteria source, open the top of one petri dish, and gently rub the swab over the entire surface of the agar. Repeat this procedure, using a new cotton swab for each of the remaining petri dishes.
3. Number the petri dishes.
4. Have volunteers cut out a few disks of filter paper that are 1 cm in diameter.
5. Dissolve antibiotic tablets or capsules in water. Soak the dishes in the antibiotic solution.
6. Using forceps that have been dipped in alcohol, open petri dish 1 and place the disk in the center of the petri dish. Repeat this procedure with petri dish 2.
7. Set petri dishes 3 and 4 aside. They will act as controls.
8. Place all the petri dishes in a warm, dark place for a few days.
9. After a few days, examine the petri dishes for evidence of bacterial growth. If bacteria are present, they will appear as white dots or clumps.
• Where did the bacteria that are growing in the petri dishes obtain their food and water? (From the nutrient agar.)
• What evidence is there that the antibiotics stopped the growth of bacteria? (The clear zone around each antibiotic disk.)
• What was the purpose of petri dishes 3 and 4? (They served as controls.)

CHAPTER 2
Viruses and Monerans

INTEGRATING SCIENCE

This life science chapter provides you with numerous opportunities to integrate other areas of science, as well as other disciplines, into your curriculum. Blue numbered annotations on the student page and integration notes on the teacher wraparound pages alert you to areas of possible integration.

In this chapter you can integrate life science and microbiology (p. 36), language arts (pp. 37, 53), life science and evolution (p. 38), life science and medicine (p. 40), computer science (p. 42), life science and cells (p. 45), life science and ecology (pp. 46, 48, 50), mathematics (pp. 46, 51), earth science and atmosphere (p. 48), and earth science and geology (p. 52).

SCIENCE, TECHNOLOGY, AND SOCIETY/COOPERATIVE LEARNING

Bacteria and other microbes are being used to clean up our world and to help prevent pollution. This process, called bioremediation, has been used for almost 100 years to process sewage in municipal sewage-treatment plants. But in the past 20 years, the use of bacteria has increased dramatically.

The best-known use of bacteria is their ability to break down oil that has been spilled into the environment. Following the *Exxon Valdez* oil spill in Alaska, fertilizers were sprayed on beaches to stimulate native bacteria to break down the spilled oil into harmless byproducts.

Bioremediation is also used in warehouses to consume spilled hazardous materials, in restaurants to consume grease and other odor-causing substances, and in rivers and lakes to help break down pollutants.

One of the most promising uses of bacteria is in the area of pollution prevention. But the challenge to researchers is not only to identify bacteria that consume or break down undesirable substances, but also to find ways to grow and culture them in the laboratory. Researchers must also find a way to present bioremediation to a

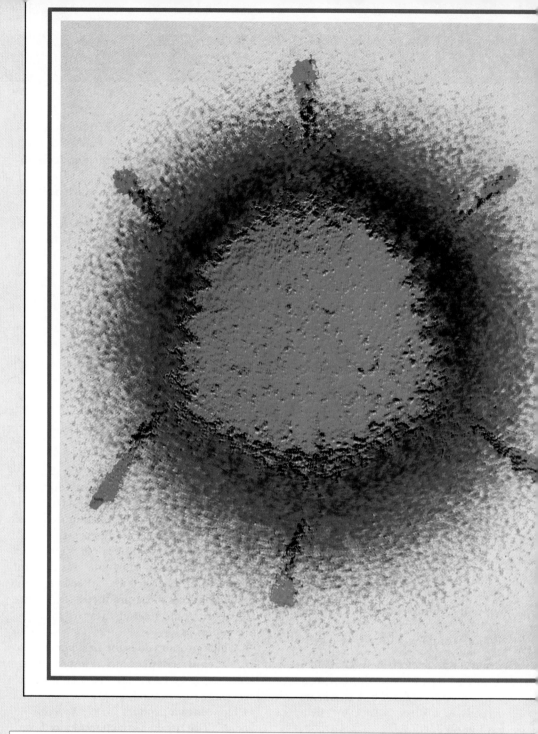

INTRODUCING CHAPTER 2

DISCOVERY LEARNING

▶ *Activity Book*

Begin teaching the chapter by using the Chapter 2 Discovery Activity from the *Activity Book*. Using this activity, students will explore different types of bacteria.

USING THE TEXTBOOK

Have students observe the picture on page B34. Without mentioning dimensions in their answers, have students answer these questions:
- **What does this picture seem to show?** (Responses might include a container of some kind, a steering wheel from a ship, or a cell.)

Have students read the caption on page B35.

Viruses and Monerans

Guide for Reading

After you read the following sections, you will be able to

2–1 Viruses
- List the parts of a virus.
- Describe how a virus reproduces and causes disease.

2–2 Monerans
- Name and describe the parts of a moneran.
- Compare autotrophic and heterotrophic monerans.
- Discuss the helpful and harmful effects of the monerans.

Achoo! And so it begins—the misery of the most common illness—the cold. Colds can be caused by any one of more than 100 kinds of viruses. One such culprit is shown on the opposite page. Because you can be infected by one kind of cold virus one month and another kind of cold virus the next month, you can get several colds every year.

Tired of catching colds? Then take heart, for help is on the way. In 1986, scientists working at the University of Virginia School of Medicine were successful in stopping the attacks of some common cold viruses. The scientists are now seeking a way to stop the cold virus before it can cause an infection. To do so, they have developed disease-fighting substances called antibodies in the laboratory. These antibodies keep the cold virus from invading unsuspecting cells. By acting like a protective shield, they guard the body from invaders.

As you read this chapter, you will learn a great deal about viruses. You will also find out about tiny one-celled organisms called monerans. So turn the page and begin your journey into a strange world of incredibly small but fascinating living things.

Journal *Activity*

You and Your World In your journal, make a list of ten questions you have about diseases. As you study this chapter, see how many of your questions are answered. Devise a plan for finding out the answers to any questions that remain unanswered.

This virus causes one kind of cold. The viruses that cause colds are remarkably small—often little more than 10 billionths of a meter across.

B ■ 35

• **How was this picture taken?** (Students should suggest with a microscope, but mention that in order to see such a tiny cell, a device more powerful than an ordinary microscope must be used.)

Point out that the pictured image of a cold virus is only one of more than 100 different viruses that can cause a cold. Although this virus may not appear particularly menacing to them, if they have the proper receptor sites in their bodies, this strange "landing-craft" virus will land on those sites and attempt to take control of their cells' internal machinery.

Have students read the chapter introduction on page B35.

Mention to students that along with viruses, they will also be learning about bacteria in this chapter and that medications can also help to cure some bacterial diseases. Also, mention that not all bacteria are harmful to people—some are actually helpful.

---

society that harbors many fears about bacteria and the dangers they might cause if they are released into our world.

Cooperative learning: Using preassigned groups or randomly selected teams, have groups complete one of the following assignments:

• Acting as the advertising department of Bacteria, Inc., a firm producing bacteria used in biomedication, create an advertising campaign that informs the public about the positive uses of biomedication in cleaning up the environment. Groups can select one of the following media, or you might randomly assign one medium to each group: magazine, newspaper, television, radio, or billboard.

• Present groups with the following scenario: A large oil company's tanker has run aground off the coast of your community. As a coastal community, you are home to thousands of species of wildlife, and tourism is a major source of community income. The oil tanker's owner has proposed the use of biomedication to clean up the oil spill. You, the city council, must make a decision about the use and possible consequences of biomedication. Each group should produce a chart showing alternatives that are available to the city council and the possible political, economic, social, and environmental consequences of each proposed alternative. An oral presentation of the decision in a mock city-council meeting could be used as a method of evaluation.

See Cooperative Learning in the *Teacher's Desk Reference.*

JOURNAL ACTIVITY

You may want to use the Journal Activity as the basis of a class discussion. Have volunteers read several questions from their lists and then have the class discuss possible answers to each question. Point out that the answers to many of the questions are very small organisms that will be explored in this chapter. Students should be instructed to keep their Journal Activity in their portfolio.

2-1 Viruses

Figure 2–1 *Viruses come in many shapes. Rabies virus is shaped like a thimble (bottom left). Tobacco mosaic virus is rod-shaped (right). Rubella (German measles) virus is spherical (top left). The viruses that cause chicken pox and related diseases are oval (center) and are the largest viruses—as much as several hundred billionths of a meter long.*

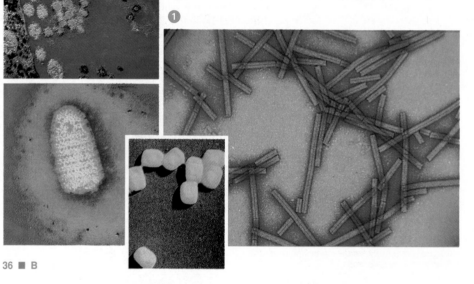

2-1 Viruses

Imagine for a moment that you have been presented with a rather serious problem. A disease is killing one of your country's most important crops. Sick plants develop a pattern of yellow spots on their leaves. Eventually, the leaves wither and fall off, and the plant dies. You must find out what is causing the disease. With this information, it may be possible to save the remaining plants.

You gather some leaves from the sick plants and crush the leaves until they produce a juice. Then you put a few drops of the juice on the leaves of healthy plants. Several days later, you discover yellow spots on the once-healthy leaves. You reason that the cause of the disease can be found in the juice from the sick plants.

You then put the juice through a filter whose holes are so small that not even cells can slip through. You figure that this should take out any disease-causing microorganisms (microscopic organisms) in the juice. In fact, when you use the best light microscope to examine the filtered juice, you find no trace of microorganisms. But to your surprise, the juice still causes the disease in healthy plants! You realize that you must have discovered a germ that is smaller than a cell—too small to be seen even with a microscope. But how could this be, you wonder. Cells are the basic unit of structure and function in living things. All living things are made of cells. There cannot be a living thing smaller than a cell! Or can there?

An additional experiment shows that the disease-causing germ can reproduce in the newly infected plants. This seems to indicate that a living thing is causing the disease. After all, the ability to reproduce is a characteristic of living things. A non-living, sickness-causing substance—a poison, for example—is not able to reproduce itself.

However, the results of other experiments seem to contradict the hypothesis that the disease-causing germ is alive. You discover that the crop-killing germ cannot be grown outside of the plants it infects. This is strange, because all living things can grow by themselves (provided they are given the correct nutrients and environmental conditions, of course). What is going on here, you wonder.

Finally, you decide to use chemical techniques to isolate the germ. After you purify enormous quantities of the juice from the infected plants, you are left with a tiny amount of whitish, needlelike crystals. These crystals show no evidence of being alive. They do not grow, breathe, eat, reproduce, or perform any other life functions. But when you inject the seemingly lifeless crystals into a healthy plant, the plant develops the disease. Clearly the mysterious crystals are the disease-causing germ itself. But what is this germ? And is it alive—or not?

This story is based on real events that started about 100 years ago and unfolded over the next forty years. The disease-causing germ is a **virus.** And whether viruses are alive or not depends on your definition of life. As you read in the story, there are good reasons to think of viruses as living things. And there are equally good reasons to think of viruses as nonliving things.

What Is a Virus?

Viruses are tiny particles that can invade living cells. Because viruses are not cells, they cannot perform all the functions of living cells. For example, they cannot take in food or get rid of wastes. In fact,

Figure 2–2 *Viruses cause a number of diseases in living things. The cherry's leaves are turning yellow and falling off as a result of a virus disease. Why are viruses considered parasites?* ❶

out that this unit is still too large for measuring viruses. Explain that the millimeter unit (which has already been divided into 1000 smaller units) would have to have each of its smaller units further divided into 1000 even smaller units in order to have a tool for measuring the size of viruses.

Be sure students understand that because of their small size, viruses cannot be seen under a light microscope. Viruses can be studied and photographed only with an electron microscope.

● ● ● ●   **Integration**   ● ● ● ●

Use the photographs of various viruses to integrate microbiology into your lesson.

CONTENT DEVELOPMENT

• **What do you think most people think about when they hear the word** *virus?* (Accept all answers. Most students will probably suggest something unfavorable because viruses cause disease.)

• **What are some diseases caused by viruses?** (Accept all answers, but mention that at this point students might suggest some nonviral diseases.)

Point out that some human diseases caused by viruses include the common cold, influenza, poliomyelitis, measles, mumps, chicken pox, hepatitis, genital herpes, and AIDS. There is also some evidence that certain types of cancer might be caused by viruses.

The word *virus* is derived from a Greek word meaning "to poison." Ask students to recall their most recent bout with a virus, particularly one involving stomach upset. This will be an unpleasant recollection, but one that will help students remember the word-root association.

WENDELL STANLEY

In 1935, Wendell Stanley became the first person to isolate a virus that caused tobacco mosaic disease. Stanley ground up more than 1000 kilograms of infected tobacco leaves. From this extract, he was able to obtain crystals of the disease-causing virus. When the crystals were placed in water and then spread on healthy tobacco leaves, the plants became infected with the tobacco mosaic disease. For this work, Stanley was awarded the Nobel prize in chemistry in 1946.

INTEGRATION

MATHEMATICS

Acquired Immune Deficiency Syndrome (AIDS) is a communicable disease caused by a virus. The spread of the AIDS virus has been explosive around the world. Have interested students research current statistics regarding the spread of this virus and the fatalities it has caused. Have students determine a fatality pattern and then extrapolate this pattern forward for the next ten years. The results should be shared with the class.

2–1 (continued)

GUIDED PRACTICE

Skills Development

Skill: Applying concepts

Distribute to each student a peanut in a shell.

• **In what way is the structure of a virus similar to the structure of a peanut?** (Accept all logical answers.)

• **What part of a virus is similar to the part of a peanut that you eat?** (The hereditary material in the core of the virus.)

• **What kinds of hereditary material make up the core of the virus?** (Lead students to suggest DNA or RNA.)

• **What is the function of the hereditary material?** (It controls the production of new viruses.)

• **What virus structure is similar to the shell of a peanut?** (The protein coat surrounding the viral nucleic acid, or the hereditary material.)

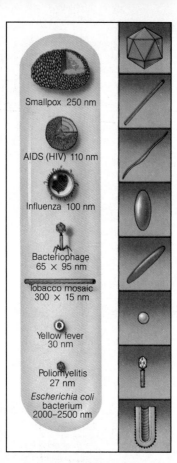

Figure 2–3 *Viruses have a wide variety of shapes and sizes. Viruses are measured in nanometers (nm). A nanometer is one-billionth of a meter. The yellow-green capsule-shaped structure represents a bacterial cell. In general, how do viruses compare to cells in terms of size? In terms of structure?* ❶

Ⓐctivity Bank

A Germ That Infects Germs, p.171

about the only life function that viruses share with cells is reproduction. However, viruses cannot reproduce on their own. They need the help of living cells. The living cells are called **hosts.** Hosts are living things that provide a home and/or food for a **parasite** (PAIR-ah-sight). A parasite is an organism that survives by living on or in a host organism, thus harming it. Because viruses harm their host cells, they are considered to be parasites.

All five kingdoms of living things—plants, animals, fungi, protists, and monerans—are affected by viruses. In fact, experts suspect that all cells are subject to invasion by some kind of virus. It is interesting to note that each type of virus can infect only a few specific kinds of cells. For example, the rabies virus infects only nerve cells in the brain and spinal cord of dogs, humans, and other mammals. So a mammal's skin cells cannot be infected with rabies. And an organism that is not a mammal—such as a frog, plant, mushroom, or protist—cannot get rabies.

The origin of viruses is unknown. Because viruses need living cells in order to reproduce, it is likely that they appeared after the first cells. Many scientists think that viruses evolved from bits of hereditary material that were lost from host cells. This may mean that a virus is more closely related to its host than it is to other viruses. Thus the virus that causes your cold may be more closely related to you than it is to the virus that causes a plant's leaves to fall off!

Structure of Viruses

A virus has two basic parts: a core of hereditary material and an outer coat of protein. The hereditary material controls the production of new viruses. Like a turtle's shell, the protein coat encloses and protects the virus. The protein coat is so protective that some viruses survive after being dried and frozen for years. The protein coat also enables a virus to identify and attach to its host cell.

With the invention of the electron microscope in the 1930s, scientists were able to see and study the shapes and sizes of certain viruses. Some viruses, such as those that cause the common cold, have 20 surfaces. Each surface is in the shape of a triangle that has equal sides. Other viruses look like fine

CONTENT DEVELOPMENT

Point out that a typical virus is composed of a core of hereditary material (nucleic acid) surrounded by a protein coat. The protein coat protects the core.

A bacteriophage is a more sophisticated type of virus that infects bacteria. A bacteriophage attaches its tail to a living cell, then injects its hereditary material, and sheds its protein coat. Once inside the cell, the hereditary material takes control of the cell and produces new bacteriophages, not allowing the cell to complete its normal functions. Eventually, the cell bursts open, and many bacteriophages are released. These bacteriophages then infect nearby bacteria.

● ● ● ● **Integration** ● ● ● ●

Use the discussion about the origin of viruses to integrate concepts about evolution into your lesson.

threads. Still others resemble tiny spheres. There are even some that look like miniature spaceships.

Reproduction of Viruses

In order to understand how a virus reproduces and causes disease, it might be helpful to examine the activities of one kind of virus known as a bacteriophage (bak-TEER-ee-oh-fayj). A bacteriophage is a virus that infects bacteria (singular: bacterium). In fact, the word bacteriophage means "bacteria eater." Bacteria are unicellular (one-celled) microorganisms that belong to the kingdom Monera.

In Figure 2–4, you can see how a bacteriophage (virus) attaches its tail to the outside of a bacterium. The bacteriophage quickly injects its hereditary material directly into the living cell. The protein coat is left behind. Once inside the cell, the bacteriophage's hereditary material takes control of all of the bacterium's activities. As a result, the bacterium is no longer in control. The bacterium begins to produce new bacteriophages rather than its own chemicals.

Figure 2–4 *The electron micrograph shows a bacterium under attack by numerous bacteriophages (inset). What stage in the diagram does this represent? What are the events in the life cycle of a bacteriophage?* ❷

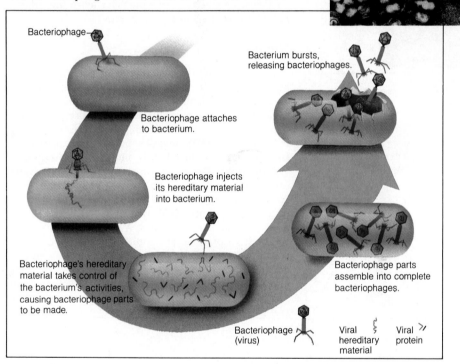

Bacteriophage

Bacterium bursts, releasing bacteriophages.

Bacteriophage attaches to bacterium.

Bacteriophage injects its hereditary material into bacterium.

Bacteriophage's hereditary material takes control of the bacterium's activities, causing bacteriophage parts to be made.

Bacteriophage parts assemble into complete bacteriophages.

Bacteriophage (virus)   Viral hereditary material   Viral protein

REINFORCEMENT/RETEACHING

▶ *Activity Book*

Students who need practice on the concept of viral discoveries should complete the chapter activity called Discovery of Viruses. They will explore the contributions made by several prominent scientists to microbiology.

ENRICHMENT

You might wish to point out that because scientists do not agree on whether viruses are living things, viruses are not classified as living things are. Traditionally, viruses have been classified according to size, shape, and the host they infect. Generally, the outer protein coats of viruses may be many-sided, spiral-shaped, or a combination of these two shapes. Viruses are also classified according to whether they infect the cells of plants, bacteria, or animals including humans. Recently, scientists have been using chemical and structural characteristics to classify viruses. For example, those viruses that contain DNA are placed in one group and those containing RNA in another. Structural characteristics are then used to subdivide viruses into smaller groups.

Viruses vary in size from approximately 20 to 400 nanometers. A nanometer is one billionth of a meter. The tobacco mosaic virus is about 300 nanometers in length, whereas the virus that causes polio is about 20 nanometers in diameter.

HOW VIRUSES ENTER A CELL

Although bacteriophages typically inject their DNA, or hereditary material, into a host cell, not all viruses invade a host cell in this manner. Many animal viruses enter the host by binding to a cell membrane, inducing the host to take in the virus particle. Other animal viruses, such as those responsible for rabies, AIDS, and influenza, leave the host cell through budding.

Figure 2–5 *Like many animal viruses, influenza viruses escape from their host cell by forming tiny bubbles at its surface.*

Figure 2–6 *A bacteriophage looks like a miniature spaceship. What are the main parts of a bacteriophage? What kind of cell is host to a bacteriophage?* ❶

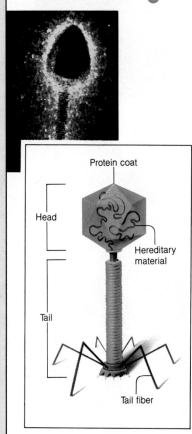

Protein coat

Head

Hereditary material

Tail

Tail fiber

Soon the bacterium fills up with new bacteriophages, perhaps as many as several hundred. Eventually, the bacterium bursts open. The new bacteriophages are released and infect nearby bacteria.

Not all viruses act like a bacteriophage. Some viruses keep their protein coat when they enter their host cell. Others, including some bacteriophages, simply join their hereditary material to that of the host cell and cause no immediate effects. A few cause disease by changing the behavior of their host cell and making it into a cancer cell. And many viruses do not cause their host cell to burst. Instead, a newly made virus particle causes the host cell to produce a tiny bubblelike structure on its edge. As you can see in Figure 2–5, this bubble eventually pinches off the side of the cell, carrying the virus with it.

Although the details of the "life" cycle vary from virus to virus, the basic pattern is the same for all viruses. **First, a virus gets its hereditary material into the host cell. Then the host cell makes more virus particles. Finally, the virus particles leave the original host cell and infect new hosts.**

Viruses and Humans

Viruses cause a large number of human diseases. Some of these diseases—such as colds, fever blisters, and warts—are simply annoying and perhaps a bit painful. Others are serious and can cause permanent damage or even death. Among the diseases caused ❶ by viruses are AIDS, measles, influenza, hepatitis, smallpox, polio, encephalitis, mumps, and herpes.

Much of the research on viruses has concentrated on ways of preventing and treating viral diseases. However, researchers have found ways of using viruses to help humans. Weakened or killed viruses are used to make some vaccines. A vaccine stimulates the body to produce antibodies (substances that prevent infection). Viruses can also be used to wage germ warfare on disease-causing bacteria and insects and on other agricultural pests. In the 1950s, a virus was used to control the population of rabbits in Australia. The virus killed about 80 percent of the

---

rabbits, enabling people to reclaim land for livestock and wildlife. Why might viruses be a better weapon against pests than poisons? ❷

Recently, researchers have learned to put hereditary material into viruses. They can then use the viruses to put the hereditary material into cells. In the not-too-distant future, scientists may be able to use viruses to replace faulty information in a person's hereditary material. This could cure diseases such as diabetes, cystic fibrosis, sickle cell anemia, and many other hereditary disorders. Scientists may also be able to use viruses to improve crops. For example, corn plants might be "infected" with hereditary material that enables them to make their own fertilizer or resist pests.

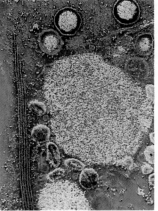

Figure 2–7 *The population of rabbits in Australia was brought under control with the help of the virus that causes the rabbit disease myxomatosis. Why are viruses good for controlling pests? ❸ Can you think of some situations in which using viruses for pest control might be a bad idea?*

Figure 2–8 *The red circles are herpes viruses inside a human cell. The colors that you see do not actually exist in real life. They are added to photographs to make structures easier to see.*

2–1 Section Review

1. How does a virus reproduce? How does this relate to how the virus causes disease?
2. Would you classify viruses as living or nonliving? Explain.
3. What is a bacteriophage?
4. Describe the structure of a virus.

Connection—*Technology*

5. Why were scientists unable to study the structure of viruses until after the electron microscope was invented?

Integration
① Computer Science

CONNECTIONS

COMPUTER VIRUSES: AN ELECTRONIC EPIDEMIC

Students who own a computer or students who have used a computer might be interested to learn of another kind of virus—the computer virus. Even though the scenario depicted in this feature is fictitious, computer viruses do exist and have caused massive losses of information. Have interested students perform additional research into computer viruses. Students might choose to uncover authentic examples of computers and computer networks infected by viruses, or they might choose to discover how large corporations or individual owners can protect themselves from these damaging viruses.

If you are teaching thematically, you may want to use the Connections feature to reinforce the themes of evolution, patterns of change, scale and structure, and unity and diversity.

Integration: Use the Connections feature to integrate computer science into your science lesson.

CONNECTIONS

Computer Viruses: An Electronic Epidemic ①

"Shuttle to mission control. Do you read me?"

"We read you loud and clear. Commence data transmission." The computer screen flickered with changing numbers and words as mission control began to receive astronomical information from the Space Shuttle. Suddenly, the screen went blank.

Before the startled technician could adjust the computer's controls, a familiar television character appeared where the data should have been.

"Cookie!" it demanded. The technician stared at the screen, horrified. She then tried to type instructions. But the computer did not respond.

"No cookie, no play. Bye-bye." With these words, the figure on the screen disappeared. And so did all the hard-won information from the Space Shuttle. Within only a few seconds, a computer virus had undone years of work!

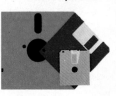

Although all the events in this story are imaginary, they are not impossible. Because many activities of modern life involve computers,

computer viruses are a matter of great concern. A computer virus is not really a virus. It is simply a program, or set of instructions to a computer. Computer viruses do not invade cells or cause diseases. They do not harm living things—at least not directly. They are made by humans and are definitely not alive. And the information they contain is in the form of magnetic or electric signals.

Some computer viruses, such as the one in the story you just read, are carried via telephone wires. Others are transmitted by computer diskettes that are "infected" with the virus program. If such diskettes are used, the virus is stored in the computer. The virus program then causes the computer to copy the computer virus onto every diskette it contacts in the future!

Don't be a victim of a virus! Whether biological or computer, viruses are enough to make you sick. A word to the wise is sufficient: Protect yourself from getting viruses and spreading them!

42 ■ B

TEACHING STRATEGY 2-2

FOCUS/MOTIVATION

Collect some common items that can be used to model some basic shapes of bacteria. For example, to illustrate spherical bacteria (cocci), use beads, marbles, or dried peas. Short pieces of dowel rods or frankfurters can be used to represent rodlike bacteria (bacilli). To make a model of spiral bacteria (spirilla), twist a pipe cleaner into a coil around a pencil. Display the models of the bacteria. Then have students use objects from the classroom to create their own models.

CONTENT DEVELOPMENT

Point out that the terms *moneran* and *bacteria* can be used interchangeably in science. Because all monerans are now considered to be bacteria, either term can be used. Most people, however, are more familiar with the term *bacteria* than the term *moneran,* so the term *bacteria* is used most often.

Bacteria are the oldest life forms on Earth and among the most numerous. Bacteria can be found above Earth's surface in the atmosphere, on the surface of the Earth, and far beneath Earth's surface.

Bacteria cells do not all look alike. Some are spherical, some are rodlike, and some are coiled, whereas others are completely shapeless. Bacteria cells also appear in many colors—ranging from reds and yellows to blues and violets. Regardless of their shape or color, bacteria cells are different from all other cells because bacteria lack a nucleus and certain other cell structures.

2-2 Monerans

Monerans are tiny organisms that consist of a single cell. As you learned in Chapter 1, moneran cells are different from all other cells because they lack a nucleus and certain other cell structures. At one time, the term bacteria was used to refer to only certain kinds of monerans. Other monerans were known as blue-green algae. Blue-green algae are now known as cyanobacteria, or blue-green bacteria. (The prefix *cyano-* means blue.) Because all monerans are now considered to be bacteria, the terms bacteria and monerans are used interchangeably. We will usually use the term bacteria in this chapter because it is more familiar and used more often than the term monerans. As you read this chapter, keep in mind that monerans and bacteria are the same thing.

As you may recall from Chapter 1, bacteria are the oldest forms of life on Earth. The first bacteria appeared about 3.5 billion years ago. Bacteria were Earth's only living things for about 2 billion years.

Bacteria are among the most numerous organisms on Earth. Scientists estimate that there are about 2.5 billion bacteria in a gram of garden soil. And the total number of bacteria living in your mouth is greater than the number of people who have ever lived!

As you might expect of such a large, ancient group, bacteria are quite varied. They may be rod-shaped like a medicine capsule, round as a marble, coiled like a stretched spring, round and stalked like a candied apple on a stick, or completely shapeless. They come in colors ranging from reds and yellows

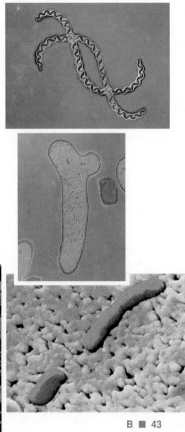

Figure 2–9 *The most common shapes of bacteria are spheres, rods, and spirals. However, some bacteria—such as Y-shaped Bifidiobacterium—have unusual shapes that do not fit into any of these categories.*

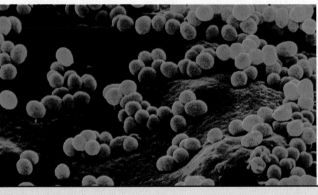

B ■ 43

2-2 Monerans

MULTICULTURAL OPPORTUNITY 2-2

Many bacteria are responsible for diseases that affect humans and animals. These diseases are most common in areas of poverty and poor sanitary conditions—areas in which health care is the poorest. Ask students to research some of the common bacterial-borne diseases including anthrax, brucellosis, salmonellosis, staphylococcosis, streptococcosis, and tuberculosis.

ESL STRATEGY 2-2

To help students assimilate terms associated with viruses and monerans, ask them to make a worksheet with two columns titled Viruses and Monerans.

Have students place each of the following terms in the appropriate category: AIDS, antibodies, antibiotics, bacteria, capsule, cell membrane, cell wall, cytoplasm, decomposers, electron microscope, endospore, flagellum, hereditary core, host cell, parasites, penumonia, protein coat, symbiosis, tiny disease-causing particles, tuberculosis, unicellular, vaccines.

Suggest that students exchange worksheets with a partner and check each other's lists.

GUIDED PRACTICE

▶ *Laboratory Manual*
Skills Development

Skills: Making observations, making comparisons

At this point you may want to have students complete the Chapter 2 Laboratory Investigation in the *Laboratory Manual* called Identifying Bacteria and How They Grow. In this investigation students will observe how bacteria grow and will identify their shapes.

REINFORCEMENT/RETEACHING

▶ *Activity Book*

Students who need practice on the sizes of viruses and monerans should complete the chapter activity called How Small Is Small? Students will relate the sizes of viruses and monerans to everyday objects.

ACTIVITY DISCOVERING

BACTERIA FOR BREAKFAST

Discovery Learning

Skills: Making observations, making inferences, using scientific instruments

Materials: yogurt, medicine dropper, glass microscope slide, coverslip, microscope

This activity allows students to observe the helpful bacteria in yogurt. The bacteria appear as dark blue capsule-shaped dots against a cloudy, pale blue background. The methylene blue serves to stain the bacteria and thus make them visible.

Have students follow up the discovery they made in this activity by doing the Activity Bank activity called Yuck! What Are Those Bacteria Doing in My Yogurt?

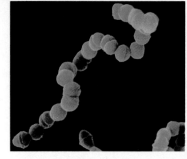
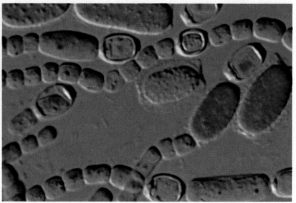

Figure 2–10 *Bacteria, such as many blue-green bacteria (right), may live in groups of cells attached to one another. The name of a bacterium can be a clue to what it looks like. For example, strepto- means chain and -coccus means a spherical bacterium. Why was this bacterium named* Streptococcus *(left)?* ❶

ACTIVITY DISCOVERING

Bacteria for Breakfast

What's inside your food? You might be surprised! Let's take a closer look at yogurt.

1. Add water to some plain yogurt to make a thin mixture.

2. With a medicine dropper, place a drop of the yogurt mixture on a glass slide.

3. With another dropper, add one drop of methylene blue to the slide.

4. Carefully place a coverslip over the slide.

5. Observe the slide under the low and high powers of a microscope.

■ Describe what you see. Why do you think you had to use methylene blue?

to blues and violets. Some bacteria live alone as single cells. Others live in groups of cells that are attached to one another.

Bacteria are found in water, air, soil, and the bodies of larger organisms. In fact, bacteria live almost everywhere—even in places where other living things cannot survive. For example, some bacteria live in volcanic vents at the bottom of the ocean. The temperature of the water in these vents can be as high as 250°C—two and one-half times the temperature of boiling water.

Bacteria are considered the simplest organisms. However, bacteria are more complex than they may appear. Each bacterial cell performs the same basic functions that more complex organisms, including you, perform.

Structure of Bacteria

One of the most noticeable features of a bacterium is the cell wall. See Figure 2–12. The cell wall is a tough, rigid structure that surrounds, supports, shapes, and protects the cell. Almost all bacteria have a cell wall. In some bacteria, there is a coating on the outside of the cell wall. This coating is called the capsule. How might the capsule provide protection for a bacterium? ❷

Lining the inside of the cell wall is the cell membrane. The cell membrane controls which substances enter and leave the cell. Within the cell membrane

---

2–2 (continued)

CONTENT DEVELOPMENT

Point out that a noticeable feature of the structure of a bacterium is its cell wall. The cell wall that surrounds most bacteria serves several functions: It supports, shapes, and protects the cell. In some bacteria, the cell wall is covered by a coating. The coating may be thin and made of a slimy material that allows the bacterium to adhere to surfaces. Or the coating may

form a thick capsule that protects the bacterium from being recognized and engulfed by white blood cells.

The cell membrane lines the inside of the cell wall, regulating which substances enter and leave the cell. Within the cell membrane is the cytoplasm. Cytoplasm forms most of the living material of a cell. In a bacterium, the hereditary material is located in the cytoplasm; it is not enclosed in a nucleus.

Some bacteria are not capable of movement; others are. Bacteria that are capa-

ble of movement have a visible structure called a flagellum. A flagellum is a long, thin, whiplike structure that helps to propel a bacterium from place to place if it is in a watery environment.

• **How are bacterial cells different from your cells?** (Bacterial cells have a cell wall.)

• **Do you think all bacteria cause disease?** (Students should infer that if that were the case, we would all be sick most of the time because there are enormous numbers of bacteria in the world.)

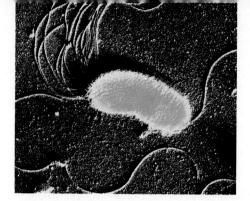

Figure 2–11 *The long, thin, whiplike structures on this soil bacterium are flagella. What is the function of flagella?* ③

is the cytoplasm. The cytoplasm is a jellylike mixture of substances that makes up most of the cell.

Unlike most other cells, the hereditary material of bacteria is not confined in a nucleus. (A nucleus is a membrane-enclosed structure that can be thought of as the "control center" of a typical cell.) ① In other words, there is no membrane separating the hereditary material from the rest of the cell in monerans.

Many bacteria are not able to move on their own. They can be carried from one place to another by air and water currents, clothing, and other objects. Other bacteria have special structures that help them move in watery surroundings. One such structure is a flagellum (flah-JEHL-uhm; plural: flagella). A flagellum is a long, thin, whiplike structure that propels a bacterium through its environment. Some bacteria may have many flagella.

Life Functions of Bacteria

Bacteria have more different ways of getting the energy they need to live than any other kingdom of organisms. In fact, bacteria obtain energy in more ways than all of the other kingdoms combined. Like most other organisms, many bacteria need oxygen in order to get energy from food. Other bacteria can thrive without oxygen. And still other kinds of bacteria will die if they are exposed to oxygen.

Many bacteria are heterotrophs. Recall from Chapter 1 that a heterotroph cannot make its own food. A heterotroph gets energy by eating food,

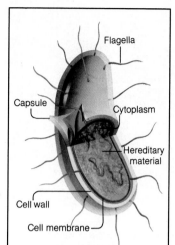

Flagella
Capsule
Cytoplasm
Hereditary material
Cell wall
Cell membrane

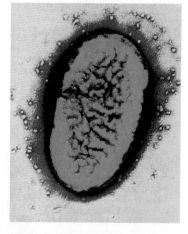

Figure 2–12 *The diagram shows the structure of a typical bacterium. Can you locate the flagella, capsule, cytoplasm, and genetic material on the photograph of the whooping cough bacterium?* ④

● ● ● ● **Integration** ● ● ● ●

Use the discussion about the nucleus of a cell to integrate life science concepts into your lesson.

INDEPENDENT PRACTICE

▶ *Activity Book*

Students who need practice identifying the various structural components of viruses and bacteria should complete the chapter activity Viruses and Bacteria.

GUIDED PRACTICE

Skills Development

Skills: Making observations, making comparisons

At this point have students complete the in-text Chapter 2 Laboratory Investigation: Examining Bacteria. In this investigation students will examine where bacteria are found.

REINFORCEMENT/RETEACHING

Remind students that bacteria are the simplest and oldest type of cells. The term bacteria is used to refer to all members of the kingdom Monera. In bacterial cells, the hereditary material is located in an area of the cytoplasm rather than enclosed in a nucleus. Like plants, bacteria have cell walls that form their outermost boundaries. Although some bacteria have flagella that aid in locomotion, many are unable to move on their own.

Figure 2–13 *If you could look at this scene with a powerful microscope, you would discover bacteria busily breaking down the remains of the dead tree. Why are some bacteria called decomposers?* ❶

usually other organisms. Some bacteria feed on living organisms. These bacteria are parasites. As you just learned, parasites are organisms that live and feed either inside or attached to the outer surface of a host organism, thus harming the host. Such bacteria ❶ cause infections in people, animals, and plants. Other bacteria feed on dead things. These bacteria are **decomposers.** Decomposers break down dead organisms into simpler substances. In the process, they return important materials to the soil and water.

Some bacteria are autotrophs. An autotroph, you will recall, makes its own food. Some food-making bacteria use the energy of sunlight to produce food. Other bacteria use the energy in certain substances that contain sulfur and iron to make food. The nauseating "sulfur" smell of mud flats or rotting food is due to the action of such bacteria.

When food is plentiful and the environment is favorable, bacterial cells grow and then reproduce by dividing into two cells. Under the best conditions, most bacteria reproduce quickly. Some types can double in number every 20 minutes. At this rate, after about 24 hours the offspring of a single bacterium would have a mass greater than 2 million ❷ kilograms, or as much as 2000 mid-sized cars! In a few more days, their mass would be greater than that of the Earth. Obviously, this does not happen. Why do you think this is so? ❷

When food is scarce or conditions become unfavorable in other ways, some bacteria form a small internal resting cell called an endospore. As you can see in Figure 2–15, an endospore consists of hereditary material, a small amount of cytoplasm, and a

Figure 2–14 *Bacteria reproduce by splitting into two cells (left). This process increases the number of bacteria. There are also processes that produce new bacteria but do not increase the number. An old bacterium is changed into a new one when genetic material is transferred via a special tube from one bacterium to another (right).*

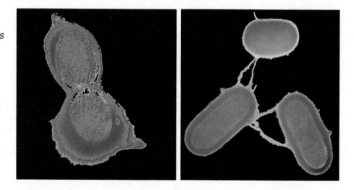

thick protective outer coat. An endospore can survive long periods in which the environment is not suitable for bacterial growth. Some endospores can survive being touched by disinfectant chemicals, blown through the atmosphere, frozen in polar ice, baked in the desert sun, boiled for an hour, and bombarded with powerful radiation. When environmental conditions improve, the endospores burst out of their protective coats and develop into active bacteria. Can you now explain why bacteria are found all over the world—and why some bacteria are extremely hard to kill? 3

Bacteria in Nature

Most types of bacteria are not harmful and do not cause disease. In fact, many types of bacteria are helpful to other living things and perform important jobs in the natural world.

FOOD AND ENERGY RELATIONSHIPS **Bacteria are an essential part of the food and energy relationships that link all life on Earth.** As you learned earlier, some bacteria (decomposers) break down dead materials to form simpler substances. These simpler substances can be used by autotrophs—such as green plants and blue-green bacteria—to make food. Small heterotrophs (organisms that cannot make their own food) such as certain protists and tiny animals feed on plants and blue-green bacteria. These small heterotrophs are food for large heterotrophs, which are

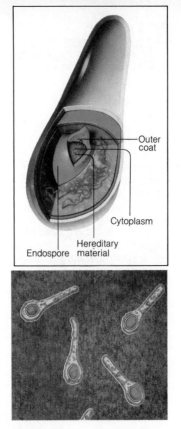

Figure 2-15 *The bacterium that causes the disease tetanus, or lockjaw, forms endospores. The structure of an endospore is shown in the diagram. What is the function of an endospore?* 4

Figure 2-16 *Flamingoes feed on blue-green bacteria and small organisms, which they filter from the water with their beaks. The structure of their filtering beak causes flamingoes to do "headstands" when feeding. Colored substances in the blue-green bacteria give flamingoes their pink color.*

B ■ 47

• **What is a heterotroph?** (A heterotroph is an organism that cannot produce its own food but obtains energy from the food it eats.)
• **Is a human an example of an autotroph or a heterotroph?** (Heterotroph.)
• **Are bacteria autotrophic or heterotrophic?** (Some bacteria are autotrophic; others are heterotrophic.)

● ● ● ● **Integration** ● ● ● ●

Use the discussion of the reproductive rate of bacteria to integrate mathematics into your life science lesson.

ENRICHMENT

▶ *Activity Book*

Students will be challenged by the Chapter 2 activity called Get the Number of That Bacterium! to determine a numerical value for various species of bacteria.

GUIDED PRACTICE

▶ *Laboratory Manual*
Skills Development
Skills: Making observations, recording data, making comparisons

At this point you may want to have students complete the Chapter 2 Laboratory Investigation in the *Laboratory Manual* called Bacteria That Dine on Vegetables. In this investigation students will explore the conditions required for the growth of bacteria.

CONTENT DEVELOPMENT

Remind students of the differences between the terms *autotroph* and *heterotroph*.
• **What do the terms *atuotroph* and *heterotroph* refer to?** (The terms refer to the way an organism obtains food, or energy.)
• **What is an autotroph?** (An autotroph is an organism that is able to produce its own food from raw materials.)

Discovery Learning

Skills: Making observations, inferring

Materials: 2 small glass tumblers, milk

This activity allows students to observe and compare the effects of heat and refrigeration on milk spoilage. Students should observe and note that the contents of both glasses differ. The glass of milk that was kept cold shows no visible signs of deterioration. The glass of milk which remained warm for several days, however, shows signs of deterioration both by sight and by smell.

The changes were caused by the action of bacteria (specifically, by the bacteria breaking down components of the milk for food and releasing waste products such as curd-causing lactic acid into the milk). Such changes can be prevented by refrigeration, which slows the growth of bacteria.

2–2 (continued)

CONTENT DEVELOPMENT

Not all types of bacteria are harmful to living things. Point out that many types of bacteria are helpful to living things.

• **Why are bacteria an essential part of the food and energy relationships of Earth?** (Bacteria break down dead material into simpler substances, which are used by autotrophs to make food. Small heterotrophs feed on these autotrophs. Small heterotrophs are food for large heterotrophs. When heterotrophs and autotrophs die, they are food for decomposing bacteria. The cycle continues in this fashion without ceasing.)

• **Name one example of how bacteria produce oxygen for our environment.** (The food-making process of blue-green bacteria produces oxygen as well as food.)

Food Spoilage

1. Obtain two small glass tumblers.

2. Place a small amount of milk in each glass.

3. Place one glass of milk in the refrigerator. Place the other glass of milk in a warm place.

4. After several days, examine both glasses. How are the contents of the glasses different?

■ What caused these changes?

■ How can such changes be prevented?

Figure 2–17 *Certain sea squirts are involved in a symbiosis with food-making bacteria. What is symbiosis?* ❶

48 ■ B

---

in turn food for still larger heterotrophs. When heterotrophs and autotrophs die, they become food for decomposers, such as certain bacteria. Thus the cycle continues.

OXYGEN PRODUCTION The food-making process of blue-green bacteria produces oxygen as well as food. ❶ Billions of years ago, this resulted in a dramatic change in the composition of the Earth's atmosphere. The percentage of oxygen gas increased from less than 1 percent to about 20 percent. This made it possible for oxygen-using organisms such as protists, plants, animals, and fungi to evolve.

CHANGING ENVIRONMENTS Bacteria continue to change environments in ways that make them suitable for other organisms. For example, bacteria are among the first organisms to grow on the bare rock created by the action of volcanoes. As they grow, the bacteria break down the rock and help create soil. Soon small plants can take root in the developing soil. The bacteria and small plants provide food and homes for tiny animals, fungi, and protists. These new arrivals further change the environment, making it possible for larger organisms to survive. Eventually, a variety of large and small organisms live on what was once bare, lifeless rock.

SYMBIOSES Some bacteria help other organisms ❷ by forming a partnership, or **symbiosis** (sihm-bigh-OH-sihs; plural: symbioses). Symbiosis is a relationship in which one organism lives on, near, or even inside another organism and at least one of the organisms benefits. Here is an example. Food-making

---

• **Name one example of how bacteria change the environment in ways that are beneficial to other organisms.** (Responses might include bacteria are the first organisms to grow on the bare rock created by volcanoes. They break down the rock and create soil, which allows plants to root. The bacteria and small plants provide food and homes for tiny animals, who make it possible for larger organisms to survive.)

• **Describe an example of a symbiotic relationship between two organisms.** (Accept all reasonable responses. Possible responses include certain bacteria live in, and are protected by, the stomach of a cow. In return, the bacteria help the cow digest plant materials that otherwise would not be able to be digested.)

bacteria live in the bodies of sea squirts and deep-sea tube worms. The bacteria are protected by the body of their host. In return, they provide their host with food. In another example of symbiosis, certain bacteria that live in the intestines of animals such as cows, termites, horses, and humans break down plant cell walls. This enables the host to digest plant materials such as wood, grass, and fruit. Bacteria called nitrogen-fixing bacteria turn nitrogen gas, which plants cannot use, into nitrogen compounds that plants can use to make biologically important substances. Nitrogen-fixing bacteria also help replace the nitrogen compounds in the soil. Without such nitrogen-fixing bacteria, most nitrogen compounds in the soil would be quickly used up and plants could no longer grow. Some nitrogen-fixing bacteria live as individual cells in soil. Others form strandlike colonies in water. Still others live in lumps on the roots of plants such as alfalfa, soybean, and clover. See Figure 2–18.

Bacteria and Humans

Bacteria help humans in many ways. Bacteria are involved in the production of food, fuel, medicines, and other useful products. Some are used in industrial processes. Others help break down pollutants, which are substances such as waste materials or harmful chemicals that dirty the environment.

Although most bacteria are either helpful or harmless, a few can cause trouble for humans. The trouble comes in a number of forms. Some harmful bacteria spoil food or poison water supplies. Others damage property or disrupt manufacturing processes. Still others cause diseases in people, pets, livestock, and food crops.

FOOD Many food products, especially dairy products, are produced with the help of bacteria. Bacteria (and products made by bacteria) are used to make cheeses, butter, yogurt, sour cream, pickles, soy sauce, vinegar, and high-fructose corn syrup. Nitrogen-fixing bacteria provide substances that crops need to grow. In flooded fields such as those used to grow rice, nitrogen-fixing blue-green bacteria fertilize the crops naturally. But bacteria do more

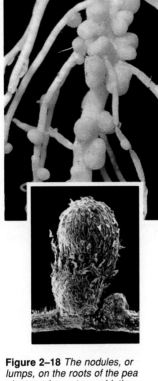

Figure 2–18 *The nodules, or lumps, on the roots of the pea plant are home to symbiotic nitrogen-fixing bacteria (top). If you were to break open a nodule such as this, you would discover the bacteria housed inside (bottom). How does this symbiosis help the pea plant and the bacteria?* 2

Activity Bank

Yuck! What Are Those Bacteria Doing in My Yogurt?, p. 172

B ■ 49

ANNOTATION KEY

Answers

1 Symbiosis is a close relationship between two species in which at least one species benefits. (Applying definitions)

2 The pea plant provides a protective environment for the nitrogen-fixing bacteria to live, and the nitrogen-fixing bacteria replace nitrogen compounds in the soil that are needed by the pea plant. (Relating cause and effect)

Integration

1 Earth Science: Atmosphere. See *Exploring Planet Earth*, Chapter 1.

2 Life Science: Ecology. See *Ecology: Earth's Living Resources*, Chapter 1.

ENRICHMENT

• **Why are flamingos pink?** (Answers will vary.)

Explain that flamingos are pink because of the bacteria they eat. The blue-green bacteria they eat contain a number of red and orange carotenoid pigments. The flamingo's body uses these pigments to color its feathers.

A flamingo's color plays an important role in reproduction. A captive flamingo will become whitish in color and fail to breed unless carotenoid pigments are added to its diet.

The crop milk that flamingos feed to their chicks is rich in a red carotenoid pigment. This fact probably explains the folklore that flamingos feed their blood to their offspring.

● ● ● ● **Integration** ● ● ● ●

Use the discussion of blue-green bacteria's production of oxygen to integrate earth science concepts into your lesson.

Use the discussion of symbiosis to integrate ecology into your lesson.

INDEPENDENT PRACTICE

▶ *Activity Book*

Students who need practice on the relationship of bacteria and food should complete the chapter activity called Cheesemaking. In this activity students explore the role of bacteria in the manufacture of various kinds of cheeses.

Figure 2–19 *Some bacteria harm humans and livestock when they grow in and poison water supplies.*

than help to make foods. They can serve as the food itself. Blue-green bacteria, which may grow in water as large masses of strands, have long been used as food. True, blue-green bacteria might not sound very appetizing, but they are quite nutritious. They are about 70 percent protein, rich in vitamins and other nutrients, and easy to digest.

Some helpful bacteria break down food to make tasty or useful products. For example, one kind of helpful bacterium breaks down milk to produce yogurt. Unfortunately, harmful bacteria may also break down food, making smelly, bad-tasting, or even poisonous products. In other words, some harmful bacteria may cause food to spoil.

Food spoilage can be prevented or slowed down by heating, drying, salting, cooling, or smoking foods. Each of these processes prevents or slows down the growth of bacteria. For example, milk is heated to 71°C for 15 seconds before it is placed in containers and shipped to the grocery or supermarket. This process, called pasteurization, destroys most of the bacteria that would cause the milk to spoil quickly. Heating and then canning foods such as vegetables, fruits, meat, and fish are also used to prevent bacterial growth. But if the foods are not sufficiently heated before canning, bacteria can grow inside the can and produce poisons called toxins. Toxin-producing bacteria may also produce a gas

Figure 2–20 *Although the 1990 oil spill in Huntington Beach, California, was not particularly large, it still took a great deal of effort to clean it up. Oil-eating bacteria are now being used on an experimental basis to help clean up such spills.*

2–2 (continued)

GUIDED PRACTICE

Skills Development

Skill: Identifying relationships

Divide students into groups of four to six students per group. Instruct each group to write down as many examples as possible of how bacteria benefit humans and the environment. Also, instruct groups to write down as many examples as possible of how bacteria are harmful to

humans and the environment. After five or ten minutes, have volunteers from each group discuss their results with the class.

ENRICHMENT

▶ *Activity Book*

Students will be challenged by the Chapter 2 activity called How Large Is Large? to explore the mathematical system of scientific notation.

CONTENT DEVELOPMENT

Remind students that bacteria can be both helpful and harmful to humans and the environment.

• **Name several ways in which bacteria are beneficial to food.** (Responses might include many dairy products are produced with the help of bacteria, and bacteria replenish nitrogen supplies in the soil, which helps crops grow.)

• **Name several ways in which bacteria are harmful to food.** (Students might sug-

that causes the can to bulge. You should never eat food from such a can.

Harmful bacteria can also spoil water supplies. When environmental conditions are changed by pollution or other causes, blue-green bacteria and other bacteria in a body of water may multiply rapidly. The huge numbers of bacteria form an ugly, often smelly, scum and poison the water. Humans and livestock can get very sick from drinking the poisoned water.

FUEL Certain bacteria break down garbage such as fruit rinds, dead plants, manure, and sewage to produce methane. Methane is a natural gas that can be used for cooking and heating. Over millions and millions of years, heat and pressure within the Earth changed the remains of ancient blue-green bacteria, among other organisms, into an oily mixture of chemicals. This mixture of chemicals is called petroleum. Petroleum is the source of heating oil, gasoline, kerosene, and many other useful substances.

ENVIRONMENTAL CLEANUP Bacteria clean up the environment in many ways. Some are used to treat sewage; others cause garbage to decompose, or rot. A few types of bacteria are able to break down the oil in oil spills. Still others break down complex chemicals such as certain pesticides and a few types of plastic.

HEALTH AND MEDICINE Some bacteria help to keep you healthy. For example, bacteria that live inside

ACTIVITY CALCULATING

Bacterial Growth

If a bacterium reproduces every 20 minutes, how many bacteria would there be after one hour? After two hours? After four hours? After eight hours? (Assume all bacteria survive and reproduce.)

If each bacterium is 0.005 mm long, how many bacteria, laid end to end, would equal 1 mm in length?

Using a metric ruler, determine the length of your thumbnail in millimeters. How many 0.005 mm-long bacteria laid end to end would equal the length of your thumbnail?

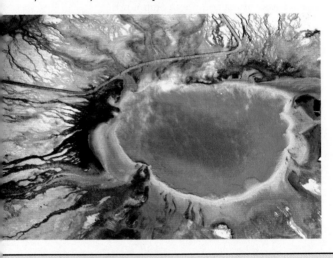

Figure 2–21 *Ancient blue-green bacteria, similar to the modern ones that give the hot springs in Yellowstone National Park their beautiful colors, were the source of the fuel called petroleum.*

B ■ 51

ACTIVITY CALCULATING
BACTERIAL GROWTH

Skills: Making computations, calculator

This activity helps to make students aware of the reproductive capacity of some bacteria. Students should determine that after one hour, there would be 8 bacteria; after two hours, there would be 64 bacteria; after four hours, there would be 4096 bacteria; and after eight hours, there would be 16,777,216 bacteria.

Students should also determine that 200 bacterium, each 0.005 mm in length, would equal 1 mm if laid end to end.

The number of bacteria that would stretch across each student's thumbnail will vary from student to student, depending on the length of the thumbnail.

Integration: Use this Activity to integrate mathematics into your science lesson.

● ● ● ● **Integration** ● ● ● ●

Use the photograph of the Huntington Beach, California, oil spill to integrate ecology into your science lesson.

ENRICHMENT

Students might be interested to know that perhaps the most powerful poison known is produced by the bacterium *Clostridium botulinum.* This bacterium can thrive in improperly sealed canned goods. People who become infected with this bacterium are said to have botulism, or food poisoning. Botulism is a serious, often fatal disease.

gest that bacteria can cause food to spoil and can spoil water supplies, which in turn spoil crops grown with that water.)

• **Name five processes that prevent or slow down food spoilage.** (Heating, drying, salting, cooling, smoking.)

• **Hundreds of years ago, when explorers traveled the oceans, they had to carry large quantities of food. How do you think they kept the food from spoiling?** (Accept all logical responses.)

• **Name several ways in which bacteria benefit the environment.** (Responses might include bacteria are used to treat sewage, decompose garbage, and break down oil from spills and complex chemicals such as certain pesticides and plastics.)

• **Some people are afraid of the use of bacteria. Give an example of what these people believe might happen if a harmful strain of bacteria were put into the environment.** (Answers will vary. Lead students in a discussion of a war tactic known as biological warfare if they have difficulty responding to the question.)

CAREERS

Bacteriologist

Many **bacteriologists** specialize in identifying unknown microorganisms from the thousands of known types of bacteria. Others try to devise methods to combat harmful bacteria. Still others study how disease-causing bacteria may be spread in our environment.

For further information on this career, write to the American Society for Microbiology, 1325 Massachusetts Avenue NW, Washington, DC 20005.

the human intestines help you to digest food. They also make vitamins that the body cannot make on its own. And they may help to prevent disease by crowding out harmful bacteria.

Certain helpful bacteria produce substances that are used to fight disease. Some bacteria naturally produce chemicals that have been found to destroy or weaken other bacteria. Such chemicals are known as **antibiotics.** Why would the production of antibiotics be useful to such bacteria? Recently, scientists have developed ways to change the hereditary material in bacteria. By putting new information into the bacteria's hereditary material, scientists cause the bacteria to produce medicines and other useful substances.

Some human diseases caused by bacteria include strep throat, certain kinds of pneumonia, diphtheria, cholera, tetanus, tuberculosis, bubonic plague, and Lyme disease. Bacteria that live in the mouth cause tooth decay and gum disease. Some of these diseases can be prevented by proper hygiene or immunization shots. Others can be treated with antibiotics.

INDUSTRY Helpful bacteria have long been used in the process of tanning leather. Recently, people have begun to use bacteria to extract copper, gold, and other useful and valuable metals from rock. Interestingly, some types of metal-rich rocks, or ores, may have been created by the actions of bacteria that lived millions of years ago. Scientists hope to develop an inexpensive way to use bacteria to make plastics and other complex compounds.

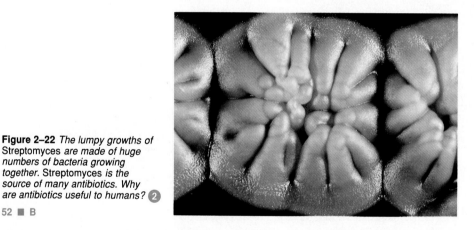

Figure 2–22 *The lumpy growths of* Streptomyces *are made of huge numbers of bacteria growing together.* Streptomyces *is the source of many antibiotics. Why are antibiotics useful to humans?* ②

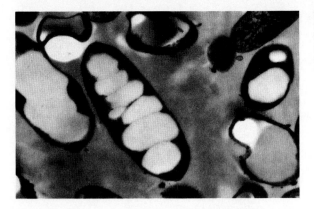

Although the time is not yet right for bacterial plastics, other substances made by bacteria are manufactured commercially. These substances are used for many different purposes—coloring food and cosmetics, tenderizing meat, removing stains, processing paper and cloth, and changing one chemical to another, to name a few.

Harmful bacteria can disrupt industrial processes. They damage leather during the tanning process, ruin paper pulp, and turn fruit juice and wine into vinegar. Some harmful bacteria break down asphalt and thus damage roads, parking lots, and other paved surfaces. And bacteria cause millions of dollars of damage each year to oil-drilling machinery, water and gas pipes, and supplies of petroleum.

ACTIVITY
WRITING

Benefiting From Bacteria

Write a 200-word essay on some of the useful ways bacteria contribute to nature and humankind. Give specific examples. ②

2–2 Section Review

1. Describe two ways in which bacteria are helpful and two ways in which they are harmful.
2. What is an endospore? How does it help bacteria survive?
3. Describe the structure of a bacterium.

Critical Thinking—*Applying Concepts*
4. Many antibiotics work by damaging a bacterium's cell wall. Explain why such antibiotics are not effective against viruses and certain kinds of bacteria that lack cell walls.

Laboratory Investigation

EXAMINING BACTERIA

BEFORE THE LAB

1. If you are using glass petri dishes, be sure they are scrubbed clean and dried before pouring the agar. This agar need not be sterile, as the poured plates should be autoclaved as a unit. If you are using sterile plastic dishes, the nutrient agar must be sterile (autoclaved 20 to 30 minutes, 240 degrees, 15-lb pressure), as these dishes cannot be autoclaved.

2. Provide a clean environment for students' work and stress aseptic procedures. Most important, stress that bacteria are everywhere and that our bodies do a good job of preventing the bacteria from infecting us. A word of caution: Because students are going to be working with concentrated colonies of potentially hazardous organisms, be sure to note that once the dishes have been contaminated, according to the instructions, they are never to be opened again until they are autoclaved. Even turning the dishes over several times could release drops of contaminated condensation from the dish, so handling should be kept to a minimum. Taping the dishes closed is a possibility you might wish to consider.

3. Be sure to provide a germicidal soap for all students to use after handling the incubated dishes.

PRE-LAB DISCUSSION

Have students read the complete laboratory procedure.

• **Can you see any bacteria on your hand?** (No.)

• **What do you think bacteria will look like in the petri dish?** (Dots.)

• **How will you be able to identify the different kinds of bacteria that are present?** (The color of the bacteria and a microscopic analysis of shape would help to identify the bacteria, if needed.)

• **Will there be things other than bacteria in the dish?** (Growths of mold will more than likely appear.)

• **How does your contamination procedure illustrate good scientific method?** (Only one variable is being considered for each dish.)

• **Why is one dish left uncontaminated?** (It is the control to show that there is nothing in the dishes that will cause anything to grow.)

Laboratory Investigation

Examining Bacteria

Problem

Where are bacteria (monerans) found?

Materials *(per group)*

```
5 petri dishes with sterile nutrient agar
glass-marking pencil
pencil with eraser
soap and water
```

Procedure 🧪

1. Turn each petri dish bottom side up on the table. **Note:** *Be careful not to open the petri dish.*
2. With the glass-marking pencil, label the bottom of the petri dishes containing the sterile nutrient agar A, B, C, D, and E. Turn the petri dishes right side up.
3. Remove the lid of dish A and lightly rub a pencil eraser across the agar in the petri dish. Close the dish immediately.
4. Remove the lid of dish B and leave it open to the air until the class period ends. Then close the lid.
5. Remove the lid of dish C and lightly run your index finger over the surface of the agar. Then close the lid immediately.
6. Wash your hands thoroughly. Remove the lid of dish D and lightly rub the same index finger over the surface of the agar. Then close the lid immediately.
7. Do not open dish E.
8. Place all five dishes upside down in a warm, dark place for three or four days.
9. After three or four days, examine each dish. **CAUTION:** *Do not open dishes.*
10. On a sheet of paper, construct a table similar to the one shown here. Fill in the table.

11. Return the petri dishes to your teacher. Your teacher will properly dispose of them.

Observations

1. How many clusters of bacteria appear to be growing on each petri dish? Are there different types of clusters?
2. Which petri dish has the most bacterial growth? Which has the least?

Petri Dish	Source	Description of Bacterial Colonies
A		
B		
C		
D		
E		

Analysis and Conclusions

1. Which petri dish was the control? Explain.
2. Did the dish that you touched with your unwashed finger contain more or less bacteria than the one that you touched with your washed finger? Explain.
3. Explain why the agar was sterilized before the investigation.
4. Design an experiment to show if a particular antibiotic will inhibit bacterial growth in a petri dish.
5. Suggest some methods that might stop the growth of bacteria.
6. What kinds of environmental conditions seem to influence where bacteria are found?

TEACHING STRATEGY

1. Make sure students follow aseptic procedures in their work.

2. Discuss with students the various places in the classroom environment that bacteria could be found. Then have them predict which petri dish will show the most colonies of bacteria.

Study Guide

Summarizing Key Concepts

2–1 Viruses

▲ Viruses are tiny noncellular particles that infect living cells.

▲ Viruses are parasites that probably affect all types of cells. Each type of virus can infect only a few specific kinds of cells.

▲ Viruses cannot carry out any life functions unless they are in a host cell. Viruses may have evolved from bits of hereditary material that were lost from host cells.

▲ The development of the electron microscope made it possible to see viruses.

▲ Viruses consist of a core of hereditary material and an outer coat of protein.

▲ The basic "life" cycle is the same for all viruses. First, a virus gets its hereditary material into the host cell. Then the host cell makes more virus particles. Finally, the virus particles infect new hosts.

▲ Viruses cause a number of diseases that range from annoying to serious to fatal.

▲ Viruses cause human diseases such as colds, chicken pox, rabies, polio, and AIDS.

▲ Weakened or killed viruses are used to make some vaccines.

▲ Viruses are used to kill pests. They are also used to change the hereditary material of cells in specific ways.

2–2 Monerans

▲ Monerans are single-celled organisms that lack a nucleus and many other cell structures.

▲ The terms monerans and bacteria are interchangeable.

▲ Bacteria are the oldest forms of life on Earth. They are also among the most numerous and varied.

▲ Some bacteria live in colonies. A colony is a group of organisms that live together in close association. Some members of these colonies may be specialized for specific functions.

▲ Some bacteria are heterotrophs. Some heterotrophic bacteria are parasites. Others are decomposers.

▲ Some bacteria are autotrophs. Some autotrophic bacteria produce oxygen.

▲ Bacteria fit into the world in many ways. They are involved in many food and energy relationships with other organisms. Some change environments in ways that make them suitable for other organisms—by making soil, oxygen, or nitrogen compounds, for example. Others are involved in symbioses with other organisms.

▲ Some bacteria are helpful to humans. Others are harmful.

Reviewing Key Terms

Define each term in a complete sentence.

2–1 Viruses
virus
host
parasite

2–2 Monerans
decomposer
symbiosis
antibiotic

B ■ 55

OBSERVATIONS

1. Answers will vary, but students should be able to distinguish between the clumps of bacterial colonies. Some may be shaped or colored differently.

2. There should be more colonies on dishes A, B, and C than on the others. Dish E should have the least growth.

ANALYSIS AND CONCLUSIONS

1. Dish E. Explanations will vary.

2. The dish touched by the unwashed finger should show more growth. Washing removed bacteria.

3. To kill any bacteria present on and in the agar.

4. Experiments will vary but should include a control in which no antibiotic was used and an experimental setup in which an antibiotic disk was used.

5. Washing, sterilization, and so on.

6. Bacteria can be found wherever they find food and suitable temperature. Most bacteria require oxygen as well.

GOING FURTHER: ENRICHMENT

Part 1

Students can develop other ideas for contaminating sterile dishes.

Part 2

Students can develop a list of contamination sources, both contact and exposure, at home and at school.

DISCOVERY STRATEGIES

Discuss how the investigation relates to the chapter by asking open questions similar to the following.

• **Why is it suggested that people wash their hands before eating?** (Accept all logical answers—relating, analyzing.)

• **Name five locations in the average home where you would be likely to find relatively high concentrations of bacteria.** (Answers might include door knobs, dirty dishes, corners, kitchen counters, telephones, and so forth—observing, predicting, analyzing.)

• **Name five locations in the average home where you would be likely to find relatively low concentrations of bacteria.** (Answers might include freezers, ovens, lights, clean dishes, furnace, and so forth—observing, predicting, analyzing.)

Chapter Review

ALTERNATIVE ASSESSMENT

The *Prentice Hall Science* program includes a variety of testing components and methodologies. Aside from the Chapter Review questions, you may opt to use the Chapter Test or the Computer Test Bank Test in your *Test Book* for assessment of important facts and concepts. In addition, Performance-Based Tests are included in your *Test Book*. These Performance-Based Tests are designed to test science process skills, rather than factual content recall. Since they are not content dependent, Performance-Based Tests can be distributed after students complete a chapter or after they complete the entire textbook.

CONTENT REVIEW

Multiple Choice
1. b
2. d
3. b
4. d
5. c
6. a
7. d
8. a

True or False
1. F, decomposers
2. T
3. F, an endospore
4. T
5. F, Viruses

Concept Mapping
Row 1: Particles
Row 2: Cells, Hosts

CONCEPT MASTERY

1. Viruses are much smaller and simpler than cells. Unlike cells, viruses cannot perform any life functions on their own.
2. First, the bacteriophage attaches to a bacterium. Then the bacteriophage injects its hereditary material, which takes control of the cell and causes the cell to create more virus particles. Finally, the host cell bursts, releasing virus particles that can infect other bacteria.
3. Weakened or killed viruses are used to make vaccines; viruses selectively kill pests; viruses can be used to insert new hereditary material into cells, thus "correcting" faulty hereditary material.

Content Review

Multiple Choice

Choose the letter of the answer that best completes each statement.

1. An example of a disease caused by a virus is
   a. bubonic plague.   c. strep throat.
   b. measles.   d. tetanus.
2. Which of the following statements is true?
   a. Because they break down wastes and dead organisms, viruses are called decomposers.
   b. Unlike other cells, viruses lack a nucleus.
   c. Viruses consist of a core of protein surrounded by a coat of hereditary material.
   d. To perform their life functions, viruses require a host cell.
3. Almost all bacteria are surrounded and supported by a tough, rigid protective structure called the
   a. cell membrane.   c. protein coat.
   b. cell wall.   d. capsule.
4. Which of the following is not found in bacteria?
   a. hereditary material   c. cytoplasm
   b. cell membrane   d. nucleus
5. Viruses are best described as
   a. autotrophs.   c. parasites.
   b. decomposers.   d. lithotrophs.
6. Bacteria cause a number of human diseases, including
   a. tuberculosis.   c. AIDS.
   b. influenza.   d. rabies.
7. Humans use viruses to
   a. make antibiotics.
   b. make fuels such as methane.
   c. break down sewage.
   d. put new hereditary material into cells.
8. Monerans are also known as
   a. bacteria.   c. viruses.
   b. bacteriophages.   d. fungi.

True or False

If the statement is true, write "true." If it is false, change the underlined word or words to make the statement true.

1. Organisms that break down wastes and the remains of dead plants and animals are called <u>producers</u>.
2. Some bacteria use whiplike structures called <u>flagella</u> to propel them through their environment.
3. When conditions become unfavorable, some bacteria produce a small internal resting cell called a <u>capsule</u>.
4. <u>Symbiosis</u> is a relationship in which one organism lives in close association with another organism and at least one of the organisms benefits.
5. <u>Bacteria</u> can be used by scientists to insert hereditary material into cells.

Concept Mapping

Complete the following concept map for Section 2–1. Refer to pages B6–B7 to construct a concept map for the entire chapter.

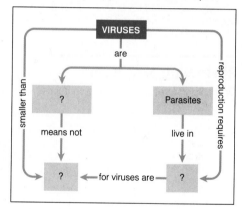

4. An autotroph makes its food from simple raw materials. A heterotroph cannot make its food—it obtains food by eating other organisms or the remains of dead organisms.
5. Symbiosis is a relationship in which one organism lives on, near, or inside another organism, with at least one organism benefiting from the relationship. Student examples will vary.
6. The cytoplasm is surrounded by the cell membrane, which is in turn surrounded by the cell wall. The cell wall may be covered by a coating called the capsule. The hereditary material is spread throughout the cytoplasm. Many monerans, or bacteria, have one or more flagella. The most important difference is that a moneran's cell lacks a nucleus and a number of the structures found in cells from the other four kingdoms.
7. Check students' responses for accuracy. Possible answers: Production—blue-green bacteria are used as food; nitrogen-fixing bacteria provide nitrogen compounds that crops need to grow; some cause dis-

Concept Mastery

Discuss each of the following in a brief paragraph.

1. How are viruses different from cells?
2. Describe how a typical bacteriophage reproduces.
3. Identify three ways in which viruses help people.
4. How do autotrophs and heterotrophs differ in the ways they obtain food?
5. What is symbiosis? Give one example of symbiosis.

6. Briefly describe the structure of a typical bacterial cell. What is the most important structural difference between the cells of bacteria and the cells of organisms in the other four kingdoms?
7. How do bacteria affect food production, food processing, and food storage?
8. Can viruses be grown in the laboratory on synthetic material? Explain.

Critical Thinking and Problem Solving

Use the skills you have developed in this chapter to answer each of the following.

1. **Applying definitions** Why is the relationship between a parasite and a host considered to be a form of symbiosis?
2. **Making inferences** When a cell is placed in water that contains a large quantity of dissolved substances such as sugar or salt, the cell will shrivel up and die. Explain why food can be preserved by putting it into honey or brine (very salty water).
3. **Interpreting a graph** The accompanying graph shows the growth of bacteria. Describe the growth of the bacteria using the numbers given for each growth stage. Why do you think the growth leveled off in stage 3 and then fell in stage 4?

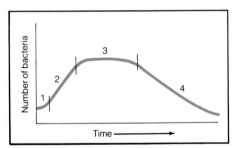

4. **Designing an experiment** Design an experiment to test the effect of temperature on the growth of bacteria. Be sure you have a control and a variable in your experiment. What results do you expect to get in this experiment? How can you apply these results in order to control food spoilage?
5. **Applying concepts** Every few years, a farmer plants a field with alfalfa or clover. At the end of the growing season, the farmer does not harvest these plants. Instead, the alfalfa or clover is plowed under, and the field is replanted with grain. Explain how the actions of bacteria account for the farmer's behavior.
6. **Using the writing process** In *The War of the Worlds,* a wonderful book by H. G. Wells, the Earth is invaded by Martians. Human weapons are useless against the invaders, and the Earth seems doomed. The Earth is saved, however, when the invaders die from diseases they contract here. Many other science-fiction writers have written stories that involve diseases. Now it is your turn. Write a science-fiction story in which a disease plays an important part.

---

eases in crops and livestock; some harm livestock by poisoning water supplies. Processing—used to make many foods, including vinegar, yogurt, and soy sauce; food is heated in pasteurization and in canning to kill harmful bacteria; food is processed in many other ways to prevent or slow bacterial growth and prevent spoilage. Storage—food is refrigerated or frozen to slow bacterial growth; food is placed in sealed containers such as cans to keep out bacteria.

8. No. Viruses require living cells in order to grow.

CRITICAL THINKING AND PROBLEM SOLVING

1. Symbiosis is a relationship in which two organisms live in close association with each other and at least one organism benefits. In the parasite-host relationship, the parasite lives in or on the host and benefits from the relationship.
2. The high concentrations of sugar in honey and salt in brine kill bacteria that cause food to spoil. Because the concentration of solutes in the environment is higher than in cells, the bacteria lose too much water via osmosis and die.
3. In stages 1 and 2, the bacteria population is growing because conditions are favorable. In stage 3, the population growth has leveled off because the conditions have become less favorable. In stage 4, the population is decreasing because individuals are dying off due to unfavorable conditions.
4. Accept all logical experiments and answers. Students should recognize that the appropriate variable is temperature. An appropriate control might be a culture grown in an incubator set at standard room temperature. Students will probably expect to find that the bacteria grow best in a narrow temperature range and that the cultures farthest from the optimum temperature grow at the slowest rates. The results of such an experiment would probably support the idea that refrigeration and cooking slow or prevent the growth of food-spoiling bacteria.
5. Both alfalfa and clover have lumps on their roots that contain nitrogen-fixing bacteria. The nitrogen-fixing bacteria replace the nitrogen compounds in the soil. When the alfalfa or clover is plowed under, bacteria known as decomposers break down the dead plants. This further enriches the soil.
6. Stories will vary but should contain a scenario in which a disease plays an important role.

KEEPING A PORTFOLIO

You might want to assign some of the Concept Mastery and Critical Thinking and Problem Solving questions as homework and have students include their responses to unassigned questions in their portfolio. Students should be encouraged to include both the question and the answer in their portfolio.

ISSUES IN SCIENCE

The following issue can be used as a springboard for discussion or given as a writing assignment:

1. What do you think would happen to life on Earth if all bacteria were killed? What do you think would happen if all living things except bacteria were killed?

SECTION	HANDS-ON ACTIVITIES
3–1 Characteristics of Protists pages B60–B62 Multicultural Opportunity 3–1, p. B60 ESL Strategy 3–1, p. B60	**Student Edition** ACTIVITY (Doing): Protist Models, p. B62 **Activity Book** CHAPTER DISCOVERY: Observing Protists, p. B71 ACTIVITY: Making a Microorganism Mobile, p. B77 **Teacher Edition** Identifying Characteristics of Living Things, p. B58d
3–2 Animallike Protists pages B63–B72 Multicultural Opportunity 3–2, p. B63 ESL Strategy 3–2, p. B63	**Student Edition** ACTIVITY (Doing): Capturing Food, p. B67 ACTIVITY BANK: Putting the Squeeze On, p. B174
3–3 Plantlike Protists pages B73–B77 Multicultural Opportunity 3–3, p. B73 ESL Strategy 3–3, p. B73	**Student Edition** ACTIVITY BANK: Shedding a Little Light on Euglena, p. B176
3–4 Funguslike Protists pages B77–B79 Multicultural Opportunity 3–4, p. B77 ESL Strategy 3–4, p. B77	**Student Edition** LABORATORY INVESTIGATION: Examining a Slime Mold, p. B80 **Laboratory Manual** Comparing Protists, p. B27 **Activity Book** ACTIVITY: Collecting Protists, p. B79
Chapter Review pages B80–B83	

OUTSIDE TEACHER RESOURCES

Books

Dixon, Bernard, *Magnificent Microbes.* Atheneum.

Farmer, J. N. *The Protozoa,* Mosby.

Gortz, H. D., ed. *Paramecium,* Springer-Verlag.

Jahn, T. L. *How to Know the Protozoa.* Brown.

OTHER ACTIVITIES	MEDIA AND TECHNOLOGY
Activity Book ACTIVITY: Surface Area and Volume, p. B89 **Review and Reinforcement Guide** Section 3–1, p. B23	**English/Spanish Audiotapes** Section 3–1
Student Edition ACTIVITY: (Calculating): How Big Is Big? p. B68 ACTIVITY: (Calculating): Divide and Conquer, p. B71 **Activity Book** ACTIVITY: Mathematics and Life Science, p. B83 ACTIVITY: Protist Hall of Fame, p. B83, p. B83 **Review and Reinforcement Guide** Section 3–2, p. B25	**English/Spanish Audiotapes** Section 3–2
Student Edition ACTIVITY (Reading): The Universe in a Drop of Water, p. B75 **Activity Book** ACTIVITY: Identifying Protists, p. B85 **Review and Reinforcement Guide** Section 3–3, p. B27	**Transparency Binder** Protists **English/Spanish Audiotapes** Section 3–3
Activity Book ACTIVITY: Comparing Protist, p. B87 **Review and Reinforcement Guide** Section 3–4, p. B29	**English/Spanish Audiotapes** Section 3–4
Test Book Chapter Test, p. B49 Performance-Based Tests, p. B129	**Test Book** Computer Test Bank Test, p. B55

*All materials in the Chapter Planning Guide Grid are available as part of the Prentice Hall Science Learning System.

Audiovisuals

From Protistan to First Multicellular Animals, film, Benchmark Films, Inc.

Kingdom of the Protists, sound filmstrip, National Geographic Society
Life Story of the Paramecium, video, EBE

The Protist Kingdom, film, BFA
Protists: Threshold of Life, film, National Geographic Society

CHAPTER OVERVIEW

The kingdom Protista consists of one-celled organisms that have nuclei. As a group, the protists are not nearly as old as the monerans. The oldest fossils of protists are only 1.5 billion years old. Protists are extremely diverse and difficult to classify because they exhibit characteristics of animals, plants, and fungi. Scientists disagree about what organisms should be classified in this kingdom.

Animallike protists are divided into four main groups—the sarcodines, the ciliates, the zooflagellates, and the sporozoans. Amebas are examples of sarcodines that have pseudopods used for feeding and, in some cases, for movement. Paramecia are examples of ciliates that have numerous hairlike projections, or cilia, during some stage of their lives. The cilia are used for movement and for sweeping food into the cell. Zooflagellates move by whipping their flagella. Many animallike protists live in water; others live inside the bodies of animals. Some have no effect on their hosts, others benefit the host, and some are parasitic. All sporozoans are parasitic. The sporozoan *Plasmodium* is the cause of malaria. Sporozoans, like the other animallike protists, are heterotrophs; that is, they get energy by eating food that is already made.

Three interesting plantlike protists are euglenas, diatoms, and dinoflagellates. Most plantlike protists are autotrophs; that is, they make their own food. Plantlike protists supply about 70 percent of Earth's oxygen as a byproduct of their food-making process. Euglenas have flagella, chloroplasts, and a red eyespot sensitive to light. Diatoms have a two-part glassy shell that encases the cells. Diatomaceous earth, the remains of the diatoms' glassy shells, has a variety of uses. Dinoflagellates have flagella that propel them through water. Some forms are responsible for toxic red tides. Some have a twinkling light similar to that of fireflies.

The funguslike protists are heterotrophs. One interesting funguslike protist is the slime mold. One type of slime mold forms into a mass of cells that live as a unit. Slime mold reproduction involves fruiting bodies. How the individual cells that make up a cellular slime mold coordinate their activities has intrigued biologists for decades.

3-1 CHARACTERISTICS OF PROTISTS
THEMATIC FOCUS

The purpose of this section is to introduce students to the kingdom of Protista. In this section students define protists as one-celled organisms that have nuclei. The presence of the nucleus in individuals of this kingdom distinguishes them from the monerans studied in Chapter 2. Students learn that fossil evidence indicates that protists developed about 1.5 billion years ago.

The themes that can be focused on in this section are evolution, scale and structure, and unity and diversity.

*Evolution: Fossil evidence indicates that protists first developed about 1.5 billion years ago, 2 billion years after the first monerans. Protists may have evolved from a symbiosis among several types of bacteria. Ancient protists are the ancestors of modern protists and multicellular organisms. As a result, they represent the first step in a chain of events leading to the development of multicellular organisms.

*Scale and structure: Protists are unicellular organisms that possess a nucleus. Most are microorganisms that cannot be seen without the aid of a microscope.

*Unity and diversity: Although the protists are divided into three major groups—animallike protists, plantlike protists, and funguslike protists—the members of each group are quite diverse in appearance and habit. Yet all protists are similar in basic structure.

PERFORMANCE OBJECTIVES 3-1

1. Describe general characteristics of the protists.
2. Define autotrophs and heterotrophs.

SCIENCE TERMS 3-1

protist p. B60

3-2 ANIMALLIKE PROTISTS
THEMATIC FOCUS

The purpose of this section is to introduce the students to the four main groups of animallike protists—sarcodines, ciliates, flagelletes, and sporozoans. The types of structures used for movement and feeding are focused on, and the environments within which the protists live are identified.

The themes that can be focused on in this section are energy, patterns of change, and systems and interactions.

Energy: Animallike protists meet their energy needs by eating food that is already made. They are known, as a result, as heterotrophs.

*Patterns of change: Sporozoans have complex life cycles that involve more than one host—in each host a different pattern of change results. An example is the malaria-causing sporozoan *Plasmodium*. It thrives as a parasite in mosquitoes and humans.

Systems and interactions: Protists interact with other organisms in a variety of ways. Some live in symbiotic relationships with a host organism. In some relationships, the protist has no effect on the organism, whereas in others, it can help the organism. Some protists are parasitic and harm the host. Still other protists serve as food for other organisms.

PERFORMANCE OBJECTIVES 3-2

1. List four main groups of animallike protists.
2. Distinguish structural features of animallike protists.
3. Describe the methods of movement and feeding used by animallike protists.

SCIENCE TERMS 3-2

sarcodine p. B63
ciliate p. B63

zooflagellate p. B63
sporozoan p. B63
pseudopod p. B63
ameba p. B64
cilia p. B66
paramecium p. B66
flagellum p. B68

3–3 PLANTLIKE PROTISTS
THEMATIC FOCUS

The purpose of this section is to introduce students to three interesting groups of plantlike protists. Students identify the characteristics of euglenas, diatoms, and dinoflagellates. They identify most plantlike protists as autotrophs and learn that 70 percent of the world's oxygen supply is a byproduct of plantlike protists' food-making process.

The themes that can be focused on in this section are systems and interactions, energy, and stability.

Systems and interactions: Through their food-making process, plantlike protists produce most of Earth's supply of oxygen. As a result, they are a vital part of the balance of nature on Earth. The oxygen produced is essential to other organisms on Earth. Plantlike protists are also a rich food source for many sea animals and one part of a food chain involving fish and humans.

Energy: Plantlike protists are autotrophs. They get energy from the food they produce themselves. Because they are a food source for other organisms, they are also a source of energy for those organisms.

Stability: Plantlike organisms help to maintain the supply of oxygen in the atmosphere. Without them, the diminished oxygen supply could result in vast changes for other forms of life on Earth.

PERFORMANCE OBJECTIVES 3–3
1. Identify characteristics of plantlike protists.
2. Name three types of plantlike protists.

SCIENCE TERMS 3–3
euglena p. B74
diatom p. B74
dinoflagellate p. B74

3–4 FUNGUSLIKE PROTISTS
THEMATIC FOCUS

The purpose of this section is to introduce students to the funguslike protists. Students identify characteristics common to funguslike protists and compare slime molds to other types of protists. The themes that can be focused on in this section are patterns of change and unity and diversity.

*Patterns of change: Within their lifetimes, slime molds undergo many transformations. At one time, they exhibit amebalike qualities in which they engulf their food. At another, they develop fruiting bodies that enable them to reproduce much as the fungi do.

*Unity and diversity: Although slime molds exhibit charactersitics that are dissimilar to those of other protists, their structure and behavior are essentially the same as those of other protists.

PERFORMANCE OBJECTIVES 3–4
1. Identify characteristics of funguslike protists.
2. Distinguish slime molds from other types of protists.

SCIENCE TERMS 3–4
slime mold p. B78

TEACHER DEMONSTRATIONS MODELING

Identifying Characteristics of Living Things

Put a drop of clove oil (available from a pharmacy) about the size of a nickel into a small amount of water, made slightly blue by food coloring, in a shallow, clear glass dish. Place the dish on the overhead projector.

• **How could we tell if this thing is alive or not?** (Lead students to suggest criteria such as movement, food ingestion, breathing, and reacting to stimuli.)

Tell students that you can test for response to stimulus by putting a couple of drops of liquid next to the blob. With an eyedropper, place about five drops of rubbing alcohol at one edge of the oil. Do not name the liquid. After time for observation, ask these questions:

• **How would you describe what you observed?** (Try to get at least ten different descriptions.)

Tell students that they will be studying a type of protist, amebas, which look like this when they move.

To test for eating, place "food," pieces of paraffin and a toothpick about 2 mm long, at the edge of, but not touching, the blob. Add a few more drops of alcohol on the opposite side.

• **What information does this give us about whether or not this thing can eat?** (Students should see the blob surround the "food" in small bubbles.)

Say that many protists enclose their food in a bubble, or food vacuole, when they take it into their bodies.

• **Do you think this blob is alive or not?** (Answers will vary. Ask students for reasons to support their opinions.)

Identify the chemicals you used. Explain that chemicals can be used to imitate some, but not all, of the very complex behavior of protists.

CHAPTER 3
Protists

INTEGRATING SCIENCE

This life science chapter provides you with numerous opportunities to integrate other areas of science, as well as other disciplines, into your curriculum. Blue numbered annotations on the student pages and integration notes on the teacher wraparound pages alert you to areas of possible integration.

In this chapter you can integrate life science and evolution (pp. 61, 79), language arts (pp. 63, 75), earth science and geology (p. 63), mathematics (pp. 68, 71), life science and photosynthesis (p. 73), life science and marine biology (p. 76), and social studies (p. 77).

SCIENCE, TECHNOLOGY, AND SOCIETY/COOPERATIVE LEARNING

About 250 million people worldwide are affected by malaria at any one time, and 2 to 4 million people die each year from the disease. Because of the widespread effects of the disease, many health professionals feel that the elimination of malaria should be a major goal of the World Health Organization, a branch of the United Nations.

In the 1950s, the World Health Organization attempted to eliminate malaria from Borneo. Malaria is caused by *Plasmodium*, a protist. Mosquitoes infected by the protist infect humans by biting them. Because the elimination of the mosquito carriers is one way of controlling malaria, the environment was sprayed with DDT to kill the *Anopheles* mosquito. But the far-reaching effects of spraying with DDT were soon evident. Cockroaches and other insects that ingested the DDT, in turn, were consumed by geckoes (insect-eating lizards). Cats caught a much greater number of geckoes than usual because the geckoes' movements were slowed due to DDT-caused nerve damage. Because geckoes are natural predators of caterpillars, the diminished gecko population led to an increase in the caterpillar population.

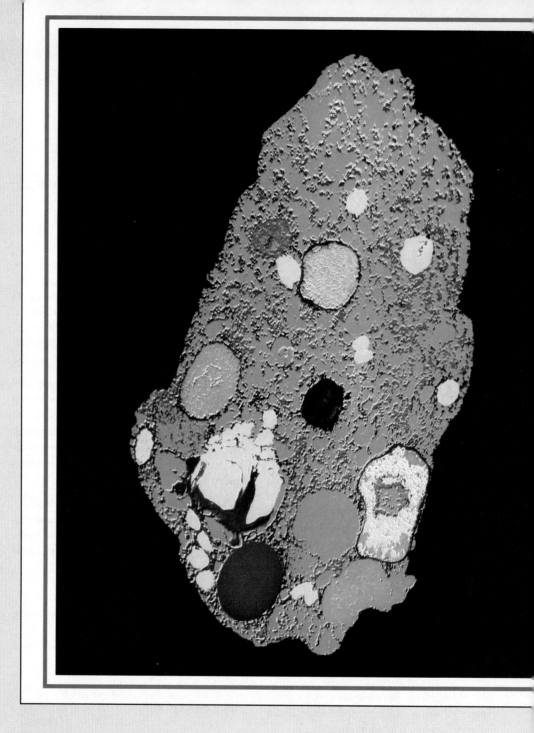

INTRODUCING CHAPTER 3

DISCOVERY LEARNING

▶ *Activity Book*

Begin teaching the chapter by using the Chapter 3 Discovery Activity from the *Activity Book*. Using this activity, students will examine some different kinds of protists and identify some of their characteristic movements, their appearance, and their behavior.

USING THE TEXTBOOK

Have students observe the photograph on page B58.

• **What do you think this photograph shows?** (Accept all answers.)

Explain that the photograph shows a single-celled microorganism from the kingdom Protista.

• **How does this microorganism differ from those you have studied in Chapter 2?** (Answers will vary, but lead students to conclude that this organism is more

Protists

Guide for Reading

After you read the following sections, you will be able to

3–1 Characteristics of Protists
- Describe the characteristics of protists.

3–2 Animallike Protists
- Identify the four groups of animallike protists.

3–3 Plantlike Protists
- Describe plantlike protists.

3–4 Funguslike Protists
- Compare slime molds to other protists.

Imagine for a moment that you have stepped into a time machine. Your destination: the world of 1.5 billion years ago. You arrive in a strange and barren place. No animals roam the land or swim in the ocean. No birds fly in the air. No trees, shrubs, or grasses grow from the soil. Earth at first seems lifeless.

A closer look at your surroundings reveals a slippery black film on the rocks and greenish threads in the water. Bacteria live here! When you use a microscope to examine the water of early Earth, you discover that a new form of life has come into being. Like the bacteria that came before it, this form of life is unicellular (one-celled). But these cells are much more complex than those of bacteria. You are surprised and delighted to recognize one large structure in the cells—the nucleus.

These first cells with a nucleus are protists. Protists represent the first step in the series of evolutionary events that eventually led to the development of multicellular (many-celled) organisms such as mushrooms, trees, fishes, birds, and humans. In this chapter, you will learn a great deal about protists—complex and fascinating organisms.

Journal *Activity*

You and Your World When did you first find out that you were surrounded by living things too small to see? In your journal, explore the thoughts and feelings you had upon learning about the unseen world around you.

◀ The numerous structures within this ameba look like tiny jewels. These structures represent an important evolutionary step toward increasing complexity in living things.

The caterpillars began eating even the thatched roofs of houses—causing many of the roofs to collapse. Even worse, rats moved in from the forest. Their predators, cats, had died from DDT poisoning. With the rats came rat fleas, which carry plague-causing bacteria. And plague can be more immediately fatal than malaria!

The World Health Organization stopped spraying DDT and parachuted a large number of cats into Borneo. Eventually, the normal balance of nature returned.

Cooperative learning: Using preassigned lab groups or randomly selected teams, have groups complete one of the following assignments:
- Illustrate the cycle of events that occurred in Borneo in the 1950s. Using your drawing and the life cycle of *Plasmodium,* propose solutions to the worldwide malaria problem. Each group's recommendations should be in the form of a chart that indicates the action to be taken, the planned outcome of the action, and any consequences of the action.
- While some health professionals think the eradication of malaria is very important, others feel this disease helps to keep the world's population in check, eradication would be a costly effort, and another disease would quickly replace malaria as a worldwide killer. Which opinion do you agree with and why? Groups could debate the issue or prepare and present a position paper.

See Cooperative Learning in the *Teacher's Desk Reference.*

JOURNAL ACTIVITY

You may want to use the Journal Activity as the basis of a class discussion. As students discuss their awareness of the unseen microorganisms, ask them to relate how they feel about the presence of these microoganisms. Help students understand that many microorganisms are helpful to humans and other forms of life. Point out that they will investigate protists in this chapter. Students should be instructed to keep their Journal Activity in their portfolio.

---

complex than the monerans studied in Chapter 2.)

Point out that a major difference between the monerans and the protists is that protists have nuclei and monerans do not. Then have students read the photograph caption and text on page B59.

• **Why do you think the protists represent the first step in the evolutionary development of multicellular organisms?** (Accept all logical answers, but point out that the protists were the first organisms that had nuclei, just as multicellular organisms have nuclei.)

• **Where do you think the protists came from?** (Students may infer that the protists evolved from the monerans.)

Explain that as with the monerans, the kingdom Protista includes a great variety of individuals. Common to each member of the kingdom is the presence of nuclei in their cells.

3–1 Characteristics of Protists

Figure 3–1 *The major difference between protists and bacteria is that protists have many complex cell structures. The most important of these is the nucleus, which looks like a string of beads in Stentor* (left), *and a thin squiggle in Vorticella* (center). *Protists also have complex external structures, such as the hairlike projections on Tetrahymena* (right).

3–1 Characteristics of Protists

The members of the kingdom Protista are known as **protists.** Like bacteria (monerans), protists are unicellular (one-celled) organisms. Also like bacteria, protists were one of the first groups of living things to appear on Earth. Although protists are larger than monerans, most cannot be seen without the aid of a microscope. Protists, however, are quite different from bacteria. The most important difference is that protists have a nucleus and a number of other cell structures that bacteria lack.

Protists can be defined as single-celled organisms that contain a nucleus. However, you will discover in this chapter that there are a few protists that do not fit this definition perfectly. In fact, as scientists develop new techniques for examining the structure and chemistry of cells, the definition of protist continues to change. And as new evidence is obtained, analyzed, and interpreted, scientists revise their ideas about which protists should be classified together— and even about which organisms should be classified as protists!

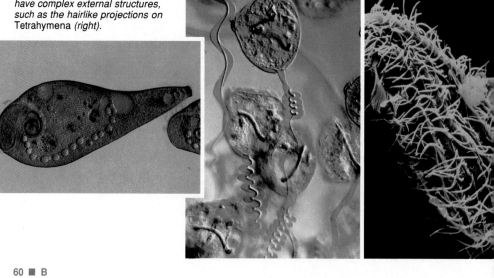

Most protists live in a watery environment. They can be found in the salty ocean and in bodies of fresh water. Some protists live in moist soil. Others live inside larger organisms. Many of these internal protists are parasites. (Parasites, you will recall, are organisms that live on or in a host organism and harm it.) But some of the protists that live inside other organisms help their hosts.

Protists generally live as individual cells. However, many protists live in colonies such as the ones shown in Figure 3–2. A protist colony consists of a number of relatively independent cells of the same species that are attached to one another. Some colonies are foamlike clusters of cells. Others are branching and treelike. A few are rings or chains of cells. And one type of colony consists of an intricate network of slime tunnels. The members of this type of colony move through the tunnels like miniature subway cars.

Evidence from fossils indicates that protists evolved about 1.5 billion years ago—about 2 billion years after bacteria. According to one hypothesis, the first protists were the result of an extremely successful symbiosis among several kinds of bacteria. Recall from Chapter 2 that a symbiosis is a close relationship of two or more organisms in which at least one organism benefits. In this case, some of the bacteria lived inside another bacterial cell. Each kind of bacterium performed activities that helped the "team" 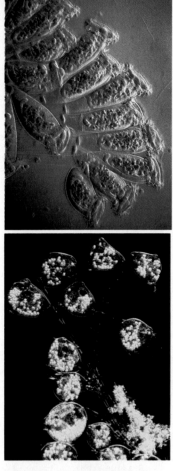 survive. Over time, the partners in the symbiosis became more specialized (suited for one particular task) and lost their independence. Traces of this early symbiosis can still be seen in protists and other cells that possess a nucleus. Within these cells are certain structures that are quite similar to bacterial cells. They even contain their own hereditary material.

Protists vary greatly in appearance and in the ways in which they carry out their life functions. For example, some protists are autotrophs. Autotrophs are able to use energy such as sunlight to make their own food from simple raw materials. Other protists are heterotrophs. Heterotrophs get their energy by eating food that has already been made. And a few protists are able to function as either autotrophs or heterotrophs, depending on their surroundings.

Figure 3–2 *Some protists, such as Epistylis (top) and Carchesium (bottom) live in colonies of attached but relatively independent cells. Colonial protists may have been the first evolutionary step toward true multicellular organisms.*

HISTORICAL NOTE
KINGDOM PROTISTA

Before the mid-1860s, microorganisms were classified as plants or animals. In 1866, a German naturalist, Ernst Haeckel, coined the word *protist* for microorganisms and established a new kingdom of living organisms, Protista. Haeckel placed the fungi, protozoa, and microscopic algae in the new kingdom. The modern classification recognizes five kingdoms—Monera, Protista, Fungi, Animalia, and Plantae. Even today, however, there is some disagreement among scientists as to what kingdom various organisms belong.

BACKGROUND INFORMATION
"FIRST ANIMALS"

Many animallike protists and some plantlike protists have been referred to as protozoa. Protozoa take their name from two Greek words that literally mean "first animal."

ENRICHMENT

▶ *Activity Book*

Students may be challenged by the chapter activity called Surface Area and Volume. In the activity students will determine and compare the ratios of surface area to the volume of cubes. The activity will enable students to better understand the relationship between surface area and volume in one-celled protists. They will also conclude that size affects the ratio.

CONTENT DEVELOPMENT

Remind students that bacteria evolved about 3.5 billion years ago.
• **Based on fossil evidence, when did the first protists appear?** (About 1.5 billion years ago.)
• **Which evolved first, bacteria or protists?** (Bacteria—about 2 billion years before protists.)
• **Is it likely that protists evolved from bacteria?** (Yes. Symbiotic bacteria may have developed into protists over time.)

● ● ● ● Integration ● ● ● ●

Use the discussion of symbiotic relationships to integrate the life science concepts of evolution into your lesson.

INDEPENDENT PRACTICE

▶ *Activity Book*

Students can gain practice in identifying characteristics of protists and different types of protists by doing the chapter activity called Making a Microorganism Mobile. Students should find that the mobiles are an effective tool for helping them remember the different types of protists and their characteristics.

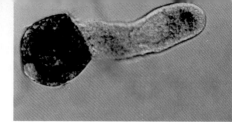

Figure 3–3 *These diatoms growing in fan-shaped colonies are autotrophs. The short, dark, thimble-shaped* Didinium *and the long, oval* Paramecium *it is eating are both heterotrophs. What is the major difference between autotrophs and heterotrophs?* ①

ACTIVITY
DOING

Protist Models

1. Using modeling clay and/or other appropriate materials, build a model of the protist of your choice. Use the diagrams in your textbook or other reference materials as a guide. You may wish to use different colors of clay or paint the model in order to show the different structures in the protist.

2. Mount the model on a piece of stiff cardboard.

3. Label each of the protist's structures.

4. Share your model and what you learned while making it with your classmates. Compare your model to the ones your classmates have constructed. How are protists different? How are they similar? What are some good characteristics in the protist models?

62 ■ B

As you might have guessed of such a large and diverse group of organisms, the kingdom Protista is divided into many phyla. Because there are many different ways to interpret what we know about protists, the classification of protists is a matter of scientific debate. Experts recognize anywhere from about nine to more than two dozen phyla. For the sake of simplicity, protists can also be grouped in three general categories. These categories, which are the ones you will read about here, are: animallike protists, plantlike protists, and funguslike protists.

3–1 Section Review

1. What are the major characteristics of protists?
2. Explain why scientists have not yet agreed on a single classification system for protists.
3. How are protists similar to monerans? How are they different?
4. When did protists first appear on Earth?

Connection—*You and Your World*
5. While standing in line at the supermarket, you notice the following newspaper headline: "Science Shocker: Alien Invaders in Every Cell of Your Body!" Explain why there may be some truth in this headline. (*Hint:* Fungi, plants, and animals evolved from protists that lived millions of years ago.)

3-2 Animallike Protists

Animallike protists are sometimes known as protozoa, which means first animals. Long ago, these organisms were classified as animals because they have several characteristics in common with animals. Their cells contain a nucleus and lack a cell wall. They are heterotrophs. Most of them can move. Animallike protists are no longer placed in a separate kingdom from more plantlike protists. Scientists have discovered that some animallike protists are so similar to certain plantlike protists that it does not make sense to place them in separate kingdoms.

Animallike protists are divided into four main groups. These four groups are the **sarcodines** (SAHR-koh-dighnz), the **ciliates** (SIHL-ee-ihts), the **zooflagellates** (zoh-oh-FLAJ-ehl-ihts), and the **sporozoans** (spohr-oh-ZOH-uhnz).

Sarcodines

Sarcodines are characterized by extensions of the cell membrane and cytoplasm known as pseudopods (SOO-doh-pahdz). The word **pseudopod** comes from the Greek words that mean "false foot" because these footlike extensions are always temporary. Pseudopods are used to capture and engulf particles of food. Some sarcodines also use psuedopods to move from one place to another.

Many sarcodines have shells that support and protect the cell. As you can see in Figure 3–4, these shells come in many forms. The shells of foraminiferans (fuh-ram-ih-NIHF-er-anz) may resemble coins, squiggly worms, the spiral burner coils of an electric stove, clusters of bubbles, and tiny sea shells. The shells of radiolarians (ray-dee-oh-LAIR-ee-uhnz) are studded with long spines and dotted with tiny holes. Because of this, radiolarian shells often look a lot like crystal holiday ornaments. When foraminiferans and radiolarians die, their shells sink to the ocean floor and form thick layers. Over millions of years, these shells are changed to rock. Some rocks that contain ancient protist shells, such as limestone and marble, are used in building. Others are used

Guide for Reading

Focus on this question as you read.

▶ *How are animallike protists similar? How are they different?*

Figure 3–4 *Many foraminiferans have beautiful shells. The shells of ancient foraminiferans help to make up limestone, marble, and chalk.*

3-2 Animallike Protists

MULTICULTURAL OPPORTUNITY 3-2

Malaria, caused by *Plasmodium,* is a serious problem in many parts of the world. At one time, malaria was coming under control through the use of DDT, an insecticide that killed the mosquito host of the *Plasmodium.* Although successful in decreasing the incidence of malaria, DDT is now banned in many countries because of a serious side effect. DDT decreases the thickness of eggshells, especially in predator birds such as the peregrine falcon. This resulted in the decline in the population of birds. Have students explore the positive and negative aspects of DDT and be prepared to discuss how concern about disease, especially common in developing nations, is balanced against concern about the extinction of several bird species.

ESL STRATEGY 3-2

As a team project, pair students who are learning English with English-speaking classmates to make models or posters that illustrate the following: how sarcodines, ciliates, and zooflagellates move about and how they obtain food and oxygen as well as eliminate waste products.

In their presentations, ask students to explain how these organisms reproduce and respond to changes in their environments.

---

and a mouse. Also display several pictures of different kinds of plants including a tree, a tulip, and a houseplant.

• **Which pictures show animals and which show plants?** (Students should be able to easily classify the pictures.)

• **What can animals do that plants cannot do?** (Answers may include eat, make sounds, move.)

• **What can plants do that animals cannot do?** (Make their own food.)

• **What do you think animallike protists have in common with animals?** (Accept all logical answers. Many students will answer that they move and eat.)

CONTENT DEVELOPMENT

Animallike protists require a moist environment. Most live in wet places, but some even survive in deserts, where they make a hard capsule around themselves and are dormant during dry times. There are virtually no vertebrates that do not host parasitic animallike protists. Most are harmless. In actual numbers, there are many more animallike protists than multicellular animals on the Earth.

● ● ● ● **Integration** ● ● ● ●

Use the information about word origins to integrate language arts into your science lesson.

Use the discussion about marble and sandstone composition to integrate concepts of geology into your lesson.

A PARASITIC AMEBA

In certain regions of the world, many people are infected with a species of ameba called *Entamoeba hystolytica,* which causes a disease known as amebic dysentery. The parasitic amebas that cause the disease live in the intestines, where they absorb food from the host. They also attack and destroy the lining of the intestine and other tissues. In humans, the symptoms of the disease include bloody feces, diarrhea, and cysts in various parts of the body. The amebas are passed out of the body in feces. In places where sanitation is poor and sewage is not treated, the amebas may find their way into the food and water supplies and can then infect other humans who use the food or water. You might ask students to consider ways that people can work together to try to eradicate this disease.

Figure 3–5 *The pores and spines of radiolarian shells make them glitter in the light like tiny glass decorations.*

Figure 3–6 *This diagram shows the structure of* Amoeba proteus, *a typical ameba. According to Greek mythology, the sea god Proteus had the magical ability to change his shape. How do amebas change their shape?* ❶

for writing and drawing—the chalk used by teachers is made of prehistoric foraminiferans!

The most familiar type of sarcodine is the bloblike **ameba** (uh-MEE-bah). Amebas use their pseudopods to move and to obtain food. An ameba first extends a thick, round pseudopod from part of its cell. Then the rest of the cell flows into the pseudopod. As an ameba nears a small piece of food, such as a smaller protist, the ameba extends a pseudopod around the food. Soon the food particle is completely surrounded by the pseudopod. As you can see in Figure 3–8, this process produces a bubblelike structure that contains the food. This structure is called a food vacuole. The food is digested (broken down into simpler materials) inside the food vacuole. The digested food can then be used by the ameba for energy and growth. The waste products of digestion are eliminated when the food vacuole joins with the cell membrane.

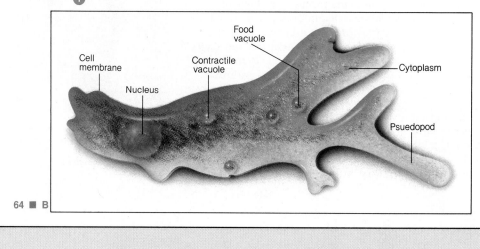

Cell membrane

Nucleus

Contractile vacuole

Food vacuole

Cytoplasm

Psuedopod

64 ■ B

CONTENT DEVELOPMENT

Amebas move by extending pseudopods. Demonstrate this movement by placing your fist on the chalkboard. Extend one finger and move your fist in the same direction. Explain that the ameba can extend pseudopods in any direction and that the rest of the cell flows into the pseudopods.

What else does the ameba use its pseudopods for? (To capture food.)

GUIDED PRACTICE

Skills Development

Skills: Manipulative, relating

Tell students that they are going to construct models demonstrating the formation of a food vacuole in an ameba. Give students two twist ties, a plastic sandwich bag, a pebble, and a pair of scissors. Have students follow these steps as you demonstrate them.

1. Put your hand inside the plastic bag. Imagine that your hand is the cytoplasm of an ameba's pseudopod and that the plastic bag is the cell membrane of the pseudopod.

2. Pick up the pebble with your bag-covered hand.

3. Twist the plastic so that the pebble is enclosed in the small "bubble." **Note:** *Do not turn the bag inside out.*

4. Tie off the bubble with one twist tie. Then tie off the bubble in a second place not far from the first tie. (In a real ameba, of course, the membrane would pinch off by itself.)

5. Cut the bag between the two twist ties.

• **What does the pebble represent?** (Food.)

• **What part of the bag represents the food vacuole?** (The bubble.)

The process just described is used to transport "large" particles, such as bits of food or solid wastes, through the cell membrane. Other substances, such as water, oxygen, and carbon dioxide do not need to be carried in and out of the cell this way. These substances simply pass right through the cell membrane.

Protists that live in fresh water, such as many kinds of amebas, must deal with a tricky problem. Water tends to come into the cell from the environment. What do you think might happen to an ameba if excess water was allowed to build up in its cell? ❷ Fortunately, protists have a special cell structure that enables them to keep the right amount of water in the cell. This structure is called the contractile vacuole. Excess water collects in the saclike contractile vacuole. When the contractile vacuole is full, it contracts (hence its name) to squirt the water out of the cell.

Amebas reproduce by dividing into two new cells. Because amebas have a nucleus, the process of cell division involves more than simply making a copy of the hereditary material and then splitting in two. (This, you should recall, is the way in which monerans reproduce.) In amebas and all other cells with a nucleus, cell division involves a complex series of events.

Amebas respond in relatively simple ways to changes in their environments. They are sensitive to bright light and move to areas of dim light. Amebas are also sensitive to certain chemicals, moving away from some and toward others. How do you think these behaviors help amebas survive? ❸

Figure 3–7 *Amebas, like most other protists, reproduce by dividing into two new cells.*

Ⓐctivity Bank

Putting the Squeeze On, p.174

Figure 3–8 *Look carefully at these images of a feeding ameba. How does an ameba capture its food?* ❹

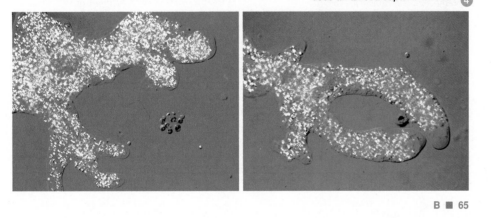

B ■ 65

BACKGROUND INFORMATION

CONTRACTILE VACUOLES

Water continually passes through the cell membrane of freshwater protists due to osmotic pressure, as it does in amebas. The osmotic pressure causes the contractile vacuole to keep filling with water and then to discharge it to the outside much as one collects water from a leaky boat in a bucket and then throws it overboard.

Most saltwater protists do not have contractile vacuoles. Because of the amount of salt dissolved in the oceans, the osmotic pressure on saltwater cells is much less. they do not need a specialized organelle to remove water.

Other materials, including O_2, CO_2, salts, and other small molecules can diffuse both into and out of the cell via the cell membrane.

• **What part represents the healed cell membrane?** (The tied portion.)
• **How do pseudopods form food vacuoles?** (By surrounding the food.)
• **What does a food vacuole consist of?** (Food and cell membrane.)
• **What is the relationship between the inner surface of the food vacuole and the outer surface of the cell membrane?** (The inner surface of the food vacuole was once the outer surface of the cell membrane.)

CONTENT DEVELOPMENT

Compare what would happen to an ameba if too much water were absorbed by it to what would happen to a balloon with too much air in it. Begin to blow up a balloon. Ask students what would happen to the balloon if too much air were put into it. (It would burst.) Ask how you could prevent the balloon from bursting. (By not adding more air or letting some air out of it.) Explain that an ameba does not have any means for preventing water from entering it.
• **How does the ameba prevent itself from absorbing too much water?** (By its contractile vacuole.)
• **How does the contractile vacuole work?** (The water collects in the contractile vacuole. When water needs to be eliminated, the contractile vacuole contracts to squirt out the water.)

Ciliates

Ciliates have small hairlike projections called cilia (SIHL-ee-uh; singular: cilium) **on the outside of their cells.** The **cilia** act like tiny oars and help these organisms move. The beating of the cilia also helps to sweep food toward the ciliates. In addition, the cilia function as sensors. When the cilia are touched, the ciliate receives information about its environment.

Cilia may cover the entire surface of a ciliate or may be concentrated in certain areas. In some ciliates, the cilia may be fused together to form structures that look like paddles or the tips of paint brushes. A few ciliates possess cilia only when they are young. As adults, they attach to a surface and lose their cilia.

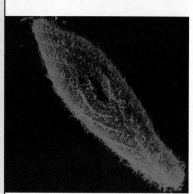

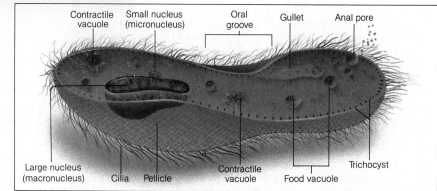

Figure 3–9 *Slipper-shaped* Paramecium *is probably the most familiar ciliate. What structures does a paramecium use in obtaining and digesting food?* ❶

One of the most interesting ciliates is the **paramecium** (par-uh-MEE-see-uhm; plural: paramecia). As you can see in Figure 3–9, a paramecium has a tough outer covering called the pellicle (PEHL-ih-kuhl). The pellicle, which consists of the cell membrane and certain underlying structures, gives the paramecium its slipper shape. The cilia of the paramecium sweep food particles floating in the water into an indentation, or notch, on the side of the paramecium. This indentation, which is called the oral groove, leads to a funnellike structure known as the gullet. At the base of the gullet, food vacuoles form around the incoming food. Then the food vacuoles pinch off into the cytoplasm. As a

66 ■ B

3–2 (continued)

GUIDED PRACTICE

Skills Development

Skills: Making observations, comparing

Tell students that they are going to use microscopes to examine another common animallike protist, but do not name the organism—a paramecium.
• **What did you observe about these protists?** (Lead students to describe shape, movement, and any structures they have seen.)

• **How is this protist different from an ameba?** (Its shape stays the same. It does not have pseudopods, is not as flat, and moves more quickly backward and forward.)

Have students adjust the light so that they can see the cilia.
• **What do you think the cilia are for?** (Accept all logical answers. Lead students to understand that the cilia provide a means of movement and food acquisition.)

CONTENT DEVELOPMENT

Food getting and digestion are important activities for all organisms and can be easily studied. As you discuss the way the paramecium gets and digests its food, have your students think about their own bodies and the digestive process as it occurs in humans.
• **In the digestive process in the paramecium, what is the function of the cilia?** (To sweep the food into the gullet.)

food vacuole travels through the cytoplasm, the food is digested and the nutrients distributed. When the work of the food vacuole is completed, it joins with an area of the pellicle called the anal pore. The anal pore empties waste materials into the surrounding water.

Paramecia, like other ciliates, have two kinds of nuclei. The large nucleus controls the life functions of the cell. The small nucleus is involved in a complicated process called conjugation. During conjugation, two ciliates temporarily join together and exchange part of their hereditary material.

As you can see in Figure 3–11, paramecia reproduce by splitting in half crosswise. During this process, many cell structures are copied, divided, broken apart, and formed anew. Now take a moment to refer to the diagram of the paramecium in Figure 3–9 on page 66. Imagine that this paramecium is about to divide. Its left half is going to become one new cell, and its right half is going to become another new cell. What structures does the left half require in order to become a fully functional paramecium? What structures does the right half require? ③

Figure 3–11 *Paramecia reproduce by dividing in half (left). This increases the number of paramecia. The process of conjugation also produces new paramecia but does not increase the number. The genetic material exchanged during conjugation may help to make one or both of the paramecia better able to deal with a changing environment (right).*

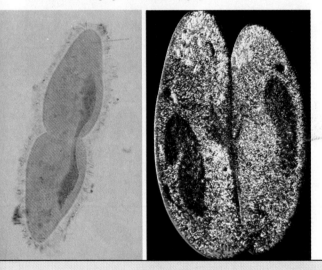

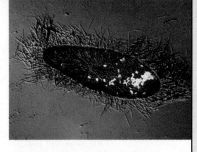

Figure 3–10 *When disturbed, a paramecium may release sharp spines known as trichocysts. How do these structures help the paramecium?* ②

ACTIVITY

Capturing Food

1. Place one drop of a paramecium culture on a microscope slide. Use a depression slide if you have one.

2. Add one drop of *Chlorella*, a green algae, to the slide.

3. Locate a paramecium under the microscope using low power. Then switch to high power. What happened to the *Chlorella*?

4. "Feed" a tiny amount of carmine red granules or India ink to the paramecia. What happens to these inedible (impossible to eat) substances? What do you think will eventually happen to them?

ACTIVITY
DOING

CAPTURING FOOD

Skills: Making observations, comparing, relating, applying

Materials: paramecium culture, microscope slide, Chlorella, carmine-red granules, microscope

The *Chlorella* is ingested by the paramecia. As a result, green-colored food vacuoles form. By adding carmine-red granules, the action of the cilia in the paramecia's oral grooves can be observed. Also the formation of a red-colored food vacuole can be seen. The movement of the cilia creates a current of water, which moves the carmine-red granules into the oral groove and then eventually into the food vacuole. The undigested carmine-red granules pass out of the paramecia through the anal pores.

• **What structure do you use for the same process?** (The arm or hand.)

• **Where does digestion occur in the paramecium?** (Inside food vacuoles within the cell.)

• **Where does digestion occur in the human body?** (Within the digestive system.)

• **Summarize the similarities and differences between digestion in the paramecium and in a human.** (Answers should focus on the complexities of the systems.)

ENRICHMENT

Point out that the paramecium's large nucleus contains multiple copies of the heredity material found in the small nucleus. Experiments have shown that if the large nucleus is removed, the paramecium dies. A paramecium, however, can live without its small nucleus, but it cannot undergo conjugation.

• **What is essential in a paramecium for conjugation?** (A small nucleus.)

• **What is essential to a paramecium for life?** (A large nucleus.)

INDEPENDENT PRACTICE

▶ *Activity Book*

Students can gain information about populations of paramecia through the chapter activity called Mathematics and Life Science. In the activity students will identify factors that affect the population growth of paramecia.

BACKGROUND INFORMATION
SPOROZOANS

Sporozoans differ from other animallike protists in several ways. The mature sporozoan is not motile. Sporozoans do not engulf their food but rather obtain their food by the diffusion of nutrients into the cell. Also, sporozoans have a complex life cycle that includes sexual and asexual reproduction.

ACTIVITY
CALCULATING

How Big Is Big? ❶

Protists range greatly in size. Some tiny flagellates are only 1 or 2 micrometers long. (A micrometer is one-millionth of a meter.) A paramecium is about 300 micrometers long. The fossil foraminiferan *Camerina* had a shell 10 centimeters wide. (A centimeter is one-hundredth of a meter.) Huge slime molds can be as large as 1 meter across.

Using a metric ruler and a calculator when necessary, find the numbers that make the following statements true:

1. I am _____ meters tall, or _____ centimeters tall, or _____ micrometers tall.
2. _____ zooflagellates 2 micrometers long placed end to end would equal 1 meter in length.
3. _____ *Camerina* shells placed end to end would equal 1 meter in length.
4. About _____ paramecia placed end to end could be placed across one *Camerina* shell.
5. _____ copies of this textbook laid end to end would equal 1 meter in length.

Figure 3–12 *Some zooflagellates cause diseases in humans. Trichomonas causes an infection of the female reproductive system (top). Giardia causes problems with the digestive system when it attaches to the walls of the small intestine (bottom).*

68 ■ B

Zooflagellates

Flagellates (FLAJ-eh-layts) are protists that move by means of flagella. Recall from Chapter 2 that a **flagellum** is a long whiplike structure that propels a cell through its environment. Flagellates may be animallike, plantlike, or funguslike. Plantlike and funguslike flagellates will be discussed later.

Animallike flagellates are called zooflagellates. (The prefix *zoo-* means animal.) **Zooflagellates usually have one to eight flagella, depending on the species.** However, some types of zooflagellates have thousands of flagella.

Many zooflagellates live inside the bodies of animals. Can you explain why such zooflagellates are said to be symbiotic? Some symbiotic zooflagellates ❶ do not have any effect on their host. Others benefit their host and may even be essential to its survival. For example, termites and wood roaches rely on zooflagellates in their intestines to digest the wood that the insects eat. Without the zooflagellates, these wood-eating insects would quickly starve. Still other zooflagellates harm their host. Do you remember what this type of symbiotic relationship is called? ❷

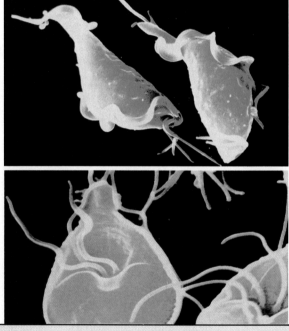

3–2 (continued)

CONTENT DEVELOPMENT

Compare flagella and cilia as modes of locomotion. Point out that the major difference between the two is the length of the cell part. Stress that the flagella of protists and bacteria differ greatly in structure. Point out the similarities between zooflagellates and plantlike flagellates. Explain that the major difference is the method of food acquisition. Zooflagel-

lates are heterotrophs, and plantlike flagellates are usually autotrophs.

Remind students of symbiotic relationships. Explain that many zooflagellates live within hosts. Many have no effect on the host, some help the host, and others are parasites. Many zooflagellates, however, are free-living—they do not live inside or attached to a host.

ENRICHMENT

Assign students to do library research on African sleeping sickness. Students

should determine what organism causes the disease, what organism transmits it, what its symptoms are, and what the treatments of the disease are.

FOCUS/MOTIVATION

Display pictures of a motorboat, a surfboard, and an inner tube.
• **Which of these can move through water on its own power?** (A motorboat.)
• **Which require outside forces to move them through the water?** (The surfboard and the inner tube.)

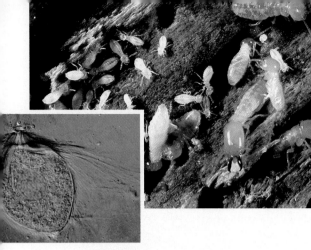

Figure 3–13 *Termites digest the wood they eat with the help of zooflagellates such as Trichonympha (inset). Why is the relationship between termites and Trichonympha an example of symbiosis?* **3**

Parasitic zooflagellates are responsible for a number of diseases in humans and animals. One parasitic zooflagellate causes African sleeping sickness, which is transmitted from one host to another through the bite of the tsetse fly. Other kinds cause various types of diseases of the intestines.

Sporozoans

All sporozoans are parasites that feed on the cells and body fluids of their host animals. Many sporozoans have complex life cycles that involve more than one kind of host animal. During these life cycles, sporozoans form cells known as spores. The spores enable sporozoans to pass from one host to another. How does this happen? A new host may become infected with the sporozoans if it eats food containing the spores. Or it may become infected if it is bitten by a tick, mosquito, or other animal that has spores in its body.

Perhaps the most famous type of sporozoan is the organism that causes the disease malaria. This organism is *Plasmodium* (plaz-MOH-dee-uhm). *Plasmodium* has two hosts: humans and the *Anopheles* mosquito. Both hosts are necessary for *Plasmodium* to complete its life cycle.

When a mosquito infected with *Plasmodium* bites a human, it injects its saliva. The saliva contains substances that keep blood flowing so that the mosquito

CAREERS

Laboratory Technician

People who perform microscopic tests in laboratories are called **laboratory technicians.** They prepare blood and other body fluid samples by adding stains or dyes to them. After this is done, they examine the slide for abnormal cell growth or the presence of a parasite. Then laboratory technicians prepare reports on their findings.

Laboratory technicians also work in veterinary hospitals and help in the detection of infectious diseases in pets and farm animals. Some may work in agriculture to help study the effects of microorganisms on farm crops.

If you are interested in this career, write to the National Association of Trade and Technical Schools, 2251 Wisconsin Avenue NW, Washington, DC 20007.

B ■ 69

• **How do these move in the water?** (They are paddled or pushed by wind, waves, or water currents, but they cannot move on their own.)

Tell students that the fourth group of animallike protists are composed of one-celled organisms that tend to move by currents. They are the sporozoans.

CONTENT DEVELOPMENT

All sporozoans are parasites that feed on the cells and body fluids of their host animals. The currents that carry the sporo-

zoans generally are in body fluids—blood or digestive tract liquids—of animals.

GUIDED PRACTICE

Skills Development

Skills: Making and reading a chart, comparing

At this time have students make a chart for identifying animallike protists. Ask them to respond to the following questions, based on the charted information:

• **How do the structures for movement**

MALARIA

About 250 million people on Earth suffer from malaria. The disease is most rampant in Africa, where it is believed about 1 million children under the age of five die each year from malaria. Malaria exhibits two forms—in the active stage, victims suffer symptoms, and in the dormant stage, they show little sign of the illness. Since 1640 when it was discovered, quinine has been a major treatment for malaria. In the mid-twentieth century, scientists developed two drugs for treatment—chloroquine for the active stage and primaquine for the dormant stage. Other drugs are also available today for treatment of the disease. Efforts to control the disease focus on mosquito abatement and drug use. Researchers are also trying to develop a vaccine for the disease.

INTEGRATION

HEALTH

Protists are the cause of many diseases in humans. Among these diseases is toxoplasmosis, which can be passed to humans through contaminated meats and house cats. Symptoms of the disease include sore throat, fever, and swelling of some internal organs. Ask students to research information about the protist that causes the disease.

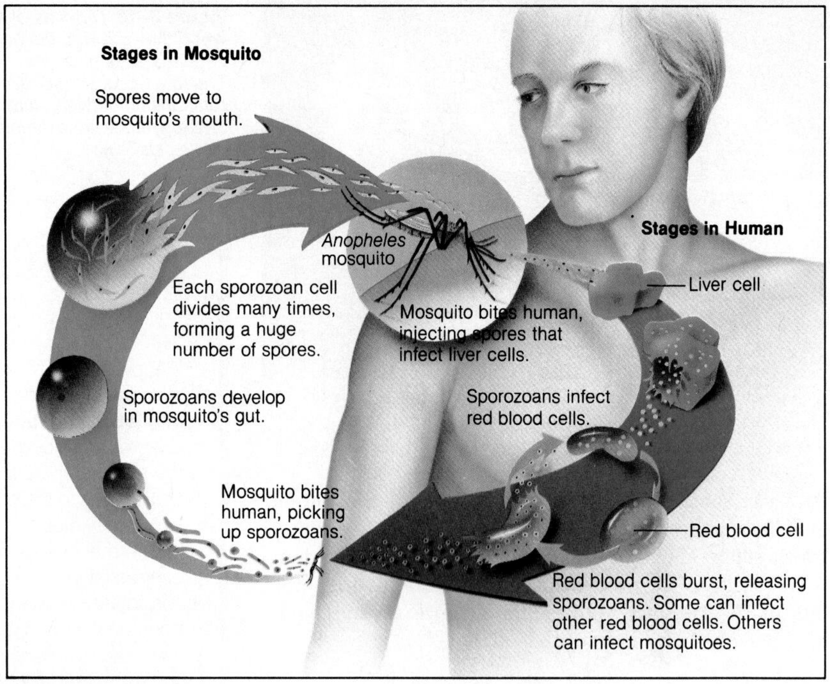

3–2 (continued)

CONTENT DEVELOPMENT

The chance of getting malaria in the United States is very small. In much of the rest of the world, however, malaria is a major human disease. Millions of people suffer from the disease caused by the sporozoan *Plasmodium*. If untreated, the disease kills.

Direct students to the text and Figure 3–14 on malaria. Outline the life cycle of *Plasmodium*. Point out that a specific type of mosquito, the *Anopheles*, injects *Plasmodium* into a person. This is not a common mosquito in the United States.

Figure 3–14 *The life cycle of* Plasmodium *is quite complex. Why is* Plasmodium *considered a parasite? What are its hosts?* ❶

can drink its fill. The saliva also contains *Plasmodium* spores. The injected spores are carried by the bloodstream to the person's liver. There the spores divide many times, producing a large number of sporozoan cells. The resulting sporozoans are a form that can infect red blood cells. After several days, these sporozoans burst out of the liver cells and invade red blood cells, where they grow and multiply. Eventually the sporozoans burst out of the red blood cells, destroying them.

Some of the sporozoans released when the red blood cells burst can infect more red blood cells. Others can infect mosquitoes. If the person with malaria is bitten by another *Anopheles* mosquito, the sporozoans along with the blood enter that mosquito. In the mosquito, the sporozoans undergo several stages of development and eventually form spores, which continue *Plasmodium's* life cycle.

The malaria-causing sporozoans inside a patient's body develop in such a way that all the millions of

• **How is a mosquito infected by the malaria-causing sporozoan?** (By biting a human infected with the disease.)
• **How is a human infected with malaria?** (By being bitten by an *Anopheles* mosquito carrying the *Plasmodium* sporozoans.)

ENRICHMENT

Tick fever, or Texas fever, nearly wiped out the cattle industry in Texas and other states at the turn of the century. Ask students to research information about the disease. What causes the disease? How is

it transmitted? How can it be controlled? Then have students consider the impact of such a disease on the economy.

INDEPENDENT PRACTICE

Section Review 3–2

1. Sarcodines are characterized by pseudopods that are used to trap food and, in some cases, move; some have shells. Ciliates have some hairlike projections on the outside of their cells at some point in their life cycle. Zooflagellates usually have one to eight flagella, or whiplike struc-

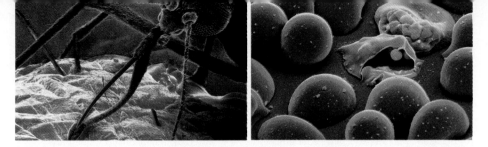

infected red blood cells burst at roughly the same time. Can you see why this is bad news for the patient? Within a few short hours, large amounts of parasites, bits of cells, and other kinds of "garbage" are dumped into the bloodstream. These released materials cause the patient with malaria to develop a high fever (about 41°C, or 105°F). As the patient's temperature climbs, the patient feels very cold and develops "goosebumps" all over. When the patient's temperature begins to return to normal, the patient feels very hot and sweats a great deal. And the disease has only just begun! The chills and fever of malaria occur again and again for weeks. In the most common kind of malaria, the chills and fever occur every 48 hours, last for about 6 to 8 hours each time, and persist for several weeks. Toward the end of that time, the chills and fever become less frequent and less severe. However, all may not be well even if the patient appears to have recovered. Several weeks after it first began, malaria often happens all over again.

3–2 Section Review

1. Briefly describe the four groups of animallike protists.
2. Compare the ways in which sarcodines, ciliates, and zooflagellates move.
3. Describe three ways in which animallike protists affect other organisms.

Critical Thinking—*Applying Concepts*
4. Explain why destroying the places where mosquitoes breed can help prevent malaria.

Figure 3–15 *Humans contract the disease malaria through the bite of an infected mosquito (left). During the course of the disease, infected red blood cells burst at regular intervals, releasing* Plasmodium *cells that can infect other red blood cells or mosquitoes (right).*

ACTIVITY
CALCULATING

Divide and Conquer ❶

The ciliate *Glaucoma* reproduces by dividing into two cells. It reproduces at the fastest rate of all protists, dividing once every three hours under the best conditions. Assuming that conditions are perfect, how many times will *Glaucoma* divide during a day? If you start off with one cell that divides at the very start of the day, how many cells will you have at the end of the day if all of the cells survive?

ACTIVITY
CALCULATING
DIVIDE AND CONQUER

Skills: Computational, calculator

Before students begin to determine the number of *Glaucoma* that will reproduce in 24 hours, ask how many three-hour periods are in one day, or 24 hours. (Eight.) Explain that *Glaucoma* will reproduce eight times in 24 hours. One *Glaucoma* at 0 hours; two at 3 hours; 4 at 6 hours; 8 at 9 hours; 16 at 12 hours; 32 at 15 hours; 64 at 18 hours; 128 at 21 hours; and 256 at 24 hours. There will be 256 *Glaucoma* cells at the end of the day if all survive.

Integration: Use this Activity to integrate mathematics into your science lesson.

tures; they are heterotrophs. Sporozoans are parasites that feed on the cells and body fluids of their hosts; they have complex life cycles and produce spores.
2. Sarcodines move by extending their pseudopods and flowing into the extension; ciliates use their cilia like tiny oars to move; and zooflagellates propel themselves by flagella.
3. Animallike protists can affect other organisms through a symbiotic relationship that is helpful or neutral to the other organism, through parasitism, or by being a food source.
4. Malaria is caused by *Plasmodium* that have mosquitoes as hosts. The mosquitoes then infect humans with the *Plasmodium*. If the mosquito breeding places are eliminated, there will be few or no mosquitoes to become infected by and to spread the malaria-causing protists.

REINFORCEMENT/RETEACHING

Monitor students' responses to the Section Review questions. If students have difficulty with any of the concepts, review the appropriate material.

CLOSURE

▶ *Review and Reinforcement Guide*
At this point students should complete Section 3–2 of the *Review and Reinforcement Guide.*

CONNECTIONS
REVENGE OF THE PROTIST

Students may not have previously considered the beneficial role protists can play in the environment. In discussing the use of *Lambornella clarki* in mosquito abatement, emphasize the benefits of using the microorganism over the use of chemical pesticides. Help students understand the dangers chemical pesticides can create in the environment. They can pollute water sources and directly affect the health of humans and other animals. The *L. clarki*, on the other hand, are not harmful to the environment. Ask students to consider any problems that the use of *L. clarki* might have in eliminating the mosquito population. (They do not affect adult mosquitoes. Mosquitoes breed in many places where standing water is available. It may not be possible to distribute *L. clarki* in all these places.)

If you are teaching thematically, use the Connections feature to reinforce the themes of energy, systems and interactions, and scale and structure.

Revenge of the Protist

Bzzzzzzz! You hear a high-pitched whine, then feel an itch on your arm. You slap at the itch. Thwack! That mosquito is never going to bite anyone again.

Swatting is a great way to get rid of a few mosquitoes. But what do you do if you want to kill lots and lots of mosquitoes? No, bug spray is not a good answer.

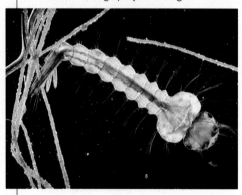

The poisons in bug spray *pollute* the environment, kill helpful insects such as bees and butterflies, and may be hazardous to your health. Although it may sound strange, a better answer is to use protists.

Scientists at the University of California at Berkeley recently discovered that a tiny ciliate called *Lambornella clarki (L. clarki)* can be a formidable foe to young mosquitoes. Young mosquitoes are worm-like, wingless, and legless creatures. They live at the surface of the water in quiet ponds, rain barrels, treeholes, and just about any other place that collects and holds water. The young mosquitoes feed on tiny particles that they filter from the water by using bristly mouthparts. These particles include small protists such as *L. clarki*.

To avoid being eaten, many microorganisms (microscopic organisms) change form. Some that are normally small and round make themselves large and flat. Others develop spines. *L. clarki* does something more amazing—it changes into a parasite that destroys the mosquitoes that feed on it.

Usually, *L. clarki* is a peaceful football-shaped ciliate that lives in treeholes and eats bacteria and other tiny bits of food. But when young mosquitoes are present, *L. clarki* cells become spherical, like a softball. These softball-shaped cells attach to the skin of young mosquitoes and then burrow into the body. The *L. clarki* cells multiply inside the body of their host, doing their deadly work. Eventually, the cells escape from the body of their dead or dying mosquito host. They can then infect other young mosquitoes. In nature, *L. clarki* cells can kill off all the young mosquitoes in a treehole. Having taken their revenge on the protist-devouring mosquitoes, the *L. clarki* cells resume life as peaceful football-shaped ciliates.

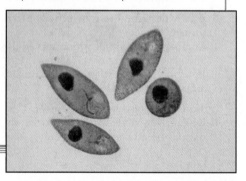

TEACHING STRATEGY 3–3

FOCUS/MOTIVATION

Bring a houseplant to class and set it on your desk or in some other visible position. Talk about the characteristics of the plant. The most obvious one, of course, is that it is green, so you will want to mention the chloroplasts and chlorophyll. You may also want to mention lack of mobility. Point out that this plant is multicellular.

• **How do plants obtain food?** (They make their own food.)

CONTENT DEVELOPMENT

Explain that students will now study plantlike protists. Ask them how plantlike protists may be similar to plants. (Accept all logical answers. Students may suggest the ability to make their own food.) Then help students compare plantlike protists to animallike protists.

• **How are plantlike protists like animallike protists?** (Answers may include they are

3-3 Plantlike Protists

Like other protists, plantlike protists are unicellular and most of them are capable of movement. **Like plants, plantlike protists are autotrophs that use light energy to make their own food from simple raw materials.** This food-making ability makes plantlike protists a vital part of the natural world. Can you see why? Many organisms rely directly on plantlike protists for food. Some of these organisms, such as animallike protists and tiny water animals, eat the plantlike protists. Others—certain animallike protists, sea anemones, corals, and giant clams, for example—are involved in symbioses with plantlike protists. The plantlike protists live inside their host's body and help to provide it with food. Still other animals rely indirectly on plantlike protists for food. For example, humans eat large fishes that eat smaller fishes that eat tiny animals that eat plantlike protists.

In addition to capturing energy and making it available to other organisms in the form of food, plantlike protists play another important role in the world. They produce oxygen as a byproduct of their food-making process. About 70 percent of the Earth's supply of oxygen is produced by plantlike protists.

Most kinds of plantlike protists are flagellates; that is, they move by means of flagella. To distinguish them from zooflagellates, plantlike flagellates are often called phytoflagellates. (The prefix

Figure 3–16 *Plantlike protists that do not belong to the three most important groups may still be rather interesting organisms. Some have elegant networks of glassy tubes that make up their skeleton (left). Others are covered with strange scales during their resting stage (top right and bottom right).*

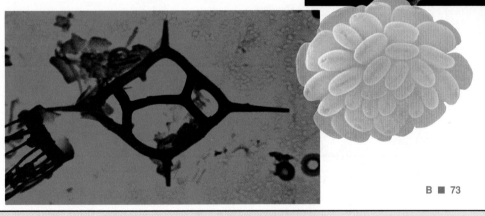

unicellular with nuclei; some live in symbiotic relationships.)

• **What is a major difference between plantlike protists and animallike protists?** (Animallike protists are heterotrophs; most plantlike protists are autotrophs throughout much of their lives. Plantlike protists have chloroplasts.)

• **Plantlike protists are most similar to what group of animallike protists? Why?** (To zooflagellates, because most plantlike protists are flagellates known as phytoflagellates.)

• **Why are plantlike protists important to humans and other animals on the Earth?** (Plantlike protists produce about 70 percent of Earth's supply of oxygen, which is essential for animal life on Earth.)

● ● ● ● **Integration** ● ● ● ●

Use the discussion of plantlike protists producing oxygen as a byproduct of food production to integrate the concepts of photosynthesis and the carbon/oxygen cycles into your lesson.

3-3 Plantlike Protists

MULTICULTURAL OPPORTUNITY 3-3

Students may be surprised at all the various places in which they find diatoms, including toothpaste, soap (especially gritty soaps for heavy-duty cleaning), fertilizers, and pesticides. Suggest that students research some ways that diatoms are used in products with which they are familiar.

ESL STRATEGY 3-3

To help students assimilate the organization and spelling of the three plantlike-protist types mentioned in this section, begin by having them unscramble the following letters to form the names of the three types:

saidotm   laeaefigdnolslt   geulesan

Then ask them to use the unscrambled names and headings for a three-column chart and place under each heading the corresponding characteristics from the following list:

pellicle   two-part glassy shell
red tide   twinkling lights
reddish eyespot   diatomaceous earth
toxin producer   taillike/belt flagella
chloroplasts   pouch with two flagella

Suggest that students work in pairs as they do this activity.

REINFORCEMENT/RETEACHING

Help students develop a food chain that shows the relationship between humans and plantlike protists from the sea. The chain should begin with the plantlike protists and then show that through a succession of plant-eating fish and fish-eating fish, the plantlike protist can be an indirect source of food for humans.

B ■ 73

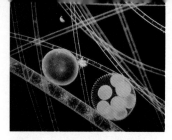

phyto- means plant.) There are many different kinds of plantlike protists. In this section, you will read about three of the more interesting groups: **euglenas** (yoo-GLEE-nahz), **diatoms** (DIGH-ah-tahmz), and **dinoflagellates** (digh-noh-FLAJ-eh-layts).

Euglenas

Euglenas come in a variety of forms. Some are long and oval. Others are shaped like triangles, hearts, or tops. Still others live in branching colonies that look like bushes with oversized leaves. And a few live in cup-shaped "houses." Although euglenas are quite varied, most share three characteristics: a pouch that holds two flagella, a reddish eyespot, and a number of grass-green structures that are used in the food-making process. Scientists call the green food-making structures chloroplasts (KLOHR-oh-plasts).

One kind of euglena is shown in Figure 3–18. Like the paramecium you read about in the previous section, euglenas have a tough outer covering called the pellicle. The pouch on one end of the euglena holds two flagella, one long and one short. The long flagellum is used in movement. In the cytoplasm near the pouch is the reddish eyespot. The eyespot is sensitive to light. Why is it important for a euglena to be able to find light? ❶

Figure 3–17 *Scientists do not agree where the dividing line between plantlike protists and plants should be placed. We classify long-stranded Spirogyra as a protist and spherical Volvox as a plant.*

Activity Bank

Shedding a Little Light on Euglena, p.176

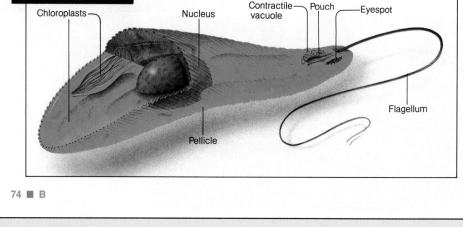

Figure 3–18 *The diagram shows the structure of a typical euglena. The pattern of grooves in a euglena's pellicle may be seen with the help of a scanning electron microscope (inset).*

Chloroplasts — Nucleus — Contractile vacuole — Pouch — Eyespot — Flagellum — Pellicle

• **Why do you think this organism is classified as a plantlike protist?** (It is an autotroph that contains chlorophyll, just as multicellular plants do.)

CONTENT DEVELOPMENT

When discussing *Euglena*, explain that ordinarily these organisms are autotrophs. Experiments have shown, however, that if they are placed in darkness in nutrient-rich solutions, their chloroplasts will disintegrate. The *Euglena* then become heterotrophs, absorbing nutrients

through their cell membranes. If returned to a lighted area, they continue to be heterotrophic.

• **Why do you think the euglena remain heterotrophic after they are returned to a lighted area?** (Lead students to conclude that because they have lost their chloroplasts, the *Euglena* no longer have the capacity to make their own food and must rely on outside food sources.)

Diatoms

How would you feel about brushing your teeth with protists? Chances are you might not be too keen on the idea. But like it or not, you are probably doing exactly this every time you brush your teeth. Why? Because a part of many toothpastes is made from plantlike protists called diatoms.

Diatoms are among the most numerous of protists. There are about 10,000 living species of these aquatic (water-dwelling) organisms. And as you can see in Figure 3–19, diatoms are among the most attractive of organisms. Each diatom is enclosed in a two-part glassy shell. The shell looks like a tiny glass box or petri dish, with one side fitting snugly into the other. The two parts of the shell are covered with beautiful patterns of tiny ridges, spines, and/or holes. Imagine how surprised people must have been when they first looked at diatoms through a microscope and discovered lacy designs like those of stained-glass windows on tiny grains of sand!

When diatoms die, their tough glassy shells remain. In time, the shells collect in layers and form deposits of diatomaceous (digh-ah-tuh-MAY-shuhs) earth. Diatomaceous earth is a coarse, powdery

ACTIVITY READING

The Universe in a Drop of Water

Protist-sized, their entire world is a puddle of fresh water. They talk to diatoms and paramecia. They are friends with the glow-in-the-dark dinoflagellate *Noctiluca*. For them, a tiny shrimp or crayfish is an incredibly huge monster. Unbelievably, they are humans!

Discover a strange world of microscopic humans in the science-fiction short story "Surface Tension," by James Blish. ❶

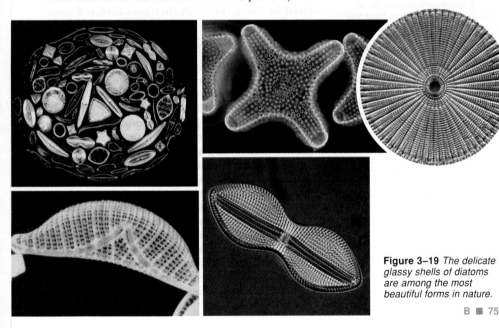

Figure 3–19 *The delicate glassy shells of diatoms are among the most beautiful forms in nature.*

B ■ 75

ANNOTATION KEY

Answers

❶ It needs light for its food-making process. (Inferring)

Integration

❶ Language Arts

ACTIVITY READING

THE UNIVERSE IN A DROP OF WATER

Skill: *Reading comprehension*

Before reading the short story, ask students to predict what problems microscopic humans may encounter in a world in which most other organisms are much larger than they are. You and the students may enjoy the short story in a class read-aloud session. Ask students to summarize the story after they have finished reading it.

Integration: Use this Activity to integrate language arts into your science lesson.

atom cell wall with a petri dish. A crumbled tissue or paper towel can be placed in the petri dish to represent the cell itself. Explain that the top and bottom halves of the petri dish are like the overlapping cell walls of the diatom. Point out that the cell walls contain silica, a compound used in making glass, and that silica produces diatomaceous earth.

GUIDED PRACTICE

🖳 Media and Technology

Skills Development

Skill: *Interpreting illustrations*

Use the transparency called Protists from the *Transparency Binder* to reinforce information about the structure and types of protists.

INDEPENDENT PRACTICE

▶ *Activity Book*

Students who need additional practice in identifying the structures of protists should complete the chapter activity called Identifying Protists. In the activity students label the structures of three different protists.

FOCUS/MOTIVATION

Begin your discussion of diatoms by displaying a few items in which diatomaceous earth may be an ingredient or may be used in their production: toothpaste, car polish, and silver polish (used as an abrasive); paint (used as a filler); and acoustical tile (used as a soundproofing material).

• **What do you think these products have to do with diatoms?** (Diatomaceous earth, derived from the glassy cell walls of diatoms, is used in making these products.)

Discuss the specific use of diatomaceous earth in each product displayed. Then demonstrate the structure of a di-

Among the funguslike protists are oomycetes, commonly known as water molds. Oomycetes can cause the disease that forms the white fuzz on aquarium fish and organic matter in water. Despite their common name, some oomycetes can live on land, but only under damp, humid conditions. Oomycetes were the cause of the Great Potato Famine in Ireland.

The cell walls of oomycetes are made of cellulose, through which the protists absorb food. Oomycetes produce motile spores that can swim to new locations to find food.

Figure 3–22 *The fuzzy white growth that covers this dead aquarium fish is the funguslike protist that killed it.*

Figure 3–23 *Unlike most protists, slime molds are visible to the unaided eye at some stages of their life cycle. In one phylum of slime molds, a single amebalike cell may grow into an enormous cell that contains many nuclei (left). In the other phylum, many amebalike cells come together to produce a large mass of cells that behaves like a primitive multicellular organism (right).*

Although it is clear that funguslike protists are important, it is not clear exactly what they are. But we do know about their characteristics. Funguslike protists are heterotrophs. Most have cell walls, although a few lack a cell wall altogether. Some are almost identical to amebas during certain stages in their life cycles. Many have flagella at some point in their lives. Because of characteristics like these, many experts classify these puzzling organisms as protists.

One of the more interesting types of funguslike protists are the **slime molds.** At one point in their life cycle, these protists are moist, flat, shapeless blobs that ooze slowly over dead trees, piles of fallen leaves, and compost heaps. A few slime molds in this stage of life are shown in Figure 3–23. Can you see why the name slime molds is appropriate for these organisms? ❶

Reproduction in slime molds involves the production of a structure called a fruiting body, which contains spores. Spores are special cells that are encased in a tough protective coating. Each spore can develop into a new organism.

Spores released by a slime mold develop into small amebalike cells. In one phylum of slime molds, each amebalike cell may develop into a single huge cell several centimeters in diameter. This large cell contains many nuclei. Eventually, the large cell settles in one place and produces fruiting bodies.

In the other phylum of slime molds, the small amebalike cells live independently for a while and reproduce rapidly. When the food supply is used up, groups of the amebalike cells gather together to produce a large mass of cells. This mass of cells begins to function as a single organism. Could this have

---

3–4 (continued)

REINFORCEMENT/RETEACHING

▶ *Activity Book*

Students who need additional practice in observing protists should complete the chapter activity called Collecting Protists. In this activity students collect protists and observe them, using a microscope. Advise students that they should not go to a pond alone and that they should inform an adult about where they are going.

GUIDED PRACTICE

Skills Development

Skills: Making observations, relating facts, drawing conclusions

At this point have students complete the in-text Chapter 3 Laboratory Investigation: Examining a Slime Mold. In the investigation students observe the growth and structure of a slime mold and draw conclusions based on their observations.

ENRICHMENT

▶ *Activity Book*

Students will be challenged by the chapter activity Comparing Protists. In the activity students research information about three protists, comparing their structures and functions.

CONTENT DEVELOPMENT

When discussing slime molds, emphasize that at various stages of their lives, slime molds exhibit characteristics of amebas and of fungi.
• **How is a slime mold similar to an ameba?** (The slime mold engulfs food and moves much as an ameba does.)
• **How is slime mold similar to a fungus?** (The slime mold reproduces in the same way as the fungus does.)

• **Do you think slime molds should be classified as funguslike protists or as fungi?** (Accept all logical opinions. Students should support their opinions with information from the text.)

● ● ● ● **Integration** ● ● ● ●

Use the discussion about the slime mold to integrate the life science concept of evolution into your lesson.

Figure 3–24 *The reproductive structures in slime molds, which are known as fruiting bodies, contain spores. What do the spores become?* ②

been a way in which many-celled organisms evolved from single-celled ones? Some scientists believe it could. This unusual behavior forces scientists to stretch the definition of protists. Protists are defined as being unicellular—but here is a group of protists acting like a primitive multicellular (many-celled) organism!

The solid mass of cells may travel for several centimeters. It then forms a fruiting body that produces spores. These spores develop into amebalike cells that continue the cycle.

The slime molds that form multicellular structures are interesting to biologists who study how cells communicate. The formation of a complex structure like the fruiting body from what was formerly a group of independent cells is an intriguing process. It has kept biologists busy for decades, and its secrets are still not fully understood.

3–4 Section Review

1. How do slime molds differ from other protists?
2. What are the characteristics of funguslike protists?
3. Draw a flowchart that shows the life cycles of the two types of slime molds.
4. How do funguslike protists affect humans?

Critical Thinking—*Expressing an Opinion*
5. Should slime molds be placed in a kingdom by themselves? Why? What sort of information would you like to have in order to make a better, more informed decision on this matter?

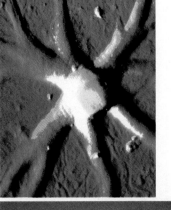

Figure 3–25 *In one phylum of slime molds, small amebalike cells come together (top) to form a mass of cells that behave like a single organism (bottom).*

B ■ 79

Laboratory Investigation

EXAMINING A SLIME MOLD

BEFORE THE LAB

1. At least one day prior to the investigation, gather enough materials for students to work in teams of two to four.
2. Cultures of the slime mold *Physarum polycephalum* can be obtained from biological-supply companies. Prepare and dispense the filter paper squares containing the slime mold at the time they are needed. To do this, use a dissecting needle or a cotton swab to transfer a small amount of mold to the small piece of filter paper.
3. The oatmeal flakes can be pulverized with a mortar and pestle.

PRE-LAB DISCUSSION

Have students read the complete laboratory investigation.
• **What is the purpose of this laboratory investigation?** (To observe some of the characteristics of a living slime mold.)
• **What do you think is the purpose of the oatmeal?** (To serve as a source of food for the slime mold.)
• **What changes do you think will take place in the slime mold over the three-day period?** (Accept all logical answers, but students will likely hypothesize that the slime mold will grow or spread.)

Laboratory Investigation

Examining a Slime Mold

Problem

What are the characteristics of a slime mold?

Materials *(per group)*

large glass beaker	crushed oatmeal flakes
small glass bowl	medicine dropper
paper towel	magnifying glass
filter paper containing slime mold	dissecting needle

Procedure 🧪 🔲

1. Wrap the small glass bowl with a paper towel so that the mouth of the bowl is covered by a smooth flat paper surface.
2. Place the covered bowl in the beaker so that the mouth of the bowl faces up.
3. Partially fill the beaker with water so that the water level is about three fourths of the way up the sides of the bowl.
4. Place the small piece of filter paper containing the slime mold in the center of the paper towel that covers the bowl.
5. Sprinkle a tiny amount of crushed oatmeal flakes next to the piece of filter paper.
6. Using the medicine dropper, add two to three drops of water to the slime mold and oatmeal flakes. Set in a cool, dark place.
7. Examine the beaker each day for three days. Record your observations.
8. After three days, remove the glass bowl from the beaker. Place the bowl on your work surface.
9. Using your magnifying glass, examine the slime mold.
10. With a dissecting needle, puncture a branch of the slime mold. **CAUTION:** *Be careful when using a dissecting needle.* Observe the slime mold for a few minutes.

Observations

1. Describe the changes that took place in the slime mold during the three-day observation period.
2. What activity did you observe in the slime mold when you examined it with the magnifying glass?
3. Describe what happened to the puncture that you made in the slime mold.

Analysis and Conclusions

1. Explain why oatmeal was sprinkled on the paper towel.
2. Is the slime mold a heterotroph or an autotroph? Explain.
3. Based on your observations, describe the characteristics of a slime mold.
4. On Your Own Design an experiment to determine the response of the slime mold to substances such as salt or sugar.

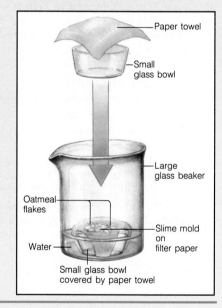

Paper towel
Small glass bowl
Large glass beaker
Oatmeal flakes
Water
Slime mold on filter paper
Small glass bowl covered by paper towel

TEACHING STRATEGY

1. Point out that the dissecting needles and magnifying glasses will not be needed on the first day of the activity.
2. Remind students to place a label on their glass beaker for identification. Glass-marking pens or masking-tape labels can be used.

DISCOVERY STRATEGIES

Discuss how the investigation relates to the chapter ideas by asking open questions similar to the following.
• **Why do you think the slime mold cultures are placed in a cool, dark storage space?** (Slime mold grows well under these conditions—inferring.)
• **If we were going to observe euglenas, what kind of food would we give them and where would we store them?** (They are autotrophs and make their own food, so

Study Guide

Summarizing Key Concepts

3-1 Characteristics of Protists

▲ Protists are microscopic, unicellular organisms that have a nucleus and a number of other specialized cell structures.

▲ According to one hypothesis, the first protists were the result of a symbiosis among several types of bacteria.

▲ Although there is much debate about their proper classification, protists can be grouped in three general categories. These are: animallike protists, plantlike protists, and funguslike protists.

3-2 Animallike Protists

▲ Like animals, animallike protists are heterotrophs that can move and that are made up of cells that contain a nucleus and lack a cell wall.

▲ Animallike protists are divided into four main groups: sarcodines, ciliates, zooflagellates, and sporozoans.

▲ Sarcodines have pseudopods.
▲ Ciliates are characterized by cilia.
▲ Flagellates move by means of flagella.
▲ Sporozoans are parasites.

3-3 Plantlike Protists

▲ Plantlike protists are autotrophs that use light energy to make their own food from simple raw materials.

▲ Many organisms rely on plantlike protists for food.

▲ About 70 percent of the Earth's supply of oxygen is produced by plantlike protists.

3-4 Funguslike Protists

▲ Some funguslike protists cause diseases in crops or in animals.

▲ Funguslike protists are heterotrophs.

Reviewing Key Terms

Define each term in a complete sentence.

3-1 Characteristics of Protists
protist

3-2 Animallike Protists
sarcodine
ciliate
zooflagellate
sporozoan
pseudopod
ameba
cilia
paramecium
flagellum

3-3 Plantlike Protists
euglena
diatom
dinoflagellate

3-4 Funguslike Protists
slime mold

B ■ 81

ANALYSIS AND CONCLUSIONS

1. To serve as a food source for the slime mold.

2. A heterotroph because it cannot make its own food.

3. A slime mold is a heterotroph that acts like an ameba, engulfing its food.

4. Experiment designs will vary but should involve placing a small amount of sugar or salt near the slime mold and watching its reaction.

GOING FURTHER: ENRICHMENT

Part 1

Have students try the experiments they designed for Analysis and Conclusions question 4. Some groups can check the effects of salt and others the effects of sugar. Suggested procedure: Students might add a small amount of sugar or salt to the oatmeal before placing it on the filter paper.

Part 2

Some students may want to perpetuate their slime mold cultures. If so, they can transfer the culture to petri dishes with agar. Some crushed oatmeal should be added to the agar surface of the dishes. The mold will spread across the surface in about three days.

food does not have to be given to them; they should be stored where they have access to light so that they can make their own food—hypothesizing.)
• **How might the investigation procedures vary if we were going to observe amebas?** (The amebas would be grown in the water, and the food source would be placed there as well; the amebas would not be punctured—hypothesizing.)

OBSERVATIONS

1. The slime mold should resemble an ameba and spread across and eat the oatmeal. The slime mold should also increase in size.

2. The slime mold should move and engulf anything in its path.

3. After a short time, the puncture should disappear.

Chapter Review

ALTERNATIVE ASSESSMENT

The *Prentice Hall Science* program includes a variety of testing components and methodologies. Aside from the Chapter Review questions, you may opt to use the Chapter Test or the Computer Test Bank Test in your *Test Book* for assessment of important facts and concepts. In addition, Performance-Based Tests are included in your *Test Book*. These Performance-Based Tests are designed to test science process skills, rather than factual recall. Since they are not content dependent, Performance-Based Tests can be distributed after students complete a chapter or after they complete the entire textbook.

CONTENT REVIEW

Multiple Choice

1. d
2. b
3. a
4. c
5. b
6. d
7. a
8. b
9. d

True or False

1. F, Animallike
2. F, chloroplasts
3. T
4. F, zooflagellates
5. T
6. F, parasites
7. F, cilia

Concept Mapping

Row 1: have; Unicellular
Row 2: Animallike; Plantlike
Row 3: Heterotrophs

CONCEPT MASTERY

1a. Sarcodines (foraminiferans is also acceptable).
b. Diatoms.
c. Dinoflagellates.
d. Sporozoans (*Plasmodium* is also acceptable).
e. Zooflagellates.
f. Euglenas.
g. Ciliates.
h. Slime molds.
i. Sarcodines.
j. Diatoms.

Content Review

Multiple Choice

Choose the letter of the answer that best completes each statement.

1. Which of the following is characteristic of most protists?
   a. They are unable to move on their own.
   b. They can be seen with the unaided eye.
   c. They lack a nucleus and many other cell structures.
   d. They are unicellular.
2. Which structure helps a freshwater protist get rid of excess water?
   a. food vacuole   c. macronucleus
   b. contractile vacuole   d. cilium
3. Malaria is caused by a type of
   a. sporozoan.   c. dinoflagellate.
   b. sarcodine.   d. zooflagellate.
4. Which of the following uses cilia to move?
   a. euglena   c. paramecium
   b. ameba   d. *Plasmodium*
5. Which of the following is considered to be an animallike protist?
   a. slime mold   c. euglena
   b. ameba   d. dinoflagellate

6. Animallike protists
   a. have a thick cell wall.
   b. produce about 70 percent of the Earth's supply of oxygen.
   c. are responsible for red tides.
   d. are heterotrophs.
7. Which is not a type of plantlike protist?
   a. zooflagellate   c. phytoflagellate
   b. dinoflagellate   d. diatom
8. Radiolarians, foraminiferans, and amebas belong to the group of protists known as
   a. ciliates.   c. sporozoans.
   b. sarcodines.   d. euglenas.
9. Slime molds
   a. are always unicellular.
   b. are autotrophs.
   c. are considered to be animallike protists.
   d. form structures called fruiting bodies.

True or False

If the statement is true, write "true." If it is false, change the underlined word or words to make the statement true.

1. <u>Funguslike</u> protists are also known as protozoa.
2. Each euglena contains one or more grass-green <u>macronuclei</u>.
3. A radiolarian captures food by using its <u>pseudopods</u>.
4. African sleeping sickness is caused by <u>ciliates</u>.
5. The first protists probably developed from a <u>symbiosis</u> among several kinds of bacteria.
6. All sporozoans are <u>producers</u>.
7. Paramecia swim using <u>flagella</u>.

Concept Mapping

Complete the following concept map for Section 3–1. Refer to pages B6–B7 to construct a concept map for the entire chapter.

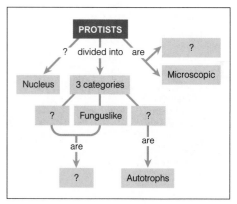

k. Funguslike protists.
l. Zooflagellates.
2. They are all animallike protists that have pseudopods. Some use their pseudopods only to capture food; others also use them for movement. Some of the have shells; others do not.
3. Particles of food such as bacteria are swept into the paramecium's oral groove by a current caused by the beating of the cilia. A bacterium travels down the oral groove into the gullet. At the end of the gullet, the bacterium is enclosed in a food vacuole. As the food vacuole travels through the cytoplasm, the bacterium is digested and nutrients move from the food vacuole into the cytoplasm. Finally, the food vacuole fuses with the cell membrane at the anal pore, releasing the undigestible parts of the bacterium into the water.
4. Malaria is caused by the sporozoan *Plasmodium*. *Plasmodium* is transmitted to a human through the bite of an infected mosquito. A mosquito becomes infected with *Plasmodium* when it takes in blood from a human with malaria.

Concept Mastery

Discuss each of the following in a brief paragraph.

1. Name the group of protists that is most closely linked with each of the following:
   a. chalk
   b. diatomaceous earth
   c. red tides
   d. malaria
   e. African sleeping sickness
   f. turning chloroplasts on and off
   g. two kinds of nuclei
   h. fruiting bodies
   i. pseudopods
   j. two-part glassy shell
   k. potato famine
   l. digesting wood
2. How are sarcodines similar to one another? How are they different?

3. Explain how a paramecium catches the bacteria on which it feeds. Then describe the cell structures a captured bacterium would encounter from the time it enters the paramecium cell to the time the digested bacterium is expelled from the cell.
4. What protist causes malaria? How is this protist transmitted from one host to another?
5. Tiny plantlike protists are found in the cells of certain radiolarians. How do these plantlike protists help their host? How might the radiolarian help its "guests"?

Critical Thinking and Problem Solving

Use the skills you have developed in this chapter to answer each of the following.

1. **Relating cause and effect** Certain kinds of chemical wastes cause plants and plantlike organisms to grow at an extremely rapid rate. With this in mind, how might pollution cause red tides?
2. **Applying concepts** Examine the protist in the accompanying photograph. In what group of protists should this organism be placed? Explain. What sort of information would help you to classify this protist?

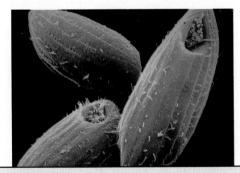

3. **Making predictions** What would happen if all the plantlike protists were to vanish from the face of the Earth? Explain.
4. **Evaluating theories** When a certain amebalike protist is treated with a bacteria-killing antibiotic, something strange happens. The small rod-shaped structures located around each of its many nuclei disappear, and the protist soon dies. Does this finding support the hypothesis that protists evolved from a symbiosis among several kinds of bacteria? Why or why not? What additional information would you like to have in order to be more sure of your answer?
5. **Using the writing process** Imagine that you are the protist of your choice. Write a letter to a potential pen pal describing yourself, where you live, your hopes and dreams, and whatever else you think is important.

---

3. Accept all logical answers. Students should recognize that plantlike protists produce most of Earth's oxygen and form the base of many food pyramids. If the plantlike protists were to disappear, most of Earth's living organisms would soon perish.

4. Accept all logical answers. Students should recognize that they probably need to know more about the rod-shaped structures and about the protist before they can be sure of the answer. The question is based on actual studies on the protist *Pelomyxa palustris*. The structures that disappeared are endosymbiotic bacteria.

5. Students' letters should be well written and reflect an accurate understanding of the protist they selected.

KEEPING A PORTFOLIO

You might want to assign some of the Concept Mastery and Critical Thinking and Problem Solving questions as homework and have students include their responses to unassigned questions in their portfolio. Students should be encouraged to include both the question and the answer in their portfolio.

ISSUES IN SCIENCE

The following issues can be used as springboards for discussion or given as writing assignments:

1. The research needed to find cures or vaccines for diseases caused by protists is costly and time-consuming. Do you think scientists should pursue such research or concentrate on research for such diseases as cancer? Explain your answer.

2. What do you think would happen to life on Earth if all protists were killed? What do you think would happen if all living organisms except protists were killed?

---

5. The plantlike protists help to make food for their host. The radiolarian may sometimes eat the plantlike protists as well. The radiolarian's relatively large size and glasslike skeletal spines may help to protect the plantlike protists.

CRITICAL THINKING AND PROBLEM SOLVING

1. If certain kinds of chemicals pollute the water, they might cause a rapid increase in the dinoflagellate population, and huge numbers of dinoflagellates form a red tide.

2. It should be classified as a ciliate because it has cilia. (The organism shown is *Tetrahymena*.) Accept all logical requests for additional information.

Chapter 4 FUNGI

SECTION	HANDS-ON ACTIVITIES
4–1 Characteristics of Fungi pages B86–B90 Multicultural Opportunity 4–1, p. B86 ESL Strategy 4–1, p. B86	**Student Edition** ACTIVITY (Discovering): Making Models of Multicellular Organisms, p. B88 ACTIVITY BANK: Spreading Spores, p. B177 **Activity Book** CHAPTER DISCOVERY: Observing Yeast, p. B97
4–2 Forms of Fungi pages B90–B94 Multicultural Opportunity 4–2, p. B91 ESL Strategy 4–2, p. B91	**Student Edition** ACTIVITY (Doing): Making Spore Prints, p. B93 LABORATORY INVESTIGATION: Growing Mold, p. B100 ACTIVITY BANK: Yeast Meets Best, p. B178 **Laboratory Manual** Examining Three Forms of Fungi, p. B31 Culturing Yeast Cells, p. B37 **Activity Book** ACTIVITY: Making Spore Prints, p. B107 ACTIVITY BANK: Psst! Wanna Make Some Dough? p. B 205 **Teacher Edition** Characteristics of Yeast, p. B84d
4–3 How Fungi Affect Other Organisms pages B94–B99 Multicultural Opportunity 4–3, p. B95 ESL Strategy 4–3, p. B95	
Chapter Review pages B100–B103	

OUTSIDE TEACHER RESOURCES
Books

Alexopoulos, C. J., and C. W. Mims. *Introductory Mycology*, Wiley.

Brodie, J. *Fungi: Delight of Curiosity*, University of Toronto Press.

McKnight, Kent H., and Vera McKnight. *A Field Guide to Mushrooms of North America*, Houghton Mifflin.

Pearson, Lorentz C. *The Mushroom Manual*, Naturegraph.

Pirozynski, K.A., and David Hawksworth, eds. *The Coevolution of Fungi With Plants and Animals*, Academic Press.

Rayner, A., and L. Boddy. *Fungal Decomposition of Wood: Its Biology and Ecology*, Wiley.

OTHER ACTIVITIES	MEDIA AND TECHNOLOGY
Review and Reinforcement Guide Section 4–1, p. B31	🎧 English/Spanish Audiotapes Section 4–1
Activity Book ACTIVITY: Identifying Fungi, p. B101 ACTIVITY: Graphing Fungi, p. B103 ACTIVITY: Mushroom Structure, p. B105 **Review and Reinforcement Guide** Section 4–2, p. B33	Transparency Binder Mushroom Development Bread Mold 🎧 English/Spanish Audiotapes Section 4–2
Student Edition ACTIVITY (Reading): A Secret Invasion, p. B94 ACTIVITY (Writing): Human Fungal Diseases, p. B96 ACTIVITY (Writing): Taking a "Lichen" to It, p. B97 **Activity Book** ACTIVITY: Relationships With Fungi, p. B109 **Review and Reinforcement Guide** Section 4–3, p. B35	🎧 English/Spanish Audiotapes Section 4–3
Test Book Chapter Test, p. B69 Performance-Based Tests, p. B129	◈ Test Book Computer Test Bank Test, p. B75

*All materials in the Chapter Planning Guide Grid are available as part of the Prentice Hall Science Learning System.

Audiovisuals

Fungi, film or video, Ward
Fungi and Man, film or video, Benchmark Films
Fungi (Diversity and Ecology), filmstrip, Ward

Fungi: The One Hundred Thousand, film or video, Coronet
The Impact of Fungi on Man and His Environment, filmstrip with cassette, Carolina Biological Supply Co.
Molds and How They Grow, film or video, Coronet

CHAPTER OVERVIEW

Like algae, the bodies of fungi are nonvascular. But unlike algae, they lack chlorophyll, so they are always heterotrophs. Because they cannot make food or move around like animals to find it, the fungi always live in or on a food source. Some live as parasites, taking their food from a living host. Others are saprophytes, obtaining their food from dead organisms or the products of organisms. Three familiar types of fungi are mushrooms, yeasts, and molds.

Some fungi, such as yeasts, are unicellular. Most fungi have multicellular bodies composed of threadlike filaments called hyphae. The hyphae produce digestive enzymes that are released into the substrate on which the fungus grows. The food, which is digested externally, is then absorbed by the hyphae. Most fungi reproduce by producing spores in structures called fruiting bodies. In many fungi, the spores can by produced by either a sexual or an asexual process.

Lichens are not a single organism but a combination of a fungus and an alga living in a symbiotic relationship. In this association, the fungus provides the alga with water and minerals absorbed from the surface on which the lichen grows. In turn, the alga makes food for the fungus and itself. Lichens are capable of living in a wide variety of environments ranging from frozen arctic regions to dry, hot deserts. They also can survive on bare rocks, on tree bark, and on mountaintops. Lichens are called pioneer organisms because they are often the first organisms to colonize a rocky, barren area. Acids released as a metabolic byproduct begin the process of soil formation, making it possible for other organisms to follow.

4-1 CHARACTERISTICS OF FUNGI
THEMATIC FOCUS

The purpose of this section is to describe the basic characteristics of fungi. Topics such as basic structure, reproduction, and method of obtaining food will be discussed.

The role of fungi as recyclers of organic material will be discussed. This recycling is important because it prevents the loss of chemical energy and it returns nutrients to the soil.

Fungi are found almost everywhere on Earth and in almost every kind of environment. An interesting aspect of fungi that will be discussed is the many ways in which fungi disperse spores in order to accomplish asexual reproduction.

The themes that can be focused on in this section are energy, scale and structure, and unity and diversity.

Energy: Fungi are heterotrophs that obtain energy by absorbing food. Some fungi obtain food from living organisms. Others are decomposers.

*Scale and structure: Students will discover that except for yeasts, which are unicellular, fungi are made up of a tangled mass of filaments called hyphae.

*Unity and diversity: Although fungi vary greatly in appearance, they are similar in the way they obtain their food, in basic structure, and in the way they reproduce.

PERFORMANCE OBJECTIVES 4-1
1. Describe the food-absorbing and reproductive structures common to most fungi.
2. Explain the functions of hyphae and spores in fungi.

SCIENCE TERMS 4-1
fungus p. B86
hypha p. B88
spore p. B89

4-2 FORMS OF FUNGI
THEMATIC FOCUS

The purpose of this section is to discuss three types of fungi—mushrooms, yeasts, and molds. Students will learn that fungi affect humans in many ways, some of which are beneficial. For example, yeasts are important to people because of their roles in baking and brewing. Mushrooms—the kind that are not poisonous—are important as a desirable food item.

The themes that can be focused on in this section are evolution and patterns of change.

*Evolution: Fungi, like other organisms, are formally classified in a way that best shows the evolutionary relationships among the members of the group.

*Patterns of change: Multicellular fungi undergo many transformations during their life cycles. The development of spores into mushrooms is a good example of the transformation process.

PERFORMANCE OBJECTIVES 4-2
1. Identify the structures of a mushroom's fruiting body.
2. Describe reproduction in yeasts.
3. Name two examples of molds.

SCIENCE TERMS 4-2
mushroom p. B91
yeast p. B92
mold p. B93

4–3 HOW FUNGI AFFECT OTHER ORGANISMS

THEMATIC FOCUS

The purpose of this section is to continue the discussion of the ecological significance of fungi and to view the effects that fungi have on human life. Another topic of discussion will be the participation of fungi in symbiotic relationships with other organisms.

Effects of fungi that are not beneficial to humans include plant diseases that damage crops. Among the plant diseases that will be discussed are chestnut blight, wheat rust, and corn smut. Students will learn that although most fungal diseases are associated with plants, some fungi cause disease in humans. Among the more familiar conditions caused by fungi are athlete's foot, ringworm, and thrush.

The themes that can be focused on in this section are systems and interactions and stability.

Systems and interactions: Fungi interact with other organisms in many ways. Some are involved in helpful symbiotic relationships. Others are parasites. A few are predators. Still others are a source of food for other organisms.

Stability: Some fungi break down dead organisms. The broken-down materials can then be used by other organisms. This keeps materials from being "locked up" in the bodies of dead organisms.

PERFORMANCE OBJECTIVES 4–3

1. Describe the structure and role of the organisms making up a lichen.
2. Explain how both fungi and plants benefit from fungus-root associations.
3. Give two examples of plant and animal diseases caused by fungi.

SCIENCE TERMS 4–3
lichen p. B97

Discovery *Learning*

TEACHER DEMONSTRATIONS MODELING

Variations in Mushrooms

Obtain several different varieties of dried mushrooms at a Chinese market or other food-supply shop. The mushroom called "Cloud Ears" works particularly well. Allow students to examine the dried mushrooms.

• **What type of organism do you think these are?** (Accept all answers.)

Tell students that the samples are dried mushrooms and that mushrooms belong to the class of organisms called fungi.

• **What do you think these mushrooms will look like if they are soaked in water?** (Accept all answers. Students should predict that the mushrooms will absorb the water and get larger.)

Soak the dried mushrooms in warm water for about 30 minutes. Then have students examine the samples. Challenge them to identify the fruiting bodies and spores in the various samples.

Characteristics of Yeast

Show the class a package of dry yeast. Point out that if conditions for growth and reproduction become unfavorable, yeasts form spores within their cells and become dormant. The dry yeast within this package is a collection of dormant yeast spores.

• **What do we need to do to cause the yeast to become active?** (Most students will be aware that the yeast can be activated by adding water to the package contents.)

Point out that in addition to moisture, yeasts also need a source of food.

• **What do yeasts usually use as a source of food energy?** (The source of energy for most yeasts is sugar.)

Prepare a culture of yeast cells by providing the conditions needed to activate the packaged yeast. To culture the yeast, mix about 5 mL of molasses and 500 mL of warm tap water in a beaker. Then add one-half package of dry yeast. Stir the mixture, cover the top loosely with aluminum foil, and allow it to stand for about 30 minutes in a warm place.

• **Why was molasses added to the yeast mixture?** (The molasses contains the sugar needed as food.)

Have the class use microscopes to view the yeast cells under low and high power. The cells might be seen better if a drop of methylene blue stain is added to the culture under the coverslip. Direct students to look for cells with buds under high power.

• **What do these buds represent?** (Yeast cells reproduce by budding; these buds will develop into new yeast cells.)

CHAPTER 4

Fungi

INTEGRATING SCIENCE

This life science chapter provides you with numerous opportunities to integrate other areas of science, as well as other disciplines, into your curriculum. Blue numbered annotations on the student page and integration notes on the teacher wraparound pages alert you to areas of possible integration.

In this chapter you can integrate life science and ecology (p. 87), life science and cells (p. 89), physical science and heat (p. 92), food science (p. 92), social studies (pp. 93, 95, 99), language arts (p. 94), mythology (p. 96), life science and symbiosis (p. 97), and earth science and geology (p. 98).

SCIENCE, TECHNOLOGY, AND SOCIETY/COOPERATIVE LEARNING

Three new breakthroughs in fungal research may prove to be beneficial for humans in a variety of ways.

In medicine, many of our most common antibiotics are produced by filamentous fungi. Some of the older antibiotics (such as penicillin), however, have become less effective against infectious diseases due to the development of resistant strains of bacteria. Years of scientific research have resulted in a new generation of antibiotics that are also produced from fungi. Cephalosporins, from the fungus *Cephalosporium,* and new varieties of penicillin are among these new antibiotics.

In agriculture, a new type of pesticide may be developed from the spores of a predatory fungus found in the rain forests of Costa Rica. When a spore from this fungus falls on an insect, it germinates, and the hyphae enter the animal's body. Growth of the hyphae interferes with the animal's nervous system, causing death. The spores germinate into mushrooms, which grow on the dead carcass and soon release more spores. Scientists are considering the idea of using these fungal spores to control crop pests. The spores would be suspended in a liquid and sprayed on crops. Insects attacking the crops would become infected with the

INTRODUCING CHAPTER 4

DISCOVERY LEARNING

▶ *Activity Book*

Begin your introduction to this chapter by using the Chapter 4 Discovery Activity from the *Activity Book.* Using this activity, students will observe some characteristics of yeast.

USING THE TEXTBOOK

Have students examine the photograph on page B84.

• **What is happening in the photograph?** (Accept all answers.)

Explain that the pig is hunting for a food called truffles. Tell students that truffles are members of a class of organisms called fungi. Another type of fungus is the mushroom. Have students read the caption to the photograph.

Fungi

Guide for Reading

After you read the following sections, you will be able to

4-1 Characteristics of Fungi
- Describe the major characteristics of fungi.

4-2 Forms of Fungi
- Compare mushrooms, yeasts, and molds.

4-3 How Fungi Affect Other Organisms
- Discuss ways in which fungi interact with other living things.

Early on a spring morning, a large pampered pig is pushed in its own wheelbarrow into an oak forest near Perigord, France. There its master gently puts the pig on a leash. The pig is now ready for the hunt!

Soon the pig catches a whiff of a wonderful odor—one that is too faint for people to smell. The pig begins to dig, but its master quickly stops it. He does not want the animal to destroy the buried treasure it has found. This "treasure" is not gold or silver. It is an ugly round black fungus known as a truffle. Because of its delicious flavor, this thick-skinned, warty cousin of the mushroom is considered a delicacy. In fact, truffles can sell for more than $1400 a kilogram!

Truffles are just one of the many different kinds of fungi. What are fungi? What do they look like? How do they affect humans and other living things? Read on to discover more about the strange world of the kingdom Fungi.

Journal *Activity*

You and Your World Have you ever eaten a mushroom? What do you remember about the first time you ate a mushroom? Has your opinion of mushrooms changed since your first tasting? In your journal, explore your thoughts and feelings about this edible fungus.

The sensitive nose of a trained pig can detect truffles buried beneath the soil. Dug up, scrubbed, and cooked, truffles become expensive taste treats.

B ■ 85

---

predatory fungus and die while the plants would remain unharmed.

Fungi may also help to provide new sources of energy. Recent experiments have led to the discovery of a common wood-rot fungus that transforms lignite coal into a liquid. Further research with this fungus has enabled scientists to isolate an enzyme that dramatically speeds up the reaction. Perhaps this enzyme can be used to liquefy coal underground so that it can be pumped to the surface like oil.

Cooperative learning: Using preassigned lab groups or randomly selected teams, have groups complete one of the following assignments:
- Select one of the three fungal breakthroughs described above and outline a 60-second news item, both audio and video portions, that shares this news with the general public. Encourage groups to be creative in their coverage of the event they select. Groups can produce a script with accompanying drawings and diagrams for their final product or actually deliver their news item to the class for evaluation.
- As the committee in charge of the "Love a Fungus" booth at an upcoming agricultural convention, prepare a display that informs farmers and ranchers of the positive roles that fungi play in our world. Groups should prepare a diagram of their booth that shows how they plan to illustrate the role of fungi in the world and outlines of any handouts they plan to distribute.

See Cooperative Learning in the *Teacher's Desk Reference.*

JOURNAL ACTIVITY

You may want to use the Journal Activity as the basis of a class discussion. In their journal entries, have students include a description of what the mushroom looked and tasted like. Students should be instructed to keep their Journal Activity in their portfolio.

- **Do you think truffles and mushrooms are plants? Why or why not?** (Accept all logical responses. Students may say that these fungi are plants because they grow in the ground and they do not move around like animals.)

Write the word *fungus* (plural: *fungi*) on the chalkboard. Have students write or express orally whatever they think they already know about these organisms.

Explain to students that at one time, organisms like truffles and mushrooms were sometimes classified with plants in the kingdom Plantae. This classification was based on some of the similarities to plants that students mentioned. Also point out that at one time, some biologists classified these organisms in the kingdom Protista. This classification was made because the organisms share certain characteristics, such as reproduction by spores, with protists.

Explain that fungi are different from plants because they do not perform photosynthesis and thus cannot make their own food.

4-1 Characteristics of Fungi

MULTICULTURAL OPPORTUNITY 4-1

As students will find out by reading the chapter on fungi, the truffle is one of the world's most valued and expensive foods. Different types of truffles grow during different times of the year. Some grow in the late fall, some in December and January, and still others in early spring. Truffles grow near trees and are found with the help of pigs or specially trained "truffle dogs."

Truffles are especially important in French cuisine, in which they are used for flavoring or as a garnish. Suggest that interested students find out about various types of mushrooms and the role they play in ethnic cooking. Following are some examples.

• The snow mushroom looks like a flower in bloom and can be purchased in Chinese markets.

• The shiitake, which grows in China, Japan, Indonesia, and Taiwan, is now sold in United States supermarkets, mostly in dried form.

• The chanterelle, a golden-colored mushroom, is valued by chefs all over Europe, Asia, and the United States.

ESL STRATEGY 4-1

Have students complete the paragraph below by using the terms given in the box. Students should choose the best term for each blank in the paragraph.

hyphae	chemicals	heterotrophs
yeast	organisms	mushrooms
molds	symbiotic	multicellular
spores	digested	unicellular

All fungi are _____. They obtain food from _____ relationships, the remains of dead _____, and by absorbing the food that _____ (released by the fungi) have _____. Many fungi reproduce by _____.

(Answers: heterotrophs, symbiotic, organisms, chemicals, digested, spores.)

Figure 4–1 When the insect-killing fungus has completed its deadly work, it produces stalked fruiting bodies that grow from the empty husks of its insect victims. What is the function of the fruiting bodies? ❶

4-1 Characteristics of Fungi

Unnoticed, a speck of dust lands on the back of an ant. But this is no ordinary dust—it is alive! Tiny glistening threads emerge from the dust and begin to grow into the ant's body. As they grow, the threads slowly devour the ant while it is still alive. Chemicals released by the threads dissolve the ant's tissues. The threads absorb the dissolved tissues and use the energy from them to spread further into the ant's body. Within a few days, the ant's body is little more than a hollow shell filled with a tangle of the deadly threads. Then the threads begin to grow up and out of the dead ant, winding together to produce a long stalk with a knobby lump at its tip. The stalked structure releases thousands of dustlike spores, which are carried by the wind to new victims.

The strange ant killer is a **fungus** (FUHN-guhs; plural: fungi, FUHN-jigh). Fungi range in size from tiny unicellular (one-celled) yeasts to huge tree fungi over 140 centimeters long. Fungi may look like wisps of gray cotton, white volleyballs, tiny brightly colored umbrellas, blobs of melted wax, stubby fingers of yellow-green slime, or miniature red bowls. (And there are many that defy description!) Although fungi come in a variety of shapes, sizes, and colors, they share many important characteristics. They are similar in the way they get their food, in their structure, and in the way they reproduce.

TEACHING STRATEGY 4-1

FOCUS/MOTIVATION

Use a freewriting activity to introduce the study of fungi. Ask the class to write for about five minutes on what they already know or think they know about fungi.

As this activity is diagnostic in nature, accept any honest ideas, even though many inaccuracies and misconceptions will be presented at this time. When finished, group students into teams of three or four and have them share their writings. Ask each team to prepare a list of questions that they would like to have answered about fungi.

Do not criticize students for their present lack of knowledge about the topic or attempt to grade their papers on the basis of accuracy. Instead, identify problem areas and clarify misinformation as you teach the sections in this chapter. The questions generated by each group can further serve as a guide to content development.

"Feeding" in Fungi

All fungi are heterotrophs (organisms that cannot make their own food). They obtain the energy and chemicals they need by growing on a source of food. **Fungi release chemicals that digest the substance on which they are growing and then they absorb the digested food.** (Animals, on the other hand, first take in—eat—their food and then digest it.)

Some fungi capture small animals for food. Oyster mushrooms, for example, release a chemical that stuns tiny roundworms in the soil. Certain threadlike soil fungi have tiny nooses that they use to snare their roundworm prey. Once a roundworm has been captured, the fungus begins to grow on it. Can you describe what happens after the fungus starts to grow on its prey? ❷

Other fungi obtain their food through symbiotic relationships. (Remember that a close relationship between two kinds of organisms in which at least one of the organisms benefits is known as a symbiosis.) Some symbiotic fungi, like the ant-killing fungus you just read about, are parasites that harm their host. Other symbiotic fungi help their host. Later in this chapter, you will read about some specific examples of harmful and helpful symbiotic fungi.

Many species of fungi get their food from the remains of dead organisms. These fungi are decomposers. Recall from Chapter 2 that decomposers ❶ break down dead plant and animal matter. The broken-down products become the foods of other living things. Why can such fungi, along with certain bacteria, be called "the Earth's cleanup crew"? ❸

Figure 4–2 *The scarlet cup fungi, purple coral fungi, and whitish earthstar fungus display just a few of the many colors and shapes of fungi.*

Figure 4–3 *An unlucky roundworm struggles in vain as the nooses of a soil fungus tighten around it in this microscopic drama.*

ACTIVITY
DISCOVERING

*Making Models
of Multicellular Organisms*

You can get a better idea of how different the structure of fungi is from that of other multicellular organisms by constructing a model. For these models you will need some sugar cubes and thin licorice whips.

1. For your model of a fungus, tangle and twist the licorice whips together to form a compact structure.

2. For your model of a plant or animal, stack the sugar cubes to produce a compact structure that is several cubes in length, height, and depth.

What do the licorice whips represent? What do the sugar cubes represent?

■ How is the basic structure of a fungus different from that of other multicellular organisms?

Structure of Fungi

A few fungi, such as the yeast used by bakers, are unicellular (one-celled). Most fungi, however, are multicellular (many-celled). The fat white toadstools at the base of a tree, the black spots of mildew on a shower curtain, and the fuzzy greenish mold on a piece of old cheese are all examples of multicellular fungi.

Multicellular fungi are made up of threadlike tubes called hyphae (HIGH-fee; singular: hypha). The **hyphae** branch and weave together in various ways to produce many different shapes of fungi. Hyphae can grow quite quickly—certain fungi can produce about 40 meters of hyphae in an hour. This is part of the reason mushrooms seem to pop up in fields and lawns overnight!

Interestingly, fungi are not multicellular in the same way that plants and animals are. All of the cells that make up the bodies of plants and animals are distinct units. Each plant or animal cell contains one nucleus (although there are some exceptions) and is enclosed by a cell membrane that separates it from other cells. The hyphae of fungi, on the other hand, are continuous threads of cytoplasm that contain many nuclei. Now can you explain why substances move more quickly and freely in fungi than in other multicellular organisms? ❶

Figure 4–4 *Fungi are made up of threadlike structures called hyphae (right). A close examination of a bread mold reveals that it is made of white, threadlike hyphae peppered with round black spore cases (left). The thick, heavy plates of shelf fungi are also made up of hyphae (center). In which kind of fungi are the hyphae more closely packed together?* ❷

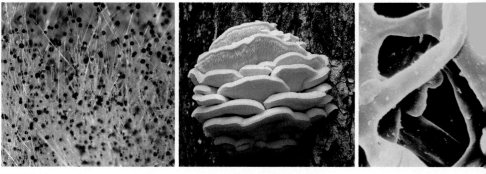

The hyphae of some fungi are divided into compartments by incomplete cross walls. Although these compartments are traditionally referred to as "cells," ❶ they are not enclosed by a cell membrane. Nuclei and other cell structures move quite freely through the openings in the cross walls. The only "real" cells are reproductive cells located at the tips of some of the hyphae.

Reproduction in Fungi

Many fungi reproduce by means of spores. Fungal **spores** are tiny reproductive cells that are enclosed in a protective cell wall. Because they are very small and light in weight, spores can be carried great distances by the wind. If a spore lands in a place where growing conditions are right, it can sprout and develop hyphae. How does this fact help to explain why fungi are found just about every place in the world? ❸

Fungi produce spores in a special structure called the fruiting body. Some fungi have simple fruiting bodies that consist of a stalk with a cluster of spores at its tip. Other fungi—mushrooms, cup fungi, and puffballs, for example—have large, complex fruiting bodies made up of many closely packed hyphae. A

Activity Bank

Spreading Spores, p.177

Figure 4–5 *Puffballs release a cloud of spores when touched (right). The girl posing beside the giant puffball (above) would be well advised to avoid touching the fungus! The spores of puffballs are spread by the wind. Some fungi spread their spores in other ways. The lacy stinkhorn produces a fluid that smells like rotting meat (top left). When a fly eats the fluid, it also takes in spores. The spores pass unharmed through the fly's body and are deposited over great distances. Tiny* Pilobolus *can throw its spore cases as much as a meter away (bottom left)! This is roughly equivalent to throwing a baseball the length of several football fields.*

B ■ 89

BACKGROUND INFORMATION

HYPHAE

Hyphae consist of tubelike structures containing cytoplasm and hundreds of nuclei. In some fungi, the hyphae are divided into sections by cross walls called septa, but there are perforations in these walls. Thus, the cytoplasm and nuclei are able to move freely throughout the hyphae. Other fungi lack septa altogether. As hyphae grow, they form a branching tangled mass called a mycelium. The mycelium forms the body of a fungus.

As a fungus matures, a part of the mycelium develops into a reproductive structure, or fruiting body. Spores are produced within fruiting bodies by either a sexual or an asexual process. Fungi can also reproduce asexually by fragmentation. In this process, fragments of broken hyphae are carried to new locations by wind or water. If conditions are favorable, the fragments will develop into new hyphae.

CONTENT DEVELOPMENT

Explain that most fungi reproduce both asexually and sexually. Asexual reproduction occurs through production of bodies called spores or by the fragmentation of the hyphae.

Point out that the spore-producing ability of mushrooms is immense. It has been determined that in an ordinary mushroom, every square millimeter of gill surface can produce about 130,000 spores that are released in five or six days. Because they are given off in such numbers, a mature mushroom cap placed on a sheet of white paper will release spores and produce a copy of its gill structure. These designs, called spore prints, are sometimes used as a means of identifying various mushrooms.

REINFORCEMENT/RETEACHING

Have students work as individuals or in small groups to arrange the following events in a logical sequence by placing a number in front of each statement. Before students begin, tell them that the statements, when properly sequenced, will show some of the events in the life cycle of a fungus. (Correct sequence for statements is shown in parentheses.)

____ Mature spores are released. (1)

____ Hyphae begin to grow from spore. (3)

____ Spores are carried by wind to a new location. (2)

____ Egg-shaped fruiting body grows upward from hyphae (4)

single fruiting body such as a large puffball may produce trillions of spores.

So why aren't we surrounded by millions of mushrooms or buried in puffballs? The answer is simple: Very few fungal spores find the proper combination of temperature, moisture, and food that they need to survive. Even fewer young fungi survive long enough to produce spores of their own. Can you now explain why fungi produce huge numbers of spores? ①

4-1 Section Review

1. Briefly discuss the basic characteristics of fungi.
2. What are hyphae?
3. How do fungi obtain food?
4. Describe how fungi reproduce.

Critical Thinking—*Relating Concepts*
5. Explain why it is important for fungi (and other organisms that cannot move) to produce offspring that can easily travel from one place to another.

4-1 (continued)

INDEPENDENT PRACTICE

Section Review 4-1

1. Fungi are heterotrophs that secrete digestive chemicals into their environment and then absorb the digested food. Fungi are made up of threadlike hyphae and typically reproduce by means of spores.
2. Hyphae are threadlike tubes that make up the body of a fungus.
3. By secreting digestive chemicals into the substance in which they are growing and then absorbing the digested materials.
4. Most fungi produce tiny reproductive cells known as spores.
5. This enables them to colonize new areas and to move away from places where the food has been depleted or where there is too much competition.

REINFORCEMENT/RETEACHING

Review students' responses to the Section Review questions. Reteach any material that is still unclear, based on students' responses.

CLOSURE

▶ *Review and Reinforcement Guide*
Have students complete Section 4-1 in the *Review and Reinforcement Guide.*

Guide for Reading

Focus on this question as you read.

▶ How are mushrooms, yeasts, and molds similar? How are they different?

Figure 4-6 *Although a morel looks, smells, and tastes like a mushroom, it is more closely related to baker's yeast than to mushrooms.*

4-2 Forms of Fungi

Fungi, like other organisms, are classified in a way that best shows the evolutionary relationships among the members of the group. As you might expect for such a strange and diverse group of organisms, the guidelines for classifying fungi are rather complex. As a result, people find it useful to group fungi according to their basic form, or shape. It is important to note that these groupings are not the formal classifications you learned about in Chapter 1. There are three basic forms of fungi: mushrooms, yeasts, and molds. **Mushrooms are shaped like umbrellas. Yeasts consist of single cells. And molds are fuzzy, shapeless, fairly flat fungi that grow on the surface of an object.**

TEACHING STRATEGY 4-2

FOCUS/MOTIVATION

Explain that about 4000 species of mushroom have been described. Tell students that of these perhaps only 30 or 40 species are known or suspected to be more or less toxic.

• **What does it mean for a substance to be "toxic"?** (A toxic substance would make a person sick or would possibly cause death.)

Explain that there are no certain outward signs that people can use to identify toxic mushrooms. The only sure way to know if a mushroom is toxic is to feed it to someone and see if they get sick. Feeding the mushroom to an animal is not a foolproof method. Some mushrooms are toxic to humans but not to animals.

Point out that the part of the mushroom we see is the fruiting body, which is only a small part of the organism. The bulk of the fungus consists of a large network of hyphae that spreads through the

Although these three categories are handy for everyday purposes, they do not make up a perfect organizational system. Why? Some fungi have very unusual shapes and cannot be placed in any of these categories. Others have more than one shape. The fungus that causes the disease known as thrush, for example, occurs both as a yeast and as a mold. Still others appear to have the correct shape for a group but are not placed in that group. Even though the morel shown in Figure 4–6 looks (and tastes) a lot like a true mushroom, it is not a mushroom.

Mushrooms

Have you ever ordered a pizza with all the trimmings? If so you ate fungi known as **mushrooms.** Figure 4–7 shows some mushrooms that might be encountered on a walk through the woods. As you can see, a mushroom has a stemlike structure called a stalk. In many types of mushrooms, the stalk is decorated with a structure called a ring, which looks somewhat like a very short skirt. On top of the stalk is the mushroom's cap. The mushroom's spores are produced on the underside of the cap. The spores are often located on thin sheets of tissue called gills, which extend from the stalk to the outer edge of the cap.

Figure 4–7 *Most familiar mushrooms produce their spores on gills. Some mushrooms, however, produce their spores in tubes and release them through pores. Others bear their spores on tiny flaps known as teeth.*

B ■ 91

soil, wood, or other substrate in which the mushroom is growing.

GUIDED PRACTICE

▶ *Laboratory Manual*

Skills Development

Skills: Observing, Comparing

At this point you may wish to have students complete the Chapter 4 Laboratory Investigation in the *Laboratory Manual* called Examining Three Forms of Fungi.

In this investigation, students compare and contrast yeasts, mushrooms, and molds.

INDEPENDENT PRACTICE

▶ *Activity Book*

Students can review what they have learned about mushroom form and function by completing the Chapter 4 activity called Mushroom Structure.

4-2 Forms of Fungi

MULTICULTURAL OPPORTUNITY 4–2

The blue and green molds that appear on damaged oranges and lemons belong to *Penicillium*, a genus with many species. The first antibiotics to be used against various infectious diseases, the pencillins, were isolated from cultures of *Penicillium notatum*.

Molds of the genus *Penicillium* are responsible for other prized products; for instance, the flavors of some of the most sought-after cheeses, such as Gorgonzola, Roquefort, and Camembert. Ask students to identify these cheeses at their local supermarkets or grocery stores and to investigate their countries of origin as they appear on the labels. Have them research how these cheeses are made.

ESL STRATEGY 4–2

Dictate the following activities and ask students to make some notes in order to be prepared to answer orally. Have them work in small groups. One student reads the directions, another student responds, and a third confirms, corrects or clarifies the answer.

1. Identify the fungi often used as a topping for pizza (Mushrooms.)

2. Identify the fungi used to make bread rise. (Yeasts.)

3. Identify the fungi responsible for penicillin. (A mold called *Penicillium.*)

4. Describe a treatment used in your native country to treat an infection. (Answers will vary.)

🖥 Media and Technology

Use the transparency in the *Transparency Binder* called Mushroom Development to illustrate the life cycle of a mushroom.

PROBLEM SOLVING
A HOT TIME FOR YEAST

Students use their skills in interpreting graphs to explore the relationship between temperature and the activity of yeast cells. The answers to the questions in the activity are

1. The rate of activity increases slightly between 0°C and 10°C, increases rapidly between 10°C and 25°C, then decreases sharply and stops at about 50°C.
2. About 25°C.
3. About 32°C.
4. Warm water helps to dissolve the yeast and provides the proper temperature for yeast activity.
5. The slightly warm oven would encourage the activity of the yeast cells and help the bread rise. Once the bread is put into a hot oven (greater than 50°C), the heat will start to kill the yeast cells.
6. Accept all logical answers. Most students would not expect to find live yeast cells because the cells stop functioning at temperatures greater than 50°C.

Integration: Use the Problem Solving feature to integrate the physical science concepts of heat into your lesson.

4–2 (continued)

ENRICHMENT
▶ *Activity Book*

Students who are interested in learning about mushroom identification will be challenged by the Chapter 4 activity called Making Spore Prints.

INDEPENDENT PRACTICE
▶ *Activity Book*

Students can explore the use of graphs to analyze data about fungi in the Chapter 4 activity called Graphing Fungi.

CONTENT DEVELOPMENT

Explain that yeasts belong to a group of fungi called sac fungi. With about 30,000 species, the sac fungi comprise the largest group of fungi. The yeasts are of enormous commercial value in the baking process and as a source of alcoholic beverages and industrial alcohol.

PROBLEM Solving

A Hot Time for Yeast ❶

Yeasts are tiny unicellular fungi. Some yeasts are used to make alcoholic beverages and foods such as bread. The accompanying graph shows the effect of temperature on the level of activity of yeast cells. Use the graph to answer the questions that follow:

Interpreting Graphs
1. How does temperature affect yeast activity?
2. At what temperature is the yeast activity the highest?
3. At what temperature does the yeast activity decrease sharply?
4. Why is yeast dissolved in warm water when it is used to make bread?
5. Explain why bread dough is sometimes placed in a slightly warm oven for

an hour or so before it is made into a loaf and baked in a hot oven.
6. Would you expect to find live yeast cells in a slice of bread? Explain.

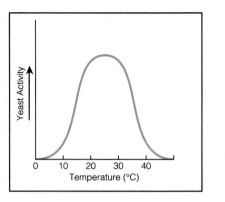

Figure 4–8 *Yeasts are used to produce bread, fuel, vitamins, chemicals, and even medicines such as the vaccine for hepatitis B. Yeasts reproduce by budding. A round scar results when a bud breaks off from its parent cell. How many buds has the larger yeast cell produced?* ❶

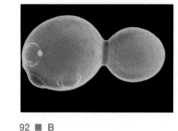

Yeasts

Most people cannot help but stop and take a deep breath when they pass a bakery. There is something about the smell of fresh bread that excites the senses. The next time you pass a bakery you might whisper a soft thank-you to another type of fungi, the **yeasts.**

In order to make soft, fluffy bread, bakers add yeast to the flour, water, sugar, salt, and other ingredients that make up bread dough. The bakers then allow the dough to sit for a while in a warm place. ❷ Bread dough is a great environment for yeast—moist, warm, and full of food. As it grows, the yeast produces carbon dioxide gas. The carbon dioxide gas forms millions of tiny bubbles in the dough. You

92 ■ B

● ● ● ● **Integration** ● ● ● ●

Use the information about uses of yeast to integrate food science concepts into your lesson.

GUIDED PRACTICE
▶ *Laboratory Manual*
Skills Development
Skills: Making observations, making comparisons, applying concepts

At this point you may want to have stu-

dents complete the Chapter 4 Laboratory Investigation in the *Laboratory Manual* called Culturing Yeast Cells. In the investigation students will explore the factors in the environment that are necessary for yeast to grow.

CONTENT DEVELOPMENT

Introduce the discussion of the discovery of the antibiotic penicillin by showing the class some examples of common molds of the genus *Penicillium*. Bluish-

see these bubbles as holes in a slice of bread. What do you think would happen if a baker forgot to put yeast in the bread dough? ❷

Unlike other fungi, yeasts may reproduce by a process known as budding. During budding, a portion of the yeast cell pushes out of the cell wall and forms a tiny bud. In time, the bud breaks away from the parent cell and becomes a new yeast.

Molds

Centuries ago people sometimes treated ❸ infections in a rather curious way. They placed decaying breads, cheeses, or fruits on the infection. Although the people did not have a scientific reason for doing this, every once in a while the infection was cured. What these people did not and could not know was that the cure was due to a type of **mold** that grows on certain foods.

In 1928, the Scottish scientist Sir Alexander Fleming found out why this treatment worked. Fleming discovered that a substance produced by the mold *Penicillium* could kill certain bacteria that caused infections. Fleming named the substance penicillin. Since that time penicillin, an antibiotic, has saved millions of lives.

Molds are used to make many foods, such as tofu (bean curd), soy sauce, and cheeses. The blue streaks in blue cheese, for example, are actually

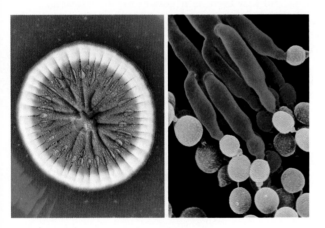

ACTIVITY
DOING

Making Spore Prints

1. Place a fresh mushroom cap, gill side down, on a sheet of white paper. Cover with a large glass jar, open end down. **CAUTION:** *If you use a wild mushroom, wash your hands thoroughly after handling it. Do not leave wild mushrooms where small children can reach them. Some wild mushrooms are poisonous.*

2. After several days, carefully remove the jar and lift off the mushroom cap. You should find a spore print on the paper.

3. Very carefully spray the paper with clear varnish or hair spray to make a permanent spore print.

4. Examine the print with a magnifying glass.

5. (Optional) Prepare a spore print using a different kind of mushroom.

What color are the spores in your spore prints? How are your spore prints similar? How are they different? How do your spore prints compare to ones prepared by your fellow classmates?

Figure 4–9 *The mold* Penicillium *(left) produces spores at the tips of tiny branches (right). What important antibiotic comes from* ❸ Penicillium?

B ■ 93

ANNOTATION KEY

Answers

❶ Three small, round bud scars are shown in the photograph, indicating that at least three other buds have been produced. (Interpreting photographs)

❷ The bread would probably be hard and dense because it would lack the air bubbles produced by the yeast. (Applying concepts)

❸ Penicillin. (Applying concepts)

Integration

❶ Physical Science: Heat. See *Heat Energy*, Chapter 1.

❷ Food Science

❸ Social Studies

ACTIVITY
DOING

MAKING SPORE PRINTS

Skills: Making observations, relating cause and effect

Materials: mushroom cap, white paper, large glass jar, clear varnish or hair spray, magnifying glass

Explain that an important characteristic used to identify mushrooms is the color of the spores. To make sure of the color, a spore print is made. Just looking at the spores is not enough—white spores on brown gills might look brown.

Students should obtain a spore print consisting of lines radiating outward like spokes.

green molds of this genus are often seen growing on citrus fruits, in Roquefort cheese, and sometimes on bread.

Ask the class to recall that antibiotics are chemicals produced by helpful bacteria that weaken or destroy disease-causing bacteria. Explain that some molds also produce antibiotics. Be sure to point out that penicillin is not actually derived from the mold on an orange or in Roquefort cheese, but from a closely related species.

● ● ● ● **Integration** ● ● ● ●

Use the historical information about the discovery of penicillin to integrate social studies into your science lesson.

REINFORCEMENT/RETEACHING

You may wish to point out to students that the term *spore* is used to refer to just about any kind of cell specialized for dispersal or survival.

• **What other organisms have you read about that produce spores?** (Bacteria, protists, fungi.)

• **How are bacterial/protist spores similar to fungus spores? How are they different? What sort of information do you need to better answer these questions?** (Accept all logical answers.)

• **Do you think that the word *spore* should be used for all these different kinds of cells? Why or why not?** (Accept all logical answers.)

ACTIVITY
READING
A SECRET INVASION

Skill: Reading comprehension

The American poet Sylvia Plath (d. 1963) was honored in 1982 with the Pulitzer Prize for poetry. She is also well known for her only novel, *The Bell Jar.*

"Mushrooms," published in 1962 in *The Colossus & Other Poems,* has been described as both "playful" and "menacing." After students have read the poem, discuss which lines might be considered "playful." (The third stanza: "Our toes, our noses/Take hold on the loam,/Acquire the air.) Which lines sound menacing? (The final stanza: "We shall by morning/Inherit the earth/Our foot's in the door.)

Integration: Use this Activity to integrate language arts into your science lesson.

---

mold. Of course, not all molds help to make foods or provide valuable medicines. Most molds are just plain, ordinary fungi. So if you discover fuzzy growths of mold on decaying breads, cheeses, or fruits, you should probably just clean out the refrigerator or take out the trash.

4–2 Section Review

1. How is a yeast different from a mushroom?
2. List five different uses for yeasts.
3. Discuss three ways in which molds affect people.

Critical Thinking—*Evaluating Classification Schemes*
4. Explain why fungi are not divided into phyla according to basic shape.

Guide for Reading

Focus on these questions as you read.

▶ How do fungi harm other organisms?

▶ How do fungi help other organisms?

A*CTIVITY*
READING

A Secret Invasion

❶ What might mushrooms say if they could speak? Are these small umbrella-shaped organisms as innocent as they seem? For one poet's answers to these questions, read the poem "Mushrooms," by Sylvia Plath.

4–3 How Fungi Affect Other Organisms

Fungi interact with other organisms in many different ways. Some fungi harm other organisms. Such fungi are often disease-causing parasites of plants and animals. Other fungi are helpful to other organisms and may even be necessary for their survival.

Fungi and Disease

Have you ever looked closely at an apple and noticed a sprinkling of small, hard brown "scabs" on its skin? Or have you ever seen round black or gray spots on the leaves of a lilac or rose bush? These scabs and spots are the result of plant diseases caused by fungi. And although the apple is still safe to eat, the lilac or rose plant might be in serious trouble.

Scabs and spots are not the only signs of fungal diseases. Some disease-causing fungi make the stem, roots, or fruit of crop plants rot, or decay. Others, such as those that cause Dutch elm disease and

---

4–2 (continued)

CONTENT DEVELOPMENT

🖳 Media and Technology

Use the transparency in the *Transparency Binder* called Bread Mold to show students the structures found in this fungi.

• **How are a mushroom and a bread mold similar in the way they obtain their food?** (Both have threadlike hyphae that grow into and absorb food.)

GUIDED PRACTICE

Skills Development

Skills: Making observations, making comparisons

At this point have students complete the in-text Chapter 4 Laboratory Investigation: Growing Mold. In the investigation students will grow mold on bread, cheese, and a piece of apple and will look for factors that influence the speed of growth and type of mold.

INDEPENDENT PRACTICE

▶ *Activity Book*

Students who need practice with the concepts of this section should be provided with the Chapter 4 activity called Identifying Fungi.

INDEPENDENT PRACTICE

Section Review 4–2

1. A yeast is unicellular and microscopic. A mushroom is multicellular, visible to

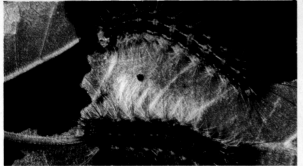

chestnut blight, kill trees that are prized for their beauty and wood. A few change kernels of growing grain into bags of useless fungal spores. Still others damage stored crops such as wheat, corn, oats, peanuts, and rice, making them unfit to eat. Can you now explain why farmers and gardeners spend millions of dollars on fungicides (fungus-killing substances) each year? ❶

In addition to damaging or even completely destroying crops, some fungi that infect plants produce toxins (poisons) that can injure or even kill humans and animals. For example, one fungus that grows on stored grain produces a toxin that is one of the most powerful cancer-causing substances. (Scientists know this toxin as aflatoxin.) In small doses, this toxin can cause liver cancer. In large doses, it is fatal. Like most fungi, the toxin-producing fungus will grow only if there is sufficient moisture. Why is it important to thoroughly dry crops such as peanuts or corn before they are stored? ❷

Another fungus replaces grains of rye with hard spiky poisonous growths known as ergot. People who eat bread or other grain products containing ergot may experience burning or prickling sensations, hallucinations, and convulsions. In extreme cases, the flow of blood to the arms and/or legs may be cut off, resulting in infections and possible loss of the affected limbs.

Ergot poisoning can be fatal—although not necessarily through the direct actions of the toxins. ❷ Some historians have suggested that the witchcraft trials in Salem, Massachusetts, may have been due in part to people's terror at the strange symptoms of ergot poisoning. Early in 1692, several girls in Salem

Figure 4–10 *Fungi affect other organisms in many ways. Leafcutter ants grow the fungi they eat on bits of plants they carry to their underground nests (top right). Gypsy moth caterpillars (left), which cause a great deal of damage to trees, can be controlled with the help of a parasitic fungus. Grains of corn are changed into bags of toxic spores by the fungus known as corn smut (bottom right).*

B ■ 95

MULTICULTURAL OPPORTUNITY 4-3

Fungal contamination of grain may damage or even completely destroy crops, resulting in widespread famine and disease. Have students research some areas of the world that suffer from famine such as India, Bangladesh, and parts of Africa. What means are taken by the governments of these countries to prevent or fight famine? Do other countries help them out? Ask the class to suggest ways of fighting famine.

ESL STRATEGY 4–3

In each group of terms, have students find "the intruder," the word that does not belong in each group.
1. aflatoxin, ringworm, mycorrhizae, ergot
2. Dutch elm disease, lichen, athlete's foot, chestnut blight

(Answers: Mycorrhizae and lichen. All the other terms are fungal poisons or diseases.)

Activity Bank

Psst! Wanna Make Some Dough? Activity Book, p. B205. This activity can be used for ESL and/or Cooperative Learning.

---

the unaided eye, and more or less umbrella-shaped.
2. Answers will vary. Possible answers are making bread rise, making alcoholic beverages, and producing fuel alcohol.
3. Answers will vary. Possible answers: Molds are the source of medicines such as penicillin; molds spoil food; molds are used to make certain foods such as soy sauce and cheeses.
4. Shape is not a good indicator of evolutionary relationships in fungi, as some fungi have more than one shape, others have misleading shapes, and still others have odd shapes that are not easily classified.

REINFORCEMENT/RETEACHING

Review students' responses to the Section Review questions. Reteach any material that is still unclear, based on students' responses.

CLOSURE

▶ *Review and Reinforcement Guide*
Have students complete Section 4–2 in the *Review and Reinforcement Guide*.

TEACHING STRATEGY 4–3

FOCUS/MOTIVATION

Arrange for students to visit a garden center, nursery, or the agricultural department of a local college or university to learn more about the role of fungi in plant growth. Of particular interest would be a visit to an orchid grower, if there is one in your area.

ACTIVITY
WRITING

HUMAN FUNGAL DISEASES

Athlete's foot is caused by the fungus *Epidermophyton floccosum*. Growth of the fungus causes itching, reddening, cracking, and scaling of the skin of the feet, especially between the toes.

Coccidioidomycosis, or San Joaquin Valley fever, is a mild lung infection caused by *Coccidioides immitis*, which grows in the desert soils of the southwestern United States and in some places in South America.

Farmer's lung is an often chronic and debilitating lung disease caused by an allergic reaction to the spores of fungi such as *Aspergillus* and *Penicillium*, which grow on grain.

Histoplasmosis is a mild lung disease caused by *Histoplasma capsulatum*, which grows in the soil and is transmitted in the droppings of chickens and other birds.

Ringworm is caused by a number of fungi, including two species of the genus *Microsporum*, which causes epidemic ringworm of the scalp in children. The itchy red patches of ringworm are sometimes round and ringlike in form, hence the name.

Sporotrichosis is caused by *Sporothrix (Sporotrichum) schenckii*, which is introduced into the body through an injury; for example, contact with a thorn. Sporotrichosis results in the formation of lumps or sores, particularly on the arms and legs, and a general infection of the lymphatic system.

Thrush (candidiasis) is caused by *Candida albicans*, which is often part of the normal flora of the body and is normally not invasive. Thrush causes sore, lumpy patches to form on the mucous membranes of the mouth and throat. Thrush is usually seen only in very young infants and in the final stages of a wasting disease. *Candida albicans* is also responsible for female "yeast infections."

Figure 4–11 *The long, curved spikes on this head of rye are ergot. What is the connection between fungi and the Salem witchcraft trials?* ①

ACTIVITY

WRITING

Human Fungal Diseases

Using reference materials in the library, prepare a brief report on one of the following diseases:

athlete's foot
coccidioidomycosis (San Joaquin Valley fever)
farmer's lung
histoplasmosis
ringworm
sporotrichosis (rose-prick fever, rose-gardener's syndrome)
thrush (candidiasis)

In your report, give the name of the fungus that causes the disease. Also include information about the transmission, symptoms, and treatment of the diseases.

began suffering from pains and odd behaviors that were thought to be caused by witchcraft. By the end of the year, 19 people had been hanged as witches as a result of the ensuing witch hunt and trials.

Animals, like plants, can become infected by fungal diseases. Fungi cause a number of severe, sometimes fatal, lung diseases in poultry (chickens, ducks, and other kinds of farm birds). They can also produce itchy or painful sores on the skin of pets such as dogs, cats, and birds. While most fungal diseases of animals are troublesome, a few have proven to be useful to humans. Plant-eating pests such as gypsy moth caterpillars, mites, and aphids can be killed by deliberately infecting them with certain fungi. These fungal pesticides are a lot more effective than chemicals—and a lot safer for the environment, too!

Although some human fungal diseases are serious, most are simply annoying. One causes the fingernails or toenails to grow in crooked and/or fall out. The diseases known as ringworm and athlete's foot cause itchy, reddened, and raw patches on the skin.

Fungus-Root Associations

As you have discovered, plants are plagued by a host of fungal diseases. However, plants get help against some of these diseases from—you guessed

Figure 4–12 *Some mycorrhizae send up mushrooms or puffballs. This sometimes produces a "fairy ring" around a tree. Folklore has* ① *it that trees surrounded by fairy rings are the favorites of fairies, so chopping down such trees is extremely bad luck.*

4–3 (continued)

CONTENT DEVELOPMENT

Emphasize to students that diseases caused by fungi attack people, plants, and animals. In order to prevent or cure fungal diseases, chemical preparations known as fungicides are used.

Point out that fungal diseases in plants are generally the most difficult to treat. In fact, experts state that most agricultural fungicides are effective only if applied

it—other fungi. Helpful fungi coat the roots of about 80 percent of the world's plants. Some of these fungi simply cover the surface of the roots. Others actually send hyphae into the roots' cells. Scientists give these fungus-root associations a fancy name: mycorrhizae (migh-koh-RIGH-zee; singular: mycorrhiza), which is Greek for "fungus roots."

The word mycorrhizae is also used to refer to the helpful fungi in fungus-root associations. The hyphae of these helpful fungi spread out into the soil in all directions, increasing their host's ability to gather nutrients by ten times or more. The fungus-root associations protect the plant against drought, cold, acid rain, and root diseases caused by harmful fungi. Currently, Australian researchers are trying to alter the hereditary material of one kind of mycorrhiza so that it produces natural insecticides. This helpful fungus would then be able to help guard its host against harmful insects as well as provide all of its normal benefits.

Lichens

Suppose someone asked you what kind of organism can live in the hot, dry desert as well as the frozen Arctic. What if the person added that this organism can also survive on bare rocks, wooden poles, the sides of trees, and even the tops of mountains? You might reply that no one organism can survive in so many different environments. In a way, your response would be right. For although **lichens** (LIGH-kuhnz) can actually live in all of these

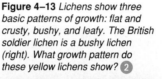

ACTIVITY WRITING

Taking a "Lichen" to It

Lichens are used in a variety of ways. Some are used to make dyes. Others are a source of food for people and livestock. Go to the library and find out the details about some of the ways lichens are used around the world. Prepare a report on your findings.

Figure 4–13 *Lichens show three basic patterns of growth: flat and crusty, bushy, and leafy. The British soldier lichen is a bushy lichen (right). What growth pattern do these yellow lichens show?* ②

B ■ 97

before a fungal disease strikes. For this reason, it is important to treat crops, trees, and garden plants with fungicides as a preventive measure.

Fungicides are also used to treat fungal diseases in humans. For example, a preparation commonly used to treat athlete's foot is called Tinactin. Some fungal infections in humans can be treated by drugs that are taken orally, such as sulfa drugs or antibiotics.

Stress that the best treatment against fungal infections is prevention, because fungi are extremely persistent and difficult to get rid of. The minimum time needed to cure a fungal infection is 10 to 14 days, and some infections take six months to a year to cure.

● ● ● ● **Integration** ● ● ● ●

Use Figure 4–12 and its caption to integrate mythology into your science lesson.

Use the discussion about fungus-root associations to integrate concepts about symbiosis into your lesson.

ECOLOGY NOTE

POLLUTION AND LICHENS

Though lichens are found in a wide variety of natural habitats, they are usually absent from areas severely affected by air pollution. Lichens absorb their needed moisture and minerals from rain water, and they are extremely susceptible to toxic materials that may be borne by rain. Thus the presence or absence of lichens in an area can serve as an indicator of that area's air quality.

4–3 (continued)

CONTENT DEVELOPMENT

Explain that even though lichens are made up of two different organisms, they are named as though they were a single organism. Presently, about 20,000 different species of lichens have been identified. Point out that the fungus part of the lichen is usually a sac fungus and that the alga is some type of unicellular blue-green or green algae.

Write the word *symbiosis* on the chalkboard and call on someone to give its definition. Expand on the definition by asking students to recall that in a symbiotic relationship, at least one of the organisms benefits. The second organism may also benefit, or it may be harmed.

• **Why is a lichen a good example of symbiosis?** (Two different kinds of organisms, an alga and a fungus, live in a close relationship.)

CAREERS

Mushroom Grower

People have been using mushrooms to flavor food for thousands of years. Usually the mushrooms were found growing wild in fields. It was not until the late 1800s that mushrooms were grown on a large scale in the United States for sale in the marketplace.

Mushroom growers must grow their crop indoors, where conditions can be controlled. The growers keep careful watch over the special devices that control temperature, humidity, and ventilation. In addition, mushroom growers must know how to identify and treat mushroom diseases.

If becoming a mushroom grower appeals to you, you can write to the American Mushroom Institute, 907 East Baltimore Pike, Kennett Square, PA 19348.

environments, they are not one organism but two. **A lichen is made up of a fungus and an alga that live together.** An alga (AL-gah; plural: algae, AL-jee) is a simple plantlike autotroph that uses sunlight to produce its food. An alga lacks true roots, stems, and leaves. Blue-green bacteria, certain protists, and a number of simple plants are considered to be algae. Combined, the two organisms (alga and fungus) in a lichen can live in many places that neither could survive in alone.

The fungus part of the lichen provides the alga with water and minerals that the fungus absorbs from whatever the lichen is growing on. The alga part of the lichen uses the minerals and water to make food for the fungus and itself. Why is the relationship between the fungus and alga considered to be one of the best examples of symbiosis? ●

Lichens are sometimes known as pioneers because they are often one of the first living things to appear in rocky, barren areas. Lichens release acids that ● break down rock and cause it to crack. Dust and dead lichens fill the cracks, which eventually become fertile places for other organisms to grow. In time, the rocky area may become a lush, green forest.

4–3 Section Review

1. Discuss three ways in which fungi harm humans.
2. How are fungal diseases useful to people?
3. Explain why fungus spores are sometimes mixed into the soil in which young trees are to be grown.
4. What is a lichen?
5. Why do agricultural inspectors check samples of grain with a microscope before permitting the grain to be ground up for food?

Connection—*Ecology*

6. Many lichens are extremely sensitive to pollution. Even when there is very little pollution, these lichens will grow poorly or not at all. How might environmental scientists use this information in their work?

• **How does the fungus benefit from the symbiosis?** (The fungus obtains food from the alga.)
• **How does the alga benefit from the symbiosis?** (The alga obtains water and minerals that are absorbed by the fungus.)

● ● ● ● **Integration** ● ● ● ●

Use the discussion of the appearance of lichen in barren areas to integrate earth science concepts of weathering and soil formation into your lesson.

REINFORCEMENT/RETEACHING

• **Based on what you have learned in this chapter, in what ways would a world without fungi be a better place?** (Accept all reasonable answers. Positive consequences of a world without fungi might include no mold on bread and no diseases such as athlete's foot.)
• **In what ways would a world without fungi be a worse place?** (Accept all reasonable answers. The negative consequences of a world without fungi might include not

CONNECTIONS

Murderous Mushrooms ❷

To avoid becoming one of the several hundred cases of mushroom poisoning in the United States each year (or worse yet, a "dear departed"), a mushroom hunter must be able to accurately identify mushrooms. The difficulty of this task is particularly well-illustrated by the mushrooms in the genus *Amanita. A. caesarae* was a favorite food of the Caesars, the rulers of ancient Rome. *A. rubescens* is also edible and delicious—but it is poisonous when it is not sufficiently cooked. *A. muscaria* is a poisonous mushroom. In fact, people in northern Europe once soaked *A. muscaria* caps in milk and set them out to attract and kill flies. If accidentally eaten by humans, *A. muscaria* causes hallucinations, sweating, wildly crazy behavior, and deep sleep. *A. verna,* which is known by the common names fool's mushroom and

destroying angel, causes cramps, severe abdominal pain, vomiting, diarrhea, liver and kidney failure, and death in over half its victims.

Interestingly, the deadly poisonous *Amanita* mushrooms taste good. (Most poisonous substances are bitter or otherwise unpleasant.) This characteristic has caused these mushrooms to be used by murderers on more than one occasion in *history.* In 54 AD, the emperor Claudius I of Rome was fed poisonous mushrooms by his wife, who wanted her son Nero to become emperor and did not want to wait for Claudius to die of natural causes. Pope Clement VII, who died in 1534, was also a victim of poisonous mushrooms.

Mushrooms in the genus Amanita *include* A. muscaria *(bottom right),* A. caesarea *(top left),* A. rubescens *(bottom left), and* A. verna *(top right).*

CONNECTIONS
MURDEROUS MUSHROOMS

Many species of mushrooms are easy to grow, taste good when properly cooked, and do not pose a danger to anyone who eats them. Wild mushrooms are a different story: Although some are edible, many are poisonous. And many species of poisonous mushrooms look almost identical to edible mushrooms. The result of eating a poisonous mushroom is severe illness and sometimes death.

In fact, it is best not to pick or eat any mushrooms found in the wild. Instead, mushroom gathering should be left to experts who can positively identify each mushroom they collect.

One field guide for mushroom hunters tells us that the safest place to hunt mushrooms is at the grocery store! Students will be able to find many interesting varieties of mushrooms at larger supermarkets as well as at gourmet food shops.

If you are teaching thematically, you may want to use the Connections feature to reinforce the themes of systems and interactions or evolution.

Integration: Use the Connections feature to integrate social studies into your science lesson.

having the extremely important drug penicillin; the breakdown of dead organic matter would not take place as completely or quickly as it needs to in order to keep energy and nutrients recycled in nature; we would not have mushrooms to eat; there would be no yeast to make bread and cake dough rise.)

ENRICHMENT

▶ *Activity Book*

Students who have mastered the concepts in this section will be challenged

by the Chapter 4 activity called Relationships With Fungi.

INDEPENDENT PRACTICE
Section Review 4–3

1. Answers will vary. Possible answers: Fungi damage or destroy crops, causing food shortages and economic losses; some fungi release toxins that poison foodstuffs; fungi cause a number of human diseases.

2. Some fungal diseases can be used as natural pesticides.

3. Certain kinds of fungi form fungus-root associations with young plants, helping them grow and thrive.

4. A special type of symbiotic relationship between an alga and a fungus.

5. To check for traces of toxin-producing molds and other impurities that might not be visible to the unaided eye.

6. Environmental scientists use lichens as pollution indicators and monitors.

REINFORCEMENT/RETEACHING

Review students' responses to the Section Review questions. Reteach any material that is still unclear, based on students' responses.

CLOSURE

▶ *Review and Reinforcement Guide*

Have students complete Section 4–3 in the *Review and Reinforcement Guide.*

Laboratory Investigation

GROWING MOLD

BEFORE THE LAB

1. Gather all materials at least one day prior to the investigation. You should have enough supplies to meet your class needs, assuming six students per group.
2. The growth of the bread mold can be hastened by using bread that does not contain preservatives.

PRE-LAB DISCUSSION

Have students read the complete laboratory procedure.

• **What is the purpose of this investigation?** (To compare the growth of mold on three substances.)

• **Have you ever seen mold on bread, cheese, or fruit? What did it look like?** (Accept all answers.)

Remind students that molds have root-like hyphae that penetrate the surface on which the mold is growing.

• **What is the purpose of the hyphae?** (To anchor the mold; to release digestive chemicals; to absorb the digested food.)

SAFETY TIPS

Some people are quite allergic to fungal spores. You may wish to have an alternative activity for students with asthma or allergies.

Students should wash their hands thoroughly after the investigation.

Laboratory Investigation

Growing Mold

Problem

How does mold grow?

Materials (per group)

large covered container
paper towel
piece of bread
piece of cheese
piece of apple
magnifying glass

Procedure 🧪

1. Line the container with moist paper towels.
2. Place the pieces of food in the container.
3. Allow the container to remain uncovered overnight. Then cover the container.
4. Store the container in a dark place. Examine its contents daily for mold.
5. After the container's contents become moldy, examine the mold with a magnifying glass. **CAUTION:** *Be careful not to accidentally inhale mold spores. Some people are allergic to mold.*
6. Make a drawing of what you observe. Record your observations next to your drawing.
7. Observe and draw the mold every two to three days for a week.

Observations

1. When did mold start to grow?
2. How do the molds appear to the unaided eye? Under a magnifying glass?
3. How do the molds change over time?
4. Did you observe fruiting bodies and spores? What did these look like?

Analysis and Conclusions

1. Do different kinds of mold grow on different kinds of substances? Why do you think this is so?
2. On which substance did the molds grow best? Develop a hypothesis to explain why. How might you test your hypothesis?
3. Why did you leave the container uncovered? What would have happened if you had covered the container immediately?
4. **On Your Own** Design an experiment to find out how light, temperature, moisture, dust, or household disinfectants affect the growth of molds. If you receive the proper permission, you may perform the experiment you have designed.

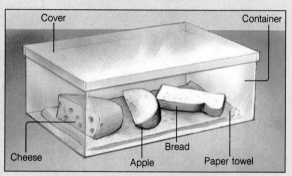

TEACHING STRATEGY

1. Remind students of your requirements for scientific drawings. Point out that unlabeled drawings are of little value.
2. Have teams follow the directions carefully as they work in the laboratory.

DISCOVERY STRATEGIES

Discuss how the investigation relates to the chapter ideas by asking open questions similar to the following.

• **On which substance do you think the mold will grow most quickly?** (Accept all logical answers—predicting.)

• **What is the purpose of the damp paper towels?** (The molds need moisture in order to grow—applying concepts, relating cause and effect.)

Study Guide

Summarizing Key Concepts

4–1 Characteristics of Fungi

▲ Fungi come in a variety of sizes, shapes, and colors.

▲ Fungi are heterotrophs. They obtain food by releasing chemicals that digest the substance on which they are growing and then by absorbing the digested food.

▲ Some fungi catch and eat tiny animals. Others are decomposers. Still others obtain food through symbiotic relationships. Some symbiotic fungi are parasites. Others help their host.

▲ Some fungi are unicellular. Most fungi are multicellular.

▲ Multicellular fungi are made up of threadlike tubes called hyphae.

▲ Fungi typically reproduce by means of spores, which are produced in structures known as fruiting bodies.

4–2 Forms of Fungi

▲ There are three basic forms of fungi: mushrooms, yeasts, and molds.

▲ Mushrooms are shaped like umbrellas. Some are good to eat. Others are poisonous.

▲ Yeasts are unicellular. They are used by bakers, brewers, industrial chemists, and medical researchers, to name a few.

▲ Molds are fuzzy, shapeless, fairly flat fungi that grow on the surface of an object. Some are used to make foods. A few are the source of important medicines.

4–3 How Fungi Affect Other Organisms

▲ Many diseases of crop and garden plants are caused by fungi.

▲ Some fungi that infect plants produce toxins that are harmful to humans and animals.

▲ Fungi cause a number of diseases in animals and humans. Some of these are simply annoying. Others are serious and even deadly.

▲ Certain fungi form fungus-root associations with most of the Earth's plants. These symbiotic relationships are extremely helpful to the plants.

▲ Lichens are produced by a symbiosis between a fungus and an alga.

Reviewing Key Terms

Define each term in a complete sentence.

4–1 Characteristics of Fungi
fungus
hypha
spore

4–2 Forms of Fungi
mushroom
yeast
mold

4–3 How Fungi Affect Other Organisms
lichen

ANALYSIS AND CONCLUSIONS

1. Some molds are very specific; others will grow on almost anything.

2. The mold probably grew best on the bread. Student hypotheses will vary. One possible response is that the bread presents the greatest amount of surface area for the mold to grow on.

3. The molds are caused by airborne fungus spores. Also, nearly all fungi require oxygen to live and grow. Covering the container might have prevented the food from being inoculated with the fungus spores. And lack of oxygen would have slowed the growth of the mold.

4. Student experiments will vary. Results may include the following: The majority of fungi grow as well in darkness as in light; most fungi grow best between 20 and 35 degrees Celsius; slow molding will occur when the relative humidity of the air reaches 70 percent; there are a great many mold spores in dust, so sprinkling food with dust will increase the likelihood of mold growth; household disinfectants will retard the growth of some molds.

GOING FURTHER: ENRICHMENT

Interested students can make a fungarium showing an assortment of different fungi. An old aquarium, no longer watertight, makes a good container. Cover the bottom with a 10-cm layer of sawdust or a mixture of soil and sand. Add enough water to make the layer moist but not sopping wet. Use a large piece of glass to cover the fungarium.

Put some of the following materials into the fungarium: paper, different kinds of fabrics, pieces of rope, various fruits or vegetables, a piece of old leather from a shoe or glove, pieces of bark or wood. Students should keep the fungarium covered and record the appearance of the different fungi that grow on the materials.

OBSERVATIONS

A variety of fungi cause molds on cheese, bread, and fruit. Results of students' experiments will vary, depending on the types of fungi that grow on their samples.

1. Mold should start to grow on the bread in a day or two. The mold on the cheese may take longer to start growing, the mold on the apple the longest.

2. Bread mold looks like dirty gray or gray-blue cotton with small specks like poppyseeds or grains of black pepper.

Decaying fruit and damp bread support fungi in the genera *Penicillium* and *Aspergillus*. Species of *Penicillium* are usually green with a white border. *Aspergillus* are often green or blue-green but can also be yellow, brown, reddish, or black.

3. The molds grow larger and may change texture or color.

4. The fruiting body should consist of a thin stalk topped with a round black spore case. Spores may be visible as tiny specks on open spore cases.

Chapter Review

Chapter Review

ALTERNATIVE ASSESSMENT

The *Prentice Hall Science* program includes a variety of testing components and methodologies. Aside from the Chapter Review questions, you may opt to use the Chapter Test or the Computer Test Bank Test in your *Test Book* for assessment of important facts and concepts. In addition, Performance-Based Tests are included in your *Test Book*. These Performance-Based Tests are designed to test science process skills, rather than factual content recall. Since they are not content dependent, Performance-Based Tests can be distributed after students complete a chapter or after they complete the entire textbook.

CONTENT REVIEW

Multiple Choice
1. b
2. c
3. a
4. d
5. a
6. d
7. b
8. b
9. d

True or False
1. F, hyphae
2. F, fungus-root associations
3. F, penicillin
4. F, mushrooms, molds, and yeasts
5. T

Concept Mapping
Row 1: classified as
Row 2: Food, eat, Spores, Hyphae

CONCEPT MASTERY

1. One reason is that the structure of fungi is fundamentally different from that of other organisms. Although most fungi are multicellular, the bulk of their body is not formed of "real" cells.
2. Fungi are heterotrophs and thus must obtain their food from other organisms. Fungi release digestive chemicals into the substance in which they are growing and then absorb the digested food.
3. The alga is an autotroph that produces food for itself and its fungal partner. The fungus protects the alga and helps to supply it with water and minerals.

Content Review

Multiple Choice

Choose the letter of the answer that best completes each statement.

1. Most fungi reproduce by means of
   a. budding.   c. binary fission.
   b. spores.   d. conjugation.
2. Yeasts consist of
   a. many long hyphae that are tightly wound together.
   b. a stemlike stalk that is topped by a cap.
   c. single cells.
   d. cup-shaped fruiting bodies.
3. Fungi that get their food from the remains of dead organisms are known as
   a. decomposers.   c. parasites.
   b. producers.   d. symbionts.
4. Fungi produce spores in structures known as
   a. buds.   c. puffballs.
   b. stalks.   d. fruiting bodies.
5. Which of the following is not necessary for fungi to grow and survive?
   a. sunlight   c. moisture
   b. proper temperature   d. food
6. A lichen is a symbiotic partnership between a
   a. yeast and a mold.
   b. fungus and a plant's roots.
   c. mold and an ant.
   d. fungus and an alga.
7. Which of the following is not a disease caused by fungi?
   a. ringworm   c. chestnut blight
   b. mycorrhiza   d. ergot poisoning
8. Any close relationship between two organisms in which at least one of the organisms benefits is known as
   a. heterotrophy.   c. autotrophy.
   b. symbiosis.   d. parasitism.
9. The month-old leftovers that you discover at the back of the refrigerator are spotted with a velvety blue-green fungus. This fungus is probably a
   a. yeast.   c. lichen.
   b. mushroom.   d. mold.

True or False

If the statement is true, write "true." If it is false, change the underlined word or words to make the statement true.

1. Most fungi are made up of threadlike structures known as <u>fruiting bodies</u>.
2. The scientific name for the organisms called <u>lichens</u> is mycorrhizae.
3. Fleming discovered that a certain mold produced an antibiotic that he called <u>cyclosporine</u>.
4. The three basic forms of fungi are <u>mushrooms, molds, and mildews</u>.
5. Fungi are classified according to <u>characteristics that show their evolutionary relationships</u>.

Concept Mapping

Complete the following concept map for Section 4–1. Refer to pages B6–B7 to construct a concept map for the entire chapter.

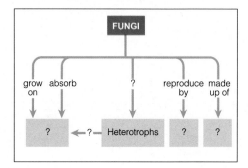

---

4. Check student answers for accuracy. Possible answers: Antibiotics such as penicillin are derived from molds, molds are used in the manufacture of certain foods, certain molds form fungus-root associations that benefit crops and desirable trees, and some molds parasitize harmful insect pests.
5. Yeast causes bread dough to rise, that is, to become fluffy with bubbles of carbon dioxide.
6. The stemlike stalk holds the cap above the surface. The cap is where the spores are produced. In many mushrooms, spores are located on gills on the underside of the cap.
7. Fungi that infect plants can destroy crops that people rely on for food or for trade. Some fungi that infect plants produce toxins that can harm or even kill humans.

CRITICAL THINKING AND PROBLEM SOLVING

1. Spores can be carried readily by the wind from place to place and from host to

Concept Mastery

Discuss each of the following in a brief paragraph.

1. Explain why fungi are placed in their own kingdom.
2. How do fungi obtain energy and nutrients?
3. How do the alga and the fungus in a lichen help each other?
4. Describe three ways in which molds help humans.
5. Why are yeast cells used in baking bread?
6. Describe the structure of a typical mushroom and explain the function of each structure.
7. Explain how fungi that infect plants can harm humans.

Critical Thinking and Problem Solving

Use the skills you have developed in this chapter to answer each of the following.

1. **Relating concepts** What is the connection between reproduction by spores and the rapid spread of crop diseases caused by fungi?
2. **Designing an experiment** Design a set of experiments to show how light, temperature, and dryness affect the growth of bread mold. Be sure to include a control in each of your experiments—and make sure that each experiment tests only one variable.
3. **Making inferences** As you can see in the accompanying photograph, a sharpei dog looks like its skin is many sizes too large. People who own a sharpei dog must rearrange its loose folds of skin every now and then. This is especially true in areas where there is a lot of moisture in the air. What might happen to a sharpei dog if its owner neglected to rearrange its folds of skin?
4. **Assessing concepts** Fungi are sometimes divided into two groups: yeasts (unicellular fungi) and molds (multicellular fungi). Is this informal classification system useful? Why or why not?

5. **Using the writing process** While looking something up in the local library's new set of encyclopedias, you notice that fungi are defined as "plants that lack chlorophyll (the green substance involved in food-making)." Write a letter to the editors of the encyclopedia explaining why this is not a good description of fungi.
6. **Using the writing process** Imagine that you are the alga in a lichen. Write a letter to an old friend describing your fungal partner.

host. Because crop plants are located close to one another in fields and one fungus produces many spores, all the plants in a field can be infected in a very short time. Spores can then be carried by the wind to infect nearby fields.

2. Student answers will vary. In describing their experimental setups, students should note how they intend to keep all variables except the experimental variable constant. For example, they should recognize that putting a mold culture on a sunny windowsill will increase its temperature as well as the amount of light it receives and will require appropriate adjustments in the control.

3. The dog might develop a fungal infection in the folds of skin.

4. Accept all logical answers. Students may or may not feel that mushrooms should be classified under "molds."

5. Student answers should explain clearly why fungi should not be classified as "plants."

SECTION	HANDS-ON ACTIVITIES
5–1 Plants Appear: Multicellular Algae pages B106–B113 Multicultural Opportunity 5–1, p. B106 ESL Strategy 5–1, p. B106	**Student Edition** ACTIVITY (Doing): That's About the Size of It, p. B108 ACTIVITY BANK: Seaweed Sweets, p. B180 **Activity Book** CHAPTER DISCOVERY: Looking at Nonvascular Plants, p. B119 ACTIVITY: Comparing Green Algae, p. B123 ACTIVITY BANK: If We Lived Near the Ocean, We'd See Seaweed, p. B211 **Teacher Edition** Introducing the Algae, p. B104d
5–2 Plants Move Onto Land: Mosses, Liverworts, and Hornworts pages B114–B118 Multicultural Opportunity 5–2, p. B114 ESL Strategy 5–2, p. B114	**Laboratory Manual** Observing Mosses and Liverworts, p. B41 **Activity Book** ACTIVITY: Comparing Mosses and Liverworts, p. B125 ACTIVITY BANK: Growing Moss, p. B213 **Teacher Edition** Land Plant Discovery Tour, p. B104d
5–3 Vascular Plants Develop: Ferns pages B119–B123 Multicultural Opportunity 5–3, p. B119 ESL Strategy 5–3, p. B119	**Student Edition** LABORATORY INVESTIGATION: Comparing Algae, Mosses, and Ferns, p. B124
Chapter Review pages B124–B127	

OUTSIDE TEACHER RESOURCES

Books

Abbe, Elfriede. *The Fern Herbal: Including the Ferns, the Horsetails, and the Club Mosses,* Cornell University Press.

Bold, H. C., and M. J. Wynne. *Introduction to the Algae,* 2nd ed., Prentice-Hall.

Conrad, Henry S., and Paul L. Redfearn, Jr. *How to Know the Mosses and Liverworts,* 2nd ed., Wm. C. Brown.

Fryxell, Greta A., ed. *Survival Strategies of the Algae,* Cambridge University Press.

Hallowell, Anne E., and Barbara Hallowell. *Fern Finder,* Nature Study.

Streams, John, ed. *Treasures of Nature: Ferns,* Crossing Press.

Taylor, Ronald J., and Alan E. Leviton, eds. *Mosses of North America,* American Association for the Advancement of Science.

Weissner, W., and D. J. Robinson, eds. *Algal Development,* Springer-Verlag.

OTHER ACTIVITIES	MEDIA AND TECHNOLOGY
Activity Book ACTIVITY: Algae Classification, p. B127 **Review and Reinforcement Guide** Section 5–1, p. B37	**English/Spanish Audiotapes** Section 5–1
Student Edition ACTIVITY (Thinking): Moss-Grown Expressions, p. B117 **Review and Reinforcement Guide** Section 5–2, p. B39	**English/Spanish Audiotapes** Section 5–2
Student Edition ACTIVITY (Reading): The Secret of the Red Fern, p. B121 **Activity Book** ACTIVITY: Life Cycle of a Fern, p. B129 **Review and Reinforcement Guide** Section 5–3, p. B41	**English/Spanish Audiotapes** Section 5–3
Test Book Chapter Test, p. B89 Performance-Based Tests, p. B129	**Test Book** Computer Test Bank Test, p. B95

*All materials in the Chapter Planning Guide Grid are available as part of the Prentice Hall Science Learning System.

Audiovisuals

The Algae, filmstrip with cassette, Biology Media

Ferns and Horsetails (Life Cycles and Ecology), filmstrip with cassette, Ward

Introduction to Algae, filmstrip with cassette, Carolina Biological Supply Co.

Mosses and Liverworts (Life Cycles and Ecology), filmstrip with cassette, Ward

Mosses, Liverworts, and Ferns, film or video, Coronet

A New Look at Algae, film or video, Ward

Simple Organisms: Algae and Fungi, rev., film or video, Coronet

CHAPTER OVERVIEW

Algae are simple plantlike autotrophs that use sunlight to produce their food. Scientists disagree about the classification of algae, but in this textbook multicellular algae and closely related unicellular algae are placed in the kingdom Plantae. Although algae are plants, they do not have the special tubes that transport water and other materials through the bodies of land plants. Algae do not have true roots, stems, leaves, or seeds. All algae live in water. The three phyla of algae get their names—brown, red, and green—from the pigments found in the cells of the algae. Green algae are believed to have shared a common ancestor with modern land plants.

To live on land, plants must make certain adaptations; they need structures for support, roots, vascular tissue, ways to minimize water loss, and methods of reproduction that do not require standing water. Mosses, liverworts, and hornworts live on land, but they are not fully adapted. They are small, live in damp places, and grow close to the ground. Ferns, on the other hand, have almost completely adapted to living on land. They have vascular tissue, a system of tiny tubes that carry water and materials throughout the plant. Vascular tissue also strengthens and supports the plant. Ferns can grow much larger and in drier environments than mosses and algae can. Ferns, too, are not completely adapted to living on land; they still need standing water in order to reproduce.

5–1 PLANTS APPEAR: MULTICELLULAR ALGAE
THEMATIC FOCUS

The purpose of this section is to introduce students to the first plants to appear on Earth, algae. Like all plants, algae contain chlorophyll and make their own food. But unlike many plants, algae do not have an internal transport system that carries water and other materials, and they do not reproduce with seeds. They must live in or near water to survive. Students will learn about the three phyla of algae—brown, red, and green—named for the predominant pigment in their cells. Students will find out about many kinds of algae and their uses. The section ends with a discussion of green algae and their similarities to land plants, which have led scientists to think that land plants evolved from green algae; that is, that modern land plants and modern species of green algae evolved from a common ancestor.

The themes that can be focused on in this section are energy, evolution, and scale and structure.

Energy: Point out to students that algae are autotrophs; that is, they are organisms that can make their own food to meet their energy needs. The chlorophyll in their cells absorbs light energy and converts it into chemical energy.

*Evolution: Be sure students understand that although land plants evolved from green algae, modern species of green algae are not the ancestors of land plants. Rather, modern species of land plants and modern species of green algae probably share a common ancestor. Review the steps in the evolution of land plants from unicellular green algae to algaelike land plants to be sure student students understand the progression.

*Scale and structure: Stress that green algae are found in many different environments and that they have evolved in many forms to adapt to those environments. They may live as single cells. Some species form a colony, in which single cells join together but continue to operate independently. Other green algae are multicellular and have specialized structures; that is, structures that perform a particular function. The important point is that the different forms of green algae provide an idea of how plants adapted to live on land.

PERFORMANCE OBJECTIVES
1. **Describe the characteristics of algae.**
2. **Identify the three algae phyla and name some characteristics of each phylum.**
3. **Describe some uses of algae.**

SCIENCE TERMS
alga p. B106
brown alga p. B107
red alga p. B107
green alga p. B107
pigment p. B107

5–2 PLANTS MOVE ONTO LAND: MOSSES,LIVERWORTS, AND HORNWORTS
THEMATIC FOCUS

The purpose of this section is to familiarize students with the adaptations that plants evolved in order to survive on land. Basically, there are five major problems that plants had to solve. Students will find that the land plants called mosses, liverworts, and hornworts solved these problems mostly by avoiding them. They live in moist places where water is plentiful. They are very small and grow very close to the ground. Thus, they are able to live on land without many special adaptations. Students will learn that mosses have many uses for people. Several species are of value to gardeners and can also be used as fuel.

The themes that can be focused on in this section are evolution and stability.

*Evolution: Point out to students that it is important to understand that land plants have evolved in ways that make them better suited to meet the challenges of life on land. It is also important to recognize what those adaptations are and to realize that if plants had not evolved to meet these challenges, they would not have survived on land.

Stability: You may wish to use the theme of stability, rather than evolution, to focus on the idea that land plants have evolved in response to the challenges of their environment. Again, review with students what adaptations plants have had to make and why. Reiterate that had plants not met the challenges of living on land, they would not have survived.

PERFORMANCE OBJECTIVES

1. Describe the adaptations that plants evolved in order to live on land.
2. Explain how mosses, liverworts, and hornworts have adapted to life on land.
3. Describe some uses of mosses.

5–3 VASCULAR PLANTS DEVELOP: FERNS

THEMATIC FOCUS

The purpose of this section is to introduce students to the concept of vascular and nonvascular plants. Vascular plants have vascular tissues, or tiny tubes that carry water and materials throughout the bodies of the plants. They have true leaves, stems, and roots, which are also parts of the transport system in plants. Algae and mosses do not have vascular tissue, so they are called nonvascular plants. Students will learn that because vascular tissue has strong cell walls and because it can transport materials quickly and efficiently to all parts of the plant, vascular plants can grow much taller and larger than nonvascular plants. Ferns, which are some of the earliest vascular plants, vary in size but can grow as tall as trees. Students will learn about the parts of ferns and the ways ferns have adapted to living in many environments. They will discover, however, that ferns are not fully adapted to life on land. Like algae and mosses, they require standing water for reproduction. Ferns are prized as decorative plants, and some species can be eaten by humans.

The themes that can be focused on in this section are evolution, scale and structure, and unity and diversity.

*Evolution: Emphasize that vascular plants, such as ferns, are better adapted for an existence on land than nonvascular plants, such as algae or mosses and their relatives. The main focus here should be that vascular tissue can move water, food, and other materials throughout plants and strengthen and support stems. Therefore, vascular plants can grow larger and in drier environments than nonvascular plants can.

*Scale and structure: Certain structures distinguish vascular plants from nonvascular plants. Vascular plants, as their name denotes, have vascular tissue, a system of tiny tubes that transport water, food, and other materials in the plants. Vascular plants also have true roots, stems, and leaves. Explain to students that the presence of these structures will enable them to identify a vascular plant.

*Unity and diversity: Point out to students that the plants they have studied in this chapter are all plants without seeds but that these plants are nevertheless a diverse group that includes unicellular, multicellular, vascular, and nonvascular forms. Encourage students to place each plant in the proper category and to list other ways that the plants differ from one another.

PERFORMANCE OBJECTIVES

1. Explain how ferns have adapted to life on land.
2. Compare ferns to mosses and algae.
3. Describe some uses of ferns.

SCIENCE TERMS

nonvascular plant p. B119
vascular plant p. B119

Discovery *Learning*

TEACHER DEMONSTRATIONS MODELING

Introducing the Algae

Both freshwater and marine algae should be readily available from a store that sells aquariums and related supplies. If possible, obtain samples of such common algae as *Spirogyra, Oedogonium, Ulva, Fucus,* and *Porphyra.* If actual specimens are not available, obtain pictures of algae from magazines such as *National Geographic* or *Omni* or refer students to the illustrations of algae that appear throughout this section.

• **What do all the algae appear to have in common?** (They are found in or near water.)
• **How do the algae obtain food?** (By using their chlorophyll and sunlight to make food.)

• **In order to carry on this food-making process, where in the water must the algae live?** (Because they need light to make food, they must live on or near the surface of the water.)

Land Plant Discovery Tour

Take students on a guided tour of the exterior of the school building. You will be looking for the many places that land plants grow. You can begin by looking in some of the obvious locations first—the lawn or garden or in the pits surrounding planted trees. Then look in some of the less obvious places—in pavement cracks, on the shady sides of the building or walls, on rocks, on trees, or near a source of standing water.

Examine each plant and have students note the following.

1. Does it have leaves?
2. Does it have veins on its leaves?
3. Where was the plant growing?
4. What is the approximate size of the plant?
5. Is the plant mosslike or does it have a green stem or a woody stem?

Use your discretion in removing some of the plant samples for additional closer observations in the classroom. Here are some activities students may enjoy.

1. Microscopic—make wet mounts from algae or moss scrapings from tree bark or walls.
2. Dissecting microscope—Examine the mosslike plants and small plants with green stems in order to better see their characteristics.
3. Tracings—Select leaves from plants with veins for students to trace. Once the outline is completed, students should draw in the pattern of the veins.
• **Which kinds of plants are more numerous?** (Plants with stems.)
• **Where are mosslike plants found?** (In shady locations, cracks in pavement, places where water collects.)
• **Did all the plants have leaves?** (Only those plants with stems had leaves.)

CHAPTER 5
Plants Without Seeds

INTEGRATING SCIENCE

This life science chapter provides you with numerous opportunities to integrate other areas of science, as well as other disciplines, into your curriculum. Blue numbered annotations on the student page and integration notes on the teacher wraparound pages alert you to areas of possible integration.

In this chapter you can integrate food science (pp. 106, 110, 111), life science and evolution (pp. 107, 112, 119), mathematics (p. 108), social studies (pp. 108, 118), earth science and astronomy (p. 112), physical science and heat (pp. 116, 123), language arts (pp. 117, 121), and life science and biogeography (p. 120).

SCIENCE, TECHNOLOGY, AND SOCIETY/COOPERATIVE LEARNING

A lake gradually turns a scummy green and begins to stink. Fish die and float on the surface of the lake amid the green scum. Eventually, the lake can no longer be used for recreation because the water is filled with algae. Seven out of eight lakes, along with coastal areas and other bodies of water, are affected by this process, which is called eutrophication. The term means "overfeeding."

Eutrophication is caused when nutrient-rich material stimulates growth of algae into an "algal bloom." As the algal bloom begins to die, there is an oversupply of detritus (dead material) for decomposers in the lake to break down. As the decomposers do their job, they use up the dissolved oxygen in the water. The depletion of the dissolved oxygen results in the death of other plants and animals—particularly, fish. Another danger of eutrophication is that some algae produce toxic substances that are taken in by other organisms in the food chain. These toxins can accumulate in some types of seafood and may result in food poisonings for consumers of fresh seafood.

Eutrophication is a natural process in all bodies of water but is aggravated by human activities. The release of sewage and other waste water containing phosphates and human waste into lakes, reservoirs,

INTRODUCING CHAPTER 5

DISCOVERY LEARNING

▶ *Activity Book*

Begin teaching the chapter by using the Chapter 5 Discovery Activity from the *Activity Book.* Using this activity, students will collect, examine, and compare non-vascular plants.

USING THE TEXTBOOK

Have students look at the photograph on page B104.
• **What do you see in the picture?** (A diver in seaweed in the ocean.)
• **What do you imagine it would be like to be in a seaweed "forest"? How would such an ocean forest compare to a land forest?** (If any students have actually dived in the ocean kelp beds, encourage them to share their experiences with the class. Otherwise, accept all answers. Point out

Plants Without
Seeds

Guide for Reading

After you read the following sections, you will be able to

5–1 Plants Appear: Multicellular Algae
- Describe the three groups of multicellular algae.

5–2 Plants Move Onto Land: Mosses, Liverworts, and Hornworts
- Describe the characteristics of mosses, liverworts, and hornworts.

5–3 Vascular Plants Develop: Ferns
- Explain how ferns are different from other kinds of plants without seeds.

The waves lapping around your face are so cold that your skin goes numb. You make a few last-minute adjustments to your wet suit, then raise your gloved hand, forming a circle with your thumb and forefinger—O.K. One of your friends on the boat gives you a thumbs-up. With a practiced movement of your arms, you descend feet first into an alien world.

Once underwater, you find yourself in a strange, dreamlike forest. Thin, vinelike "trees" stretch up toward the silver sky far above your head. Below you, the trees extend so far down into murky darkness that they seem to go on forever. Fishes dart like birds among the swaying ribbonlike leaves of the trees. A curious young harbor seal suddenly appears and stares at you with large dark eyes. Then it does a flip with a half-twist and vanishes like a ghost among the long, thin trees of seaweed.

The trees of this underwater forest are a type of seaweed known as kelp. Kelp forests—found off the coasts of Washington, Oregon, and northern California—are home to slugs, snails, fishes, seals, sea otters, and many other living things.

Kelps are just one kind of plant without seeds. In this chapter, you will read about the strange and ancient world of plants without seeds.

Journal *Activity*

You and Your World In your journal, record your thoughts about studying plants. Do you think this chapter and the one that follows will be easy? Difficult? Interesting? When you have completed Chapters 5 and 6, look back at this entry and see if the chapters were as you expected.

◀ *Long, ropy strands of kelp sway gracefully with the movement of the ocean's waves and currents as a diver explores the mysterious, dimly lit world of the seaweed "forest."*

B ■ 105

---

bays, and rivers provides algae with the nutrients they need to "bloom." Runoff from suburban and agricultural areas contains animal wastes and fertilizers that also contribute to algal blooms and eutrophication. These human activities can be controlled to prevent eutrophication: Sewage and waste water can be treated to control the amount of nutrient-rich water released into the environment; homeowners can use phosphate-free detergents; and good land management practices in both suburban and agricultural areas can reduce the runoff of water containing animal and fertilizer residues.

Cooperative learning: Using preassigned or randomly selected teams, have groups complete one of these assignments.

• Diagram the process and results of eutrophication in a lake. The groups' diagrams should show the lake as it was once used for recreation, the source(s) of the nutrient-rich water entering the lake, and the effects of that water on the lake. If possible, encourage groups to identify bodies of water in their area that have been affected by eutrophication. Have them identify the source(s) of the nutrient-rich input and describe the effects on the lake.

• Write and illustrate a "biopoem" about one of the plants without seeds—algae, mosses, liverworts, hornworts, or ferns—covered in this chapter. A biopoem has the following format:

 * Number of lines equals the number of letters in the title. (Example: "Algae"— 5 lines)
 * Write the title of the poem vertically on the page.
 * Each line must begin with the first letter (in sequence) of the spelled title.
 * Each line must include some fact about the title—classification, characteristics, examples, uses, and so on.

See Cooperative Learning in the *Teacher's Desk Reference.*

JOURNAL ACTIVITY

You may want to use the Journal Activity as the basis for a class discussion before students write in their journals. As a group, discuss students' thoughts about plants. Students should be instructed to keep their Journal Activity in their portfolio.

---

that the kelp resembles the tall trees in a land forest and that there would be animal life in both kinds of forests.)

Have students read the chapter introduction on page B105.

• **Why would the kelp provide a home for slugs, snails, fishes, seals, sea otters, and many other living things?** (It offers protection and shelter. It provides food for some animals, who in turn provide food for other animals.)

• **Sea otters have to feed constantly. Why would they find the kelp forest the ideal home?** (With all the animals that live in the kelp forest, the sea otters would be assured of a readily available food supply.)

• **Some scientists think the sea otters keep the North Pacific kelp forests in balance. Can you explain what that means?** (Sea otters help to control the populations of many animals that could overrun and destroy the kelp.)

5-1 Plants Appear: Multicellular Algae

Guide for Reading

Focus on this question as you read.

▶ How are red, brown, and green algae similar? How are they different?

Figure 5-1 *Algae come in many forms. Mermaid's cups resemble shallow drinking glasses (top). The bubblelike structures on giant kelp help to keep it afloat in its watery home (bottom right). The delicate branches of coralline algae contain so much limestone that they look and feel as if they were carved out of reddish stone (bottom left).*

5-1 Plants Appear: Multicellular Algae

Have you ever walked along a beach? If so, you may have found flat greenish-brown ribbons, brown ropes, or delicate reddish tangles of seaweed washed up on the sand. You may have discovered bits of yellowish seaweed that contain bubbles which squish with a satisfying pop. Looking into tide pools, you might have seen transparent green veils of sea lettuce, dainty mermaid's cups, or pink coralline algae.

Even if you have never been to a beach, you have probably encountered seaweed in other forms. For example, the edible blackish wrapper on certain kinds of sushi (Japanese rice rolls) is made of dried seaweed. Ground-up seaweed is mixed with water and sprayed on gardens to make plants grow better. And chemicals from seaweed are added to ice cream, jellies, candy, and many other foods to give them a smooth texture.

Seaweeds are some of the most familiar types of multicellular **algae** (AL-jee; singular: alga, AL-gah). As you learned in Chapter 4, the term alga refers to any simple plantlike autotroph that uses sunlight to produce its food. Thus the term algae is used to refer to everything from microscopic unicellular blue-green bacteria to enormous multicellular strands of kelp as long as a football field.

Scientists do not agree about the formal classification of algae. Some scientists think that all algae (except blue-green bacteria) should be classified as protists. Others think that all algae should be classified

phyll and the presence of roots, stems, leaves, and possibly flowers.)

Now show an example of a nonvascular plant. Some easily obtained specimens include *Hydrodictyon*, which is found floating on ponds during warm months. Other possibilities are a clump of moss or a piece of bark covered with *Protococcus*. This powdery-appearing green alga can be collected throughout the year.
• **This is also a plant. How is it similar to the other plants?** (Focus attention on

the fact that simple plants also possess chlorophyll.)
• **How is it different from the other plants?** (Lead students to suggest that it has a simpler structure. True roots, stems, and leaves are not present.)

CONTENT DEVELOPMENT

Point out to students that the term *algae* is used to name any of a variety of simple plantlike autotrophs (organisms that can make their own food) that use sunlight to produce food. Such organ-

as plants. Both of these views have evidence to support them. In this textbook, we have decided to use the following classification system: Blue-green bacteria belong to the kingdom Monera. Most of the other unicellular algae are assigned to the kingdom Protista. Multicellular algae and closely related species of unicellular algae are placed in the kingdom Plantae.

In Chapter 2, you learned about blue-green bacteria. In Chapter 3, you learned about plantlike protists. In this chapter, you will learn about multicellular algae and the unicellular algae that are closely related to them. For convenience and simplicity, from this point on we will refer to multicellular algae and their close relatives as algae.

Algae were the first kinds of plants to appear on Earth. The oldest fossils (preserved remains of ancient organisms) of algae are about 900 million years old. Fossil evidence indicates that land plants evolved from certain types of these ancient algae.

Although algae resemble the land plants that are familiar to you, there are some important differences. **Algae lack the special tubes that transport water and other materials through the bodies of land plants.** This means that algae do not have true roots, stems, or leaves. By definition, roots, stems, and leaves contain these special tubes. **Algae do not have seeds.** Because they lack transporting tubes and seeds, algae must live in or near a source of water in order to survive and reproduce.

Algae are divided into three phyla: **brown algae, red algae,** and **green algae.** The phyla get their names from the **pigments,** or colored chemicals, that are found within the cells of the algae.

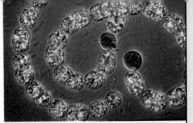

Figure 5–2 *Blue-green bacteria (top) and diatoms (bottom) are both types of algae that do not belong to the plant kingdom.*

Figure 5–3 *Multicellular algae are divided into three phyla according to the pigments they contain. Which of the algae shown here is a red alga? Which is a green alga? Which is a brown alga?* ❶

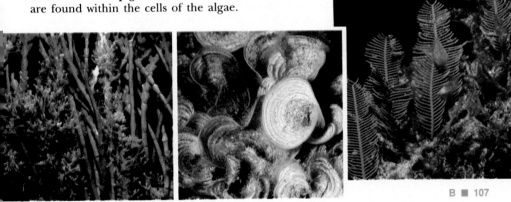

B ■ 107

BACKGROUND INFORMATION

CLASSIFICATION OF ALGAE

The question of how to classify algae can be used to demonstrate to students the difficulty of higher classification. The modern definition of species allows us to draw clear lines separating most (though not all) species of organisms. Taxonomic categories above the level of species, however, must often be made by judging the relative importance of several biological characteristics. When different people place varying weights on these characteristics, they come to different conclusions. The kingdom Protista, for example, is a hodgepodge of organisms lumped together mostly because they are unicellular. The kingdom Plantae, on the other hand, is mostly multicellular. Yet we are faced with the problem of the green algae, which clearly contain both unicellular and multicellular species. Where do we put them? We have chosen not to include multicellular organisms in the Protista and have therefore divided algae into predominately unicellular and predominantly multicellular groups. The former we have placed in the Protista, and the latter—which are also tied to higher plants by several aspects of structural and photosynthetic chemistry—we have placed in the Plantae.

isms range in size from tiny unicellular to huge multicellular. The range of organisms has caused scientists to disagree on how to classify algae. Some classify all algae as protists; others classify all algae as plants. This is a good opportunity to demonstrate to students that there are not always hard-and-fast rules in science. Many subjects are constantly debated, and some subjects that may seem closed can be reopened with the discovery of new information. This is what keeps all areas of science vital and growing.

● ● ● ● **Integration** ● ● ● ●

Use the discussion of the uses of seaweed to integrate food science into your lesson.

Use the discussion of algal fossils to integrate concepts of evolution into your lesson.

ACTIVITY DOING

THAT'S ABOUT THE SIZE OF IT

Skills: *Measuring, making computations, calculator, recording data, comparing*

Materials: *meterstick, calculator*

This activity should help students comprehend the idea of a giant kelp's incredible height. They may wish to choose one student as the measurement recorder and two or more students to do the measuring.

The number of meters in students' bakalian will depend on the number of students in the class and their respective heights. The likelihood is that one bakalian will not be 100 meters long and therefore will not equal the length of the giant kelp plant. Probably one bakalian will equal less than one-half to one-half the length of the giant kelp. The kelp's length in bakalians will also vary. For example, if the bakalian equals 50 meters, then the giant kelp would be two bakalians long.

Integration: Use this Activity to integrate mathematics into your lesson.

5–1 (continued)

CONTENT DEVELOPMENT

Be sure students understand that without a transport system, algae will be limited in size and will be confined to watery or damp environments. The simple structure of algae means that most algal cells are near the surface, where water can enter the cells directly from the environment. Plants can grow large or in dry places only if they have a way to carry water to cells that are some distance from the water source.

CONTENT DEVELOPMENT

The bodies of some brown algae consist of simple branching filaments. Others are quite large with a variety of specialized tissues. Some of them, such as *Sargassum* seen in Figure 5–4, have a branched structure called a holdfast at the base to secure

---

That's About the Size of It

1. Using a meterstick, determine the height in meters of each person in your science class (including your teacher). Record your measurements.

2. Using a calculator, add up the heights of all the people in your class (make sure you put the decimal point in the right place!). You have now defined the length of an unofficial unit of measurement known as a bakalian. How many meters long is a bakalian?

3. Giant kelps can be as long as 100 meters. How does the length of a bakalian compare to that of such a kelp plant? How long is a 100-meter kelp plant in bakalians?

In green algae, the most noticeable pigment is the green chemical chlorophyll. Chlorophyll captures light energy so that it can be used in the food-making process. Red algae and brown algae also contain chlorophyll. However, the green color of the chlorophyll is masked by other kinds of pigments. These pigments, which are known as accessory pigments, absorb light energy and transfer it to chlorophyll for use. The main accessory pigments in red algae are pink, red, reddish purple, and reddish black. What color would you expect the main accessory pigment in brown algae to be? ❶

Brown Algae

For centuries it was the subject of sailors' nightmares—a haunted sea located somewhere in the Atlantic Ocean between the African coast and the islands of the West Indies. The sailors whispered the ❷ name of this sea when they dared to say it at all—the Sargasso Sea. According to legend, if a ship was foolish or unlucky enough to sail into this sea, it would encounter seaweed that could cling to the sides of the ship like monster hands. As the ship sailed on, the seaweed would become thicker and thicker, slowing the ship to a crawl, and then to a halt. A ship trapped in the horrible seaweed of the Sargasso Sea would remain there forever, joining a fleet of dead ships guarded by skeletons.

The legends about the ghost ships of the Sargasso Sea are purely imaginary. Although there really is a Sargasso Sea, it is not an obstacle to ships. In the Atlantic Ocean between Africa and Bermuda lies the real Sargasso Sea. It is an area of calm winds and gentle waves that is a perfect home for the brown algae *Sargassum* (sahr-GAS-uhm). *Sargassum* floats on or near the ocean's surface, held up by tiny round air-filled structures that act like inflatable life preservers. These air-filled structures, or air bladders, help to ensure that the *Sargassum* stays close to the surface of the ocean, where it can get plenty of sunlight. Why is sunlight important for *Sargassum*? ❷

In most parts of the ocean, a clump of seaweed would quickly be torn apart by the action of the wind and waves. But there is little wind and wave action in the Sargasso Sea. As a result, *Sargassum* is

---

them to rocks or other objects on the sea floor. A stemlike structure called the stipe extends from the holdfast. Above the stipe is a leaflike blade with air bladders to keep it afloat. Students may mistakenly refer to the holdfast, stipe, and blade of some large brown algae as a root, stem, and leaf, respectively. Explain that though the structures resemble plant parts, they do not function as those plant parts do. Holdfasts are only anchors for the algae; they do not take in water and nutrients as roots do. The stipe and blade are not

composed of the same kind of tissue as stems and leaves and do not perform the functions that those parts do.

GUIDED PRACTICE

Skills Development

Skill: Making inferences

Ask students to look at Figure 5–4 and read the caption. Emphasize that many organisms live in beds of large brown algae.

• **Can you think of a reason why the fish and crab make their home among strands**

---

able to form enormous floating mats many kilometers long. These mats are home to fishes, crabs, jellyfish, and many other ocean animals. Can you find the crab and the fish hiding among the strands of *Sargassum* in Figure 5–4?

While the Sargasso Sea did not provide an answer to the mysterious disappearance of sailing ships, it did explain a few biological mysteries. One of these mysteries involved the snakelike fishes known as eels. For more than two thousand years, people wondered where eels in the lakes and rivers of Europe came from. No one had ever seen a baby eel. No one had ever seen eels mate or lay eggs. Aristotle, a philosopher and scientist of ancient Greece, thought that eels simply sprang out of the mud. During the eighteenth century, some people thought that eels were formed from the hairs on horses' tails.

It was not until the end of the nineteenth century that the mystery was solved: Adult eels leave the rivers and lakes and travel thousands of kilometers to the Sargasso Sea to mate and lay their eggs. The baby eels, which are leaf-shaped fish that look nothing like adult eels, live for a while among the mats of *Sargassum*. Then, over a period of several years, the baby eels gradually travel back across the ocean to Europe. As they grow, the baby eels slowly change their shape. By the time they reach the mouths of the European rivers and begin swimming upstream, the baby eels have grown up into small adult eels.

Sargassum is not the only kind of brown algae. The kelp that you read about earlier is another type of brown algae. So is the rockweed that grows on rocky coasts. As you can see in Figure 5–6 on page 110, brown algae come in a wide variety of shapes

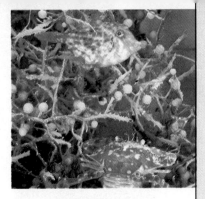

Figure 5–4 *The enormous floating mats of* Sargassum *are home to many animals. The spherical structures on the* Sargassum *are air bladders. What is the function of the air bladders?* ③

Figure 5–5 *Baby eels spend the first few years of their lives in the Sargasso Sea. Then they travel to the rivers of Europe. On their journey, they take on their adult shape (left) and develop pigments so they are no longer transparent. Silvery-gray adult European eels may wiggle across patches of dry land as they begin their journey back to the Sargasso Sea to breed (right).*

Answers
① Brown. (Making inferences)
② *Sargassum*, like all plants, needs sunlight to use to make food. (Applying concepts)
③ The air bladders keep *Sargassum* close to the surface of the ocean. (Relating facts)

Integration
① Mathematics
② Social Studies

BACKGROUND INFORMATION
BROWN ALGAE

Brown algae, or *Phaeophyta,* are classified together because they are all multicellular, contain chlorophylls *a* and *c,* the brown pigment fucoxanthin, and store their foods in the form of starches and oils. The combination of fucoxanthin and chlorophyll *c* gives these plants their yellowish-brown color.

the first to encounter the Sargasso Sea. The stories of becalmed ships and sea monsters were told and embroidered, as such stories are, by poets and writers in all seafaring nations.

● ● ● ● **Integration** ● ● ● ●

Use the discussion of the legends about the Sargasso Sea to integrate social studies concepts into your lesson.

ENRICHMENT

Interested students might like to investigate in more detail the amazing migration of both European and American eels to the Sargasso Sea to breed. Ask students to find out how a possible explanation of this behavior might be tied to Wegener's theory of continental drift.

of *Sargassum?* (Accept all logical answers, but lead students to infer that the seaweed provides an abundant source of food as well as shelter and protection.)
• **How are the fish and crab protected by** *Sargassum?* (They have acquired the colors and patterns of the seaweed, which enable them to blend in with their environment.)

CONTENT DEVELOPMENT

Explain that the area of the North Atlantic Ocean known as the Sargasso Sea is set apart from the rest of the ocean only by the vast amounts of seaweed that grow there and by the slow ocean currents that are surrounded by faster currents. The relative calm of the water and the soft, flexible nature of the seaweed's fronds allow it to live in such open sea.

Sargassum is also called gulfweed. The name *Sargassum* came from the Portuguese word for seaweed, *sargaço,* which originally meant "grape." The small air bladders on the seaweed are shaped like grapes. Portuguese sailors were some of

RED ALGAE

Red algae, or *Rhodophyta,* are classified together because they all contain chlorophylls *a* and *d* and the reddish pigments called phycobilins and because they store their food in the same form of starch.

INTEGRATION

NUTRITION

One type of red algae, *Porphyra,* is used to make a popular Japanese dish, sushi. Seasoned rice is rolled up in a sheet of dried *Porphyra,* called *nori* in Japanese, and then cut into slices like a jellyroll. A piece of raw fish is often rolled inside or served on top. *Porphyra* is such a staple of the Japanese diet that it is grown on special marine farms.

BACKGROUND INFORMATION

SLIMY ALGAE

Why do many types of algae feel slimy? The cell walls of these algae have two layers—the inner one made of cellulose and the outer one of a substance called pectose. In water, the pectose is changed to pectin, a soluble substance that makes the cells feel slimy.

5–1 (continued)

CONTENT DEVELOPMENT

Explain that algins or alginates, which are made from kelp (brown algae), are gummy substances that can be used as thickening or gelling agents. They are used in ice cream, mayonnaise, film, paint, varnish, and even buttons. They are also used in cosmetics such as lipsticks and soaps. In the United States, kelp is harvested near the coast of California. But in Asia, where kelp has long been a major food source, it is grown on special farms in the ocean. Kelp is rich in vitamins and minerals, particularly iodine, which people and animals need to ensure normal growth.

Figure 5–6 *Brown algae come in all shapes and sizes. Can you locate the air bladders on the kelp (top right)?*

and sizes. Most brown algae live in the ocean. Some, like *Sargassum,* float freely. Others, such as kelps and rockweed, are attached to the sea floor.

Brown algae have long been used as food for humans in China, Japan, Canada, Ireland, New Zealand, and many other parts of the world. They can also be used as food for livestock. For example, in some places in Scotland and Ireland, cattle and sheep graze on brown algae at low tide. Chemicals extracted from brown algae (known as algins or alginates) are added to salad dressings and other foods to make them smooth and prevent their ingredients from separating.

Red Algae

Like brown algae, most red algae are multicellular and live in the ocean. Red algae can grow to be several meters long, but they never reach the size of brown algae such as *Sargassum* or giant kelps. The shapes of red algae are just as varied as those of brown algae. Some red algae form clumps of delicate, branching red threads. Others grow in large, rounded, flat sheets. Still others produce hard, stiff branches rich in calcium carbonate (the substance that makes up the shells of foraminifers, corals, snails, and crabs).

Red algae usually grow attached to rocks on the ocean floor. Red algae can be found at depths up to 170 meters—far deeper than other kinds of algae. Very little sunlight penetrates to these extreme depths. Chlorophyll alone cannot absorb enough light energy to allow the food-making process to continue at a life-sustaining level. How, then, do deep-water red algae get the light they need to survive?

The answer involves the accessory pigments that give red algae their characteristic color and their

● ● ● ● **Integration** ● ● ● ●

Use the discussion on brown algae as food to integrate food science into your lesson.

FOCUS/MOTIVATION

Have students study the different kinds of red algae in Figure 5–7. Point out that like brown algae, red algae come in many different sizes and forms. Note that red algae are mostly small, feathery, and delicate looking.

• **What is the main difference between brown algae and red algae?** (Brown algae's accessory pigments give them a brown color; red algae's accessory pigments give them a red color.)

• **Do brown and red algae contain chlorophyll? If so, why aren't they green?** (Brown and red algae do contain chlorophyll, but their accessory pigments mask its green color.)

name. The accessory pigments are able to absorb the small amount of light energy that penetrates to deep waters. The absorbed energy is then transferred to chlorophyll. Thus the accessory pigments enable red algae in deep water to make the most of the light that reaches them. Now can you explain why red algae are able to live in deeper waters than green algae? ❶

Red algae are used by humans in a number of ways. Some species are eaten as food. Other species are harvested by the ton and added to soil to make it better for crops. Chemicals extracted from red al-❷gae are used to manufacture certain foods. The next time you eat ice cream, use a creamy salad dressing, drink chocolate milk, or frost a cake with ready-made frosting, read the list of ingredients on the package. You may find carrageenan (KAIR-uh-geen-uhn), a substance that comes from red algae, on the list. Another substance derived from red algae is agar (AH-gahr). Agar plays an important role in medical research. It is used to make the jellylike nutrient mixtures on which bacteria and other microorganisms are grown.

Green Algae

Deep in space, a silvery ship is in its second year of a four-year journey to study the planets. On board, the astronauts are about to finish dinner. Although they have brought along enough food to last the entire journey, they could not carry enough oxygen to last several years. Are the astronauts doomed to suffocate in space? Of course not.

In a tiny room near the back of the ship sits a tank of water filled with green algae. The green algae use the carbon dioxide exhaled by the astronauts

Figure 5–7 *Red algae may resemble fingers of transparent red cellophane, dark pink rock formations, or clumps of purplish-black hair.*

Activity Bank

Seaweed Sweets, p. 180

Figure 5–8 *The white streaks on the agar in this petri dish consist of millions of bacteria. Because most microorganisms cannot live on agar alone, nutrients and other substances are added to the agar. In this case, blood has changed the color of the agar from pale yellow to red.*

B ■ 111

BACKGROUND INFORMATION

VOLVOX

Large, complex colonial organisms such as *Volvox* straddle the fence between simple colonial algae and truly multicellular plants. In *Volvox*, individual cells are more or less independent but are connected to one another and move about in a coordinated way. In truly multicellular plants, individual cells become specialized and dependent on one another. In *Volvox*, the only specialized cells are those involved in reproduction.

ENRICHMENT

A sudden "bloom" of red algae near shorelines may result in a floating, decaying mass of red algae that pollutes beaches when they are washed ashore. This so-called red tide has occurred in recent years along Cape Cod in Massachusetts and along the coast of Italy, causing beaches to be closed during the height of the summer. Interested students may want to investigate the cause of these red tides and their impact on the environment.

REINFORCEMENT/RETEACHING

▶ *Activity Book*

Students can begin learning about different kinds of green algae by completing the chapter activity called Comparing Green Algae.

CONTENT DEVELOPMENT

Be sure students understand why red algae can live at greater depths in the ocean than other algae. Their red pigments can trap what little light energy does penetrate deep water and transfer that energy to the chlorophyll.

Point out that red algae is the source of a substance called carrageenan. This substance gives products a smooth texture and helps them retain moisture, which is why it is used in things like ice cream,

ready-made frosting, toothpaste, and shoe polish. Carrageenan is derived from several kinds of red algae called Irish moss, or carrageen, that grow on rocky coasts. They are harvested mostly along the eastern coasts of the United States and Canada and in Ireland and France.

● ● ● ● ● **Integration** ● ● ● ●

Use the discussion of red algae in food products to integrate food science into your lesson.

PLANT EVOLUTION

It is generally believed that green algae, or *Chlorophyta,* are the group from which land plants evolved. Among the algae, only the green algae have cellulose in their cell walls, contain chlorophylls *a* and *b,* and store their food in the form of starch, all of which are also characteristics of the land plants. Because algae do not form fossils, we do not have direct evidence for the relationship; however, one stage in the life cycle of mosses looks remarkably like a tangle of green algal filaments. This resemblance could be traced to a common algaelike ancestor for both mosses and the modern multicellular green algae.

ECOLOGY NOTE

USES OF ALGAE

In the future, it may be possible to use algae to treat sewage. Raw sewage would be diluted and run through large tanks containing algae. The algae would remove the nutrient-rich materials such as phosphates, nitrogen, and fertilizers from the sewage. The remaining water would then be safely released with less chance of causing eutrophication in lakes, rivers, and reservoirs.

Figure 5–9 *Some green algae look like exotic vines with cone-shaped leaves. Others form enormous green bubbles.*

Figure 5–10 *Hydras are freshwater relatives of jellyfishes and sea anemones. The greenish color of* Chlorohydra *is caused by the hundreds of green algal cells that live within its body. How do the algal cells help their partner?* ❶

112 ■ B

to make food. During the food-making process, the algae release oxygen, which the astronauts breathe. If you could look closely at the tank, you would see bubbles of oxygen floating toward the surface. The relationship between the breathing of the astronauts and the food-making process in the algae means that the astronauts and the green algae support each other's lives.

This scene is purely imaginary. However, science fiction may soon become science fact. Today, scientists are hard at work developing methods of growing green algae in closed environments. And someday, future space explorers may rely on algae to maintain the air supply aboard spacecraft.

For now, the only place you can find green algae is here on planet Earth. Most green algae live in fresh water or in moist areas on land. However, some green algae do live in rather unusual places. Several may live in a close partnership with fungi, forming lichens. A few can live in bodies of water many times saltier than the ocean, such as the Great Salt Lake in Utah. And some can live within the bodies of worms, sponges, and protists such as *Paramecium.*

Green algae have life cycles, pigments, and stored food supplies similar to those of complex land plants. These similarities, along with a few others, suggest that land plants evolved from green algae. Of course, modern species of green algae are not the ancestors of land plants. But in the distant past, modern species of land plants and modern species of green algae shared a common ancestor. Because the remains of algae rarely become fossils, it is unlikely that we will ever know exactly what this common ancestor was. But by examining modern species of green algae and using imagination, scientists have developed a pretty good idea of how land plants may have evolved from green algae.

Modern species of green algae that represent important steps in the development of land plants are shown in Figure 5–11. As you can see, the earliest forms of green algae were probably unicellular organisms. Later, colonies consisting of many relatively independent cells developed. The next step on the way to land plants were multicellular green algae that lived in water. After that came multicellular

---

5–1 (continued)

CONTENT DEVELOPMENT

• **How might green algae someday be used during long trips in space?** (To supply the oxygen needs of the astronauts.)
• **In what way could the astronauts and green algae support each other's lives?** (The algae could use the carbon-dioxide wastes released by the astronauts as a raw material to produce oxygen for the astronauts.)

● ● ● ● **Integration** ● ● ● ●

Use the discussion of algae in space to integrate astronomy into your lesson.

ENRICHMENT

▶ *Activity Book*

Students are challenged to use reference materials and their own knowledge to complete the chapter activity Algae Classification.

CONTENT DEVELOPMENT

Be sure students understand why scientists believe that land plants evolved from green algae. Trace the hypothetical development from unicellular to colonial to multicellular (first in water, then on

land) organisms to algaelike land plants, using Figure 5–11.

● ● ● ● **Integration** ● ● ● ●

Use the discussion on common ancestors to integrate evolution into your lesson.

INDEPENDENT PRACTICE

Section Review 5–1

1. The three phyla are brown algae, red algae, and green algae, named for the most noticeable pigment in their cells. Most brown algae and red algae live in the

green algae that lived on land. Finally, about 450 to 500 million years ago, algaelike land plants evolved from green algae ancestors. One group of early land plants evolved into mosses and their relatives. Another group evolved into ferns and other more complex land plants.

Figure 5–11 *By studying modern forms of green algae, scientists have been able to reconstruct the basic stages of the evolution of plants. Unicellular algae, such as Chlamydomonas (top left), gave rise to colonial algae such as Volvox (bottom left). Colonial algae gave rise to simple multicellular algae such as Ulva (top right), which in turn gave rise to more complex multicellular forms (bottom right). What was the next big step in the evolution of land plants?* ❷

5–1 Section Review

1. Compare the three phyla of algae.
2. Why can red algae survive in deeper water than other kinds of algae can?
3. Why are brown algae important to humans?
4. How are green algae related to land plants?

Connection—*Social Studies*

5. On his first voyage to the New World, Christopher Columbus had a lot of trouble with sailors who wanted to turn back. Some historians think that the sailors were afraid of falling off the edge of the world. Other historians disagree. Explain how brown algae may have played a part in the near-mutiny of Columbus's sailors.

B ■ 113

5–2 Plants Move Onto Land: Mosses, Liverworts, and Hornworts

MULTICULTURAL OPPORTUNITY 5–2

Have students design a plant that could adapt to various conditions. Some possible adaptations might include the following:
• a plant that could survive on very little water
• a plant that could survive in a crack in a sidewalk
• a plant that could survive next to a highway
• a plant that could survive in a polluted environment

ESL STRATEGY 5–2

Have students find pictures of land plants and choose one to illustrate and label with the five tasks that a land plant must be able to perform in order to survive in its environment.

Then ask students to compare the mosses and their relatives to the land plants in their illustration and to list reasons why these tiny plants do not have the same difficulty surviving on land as larger, more complex plants do.

TEACHING STRATEGY 5–2

FOCUS/MOTIVATION

Have students study Figure 5–12 and read the caption.
• **What do you see in the illustration?** (Plants and animals from millions of years ago.)
• **Like algae, the plants in the illustration do not have seeds. But what is the main difference between algae and these plants?** (Algae live in water; these plants lived on land.)
• **How do you think these plants were able to live on land? What did they need?** (Accept all logical answers. Lead students to consider how plants that were used to living in water could adapt to living in

soil and air. Make a list of students' suggested adaptations.)

CONTENT DEVELOPMENT

Point out that the first multicellular plants lived in water and were completely adapted to their watery environment. The water supported them physically, they used it to produce food, and it helped bring their sperm and egg cells together during reproduction. In order to make the transition to land, the ancestors of today's land plants were able to adapt to a

Guide for Reading

Focus on these questions as you read.
▶ What are some adaptations that plants need to live on land?
▶ What are the major characteristics of mosses and their relatives?

In the ocean, a kelp plant may stand as tall as a very large tree. Its leaves are held up by the water so that they are exposed to enough light. Kelp absorbs water, carbon dioxide, minerals, and all the other substances it needs from ocean water. It reproduces by releasing egg cells and sperm cells into the water, where the sperm cells can swim to the eggs and fertilize them.

Now imagine that the same kelp plant is taken from the water and planted on land. Can the plant stand upright and hold up its leaves? Can it absorb minerals and water from the new substance (air) that surrounds it? Can its sperm cells swim on dry land? Can the kelp plant survive on land? The answer to all of these questions is no.

As the example of the kelp plant indicates, it is not easy for aquatic organisms to live on dry land. The plants that invaded the land millions of years

Figure 5–12 *The first forests appeared on Earth more than 380 million years ago. Unlike the forests of today, these ancient forests were dominated by plants without seeds. What adaptations did plants require in order to successfully invade the land?* ❶

114 ■ B

dry environment. Their evolution had to include many solutions to problems of acquiring, transporting, and conserving water, as well as reproducing without water.

FOCUS/MOTIVATION

Display moss plants and a fern for students to examine.
• **How are these plants different from one another in the way that they are constructed?** (Mosses are small; ferns are much larger. Ferns have leaves and veins; mosses do not.)

ago had to evolve in ways that made them better suited to meet the challenges of their new environment. Let's look at some of the tasks that plants need to perform in order to survive on land.

Land plants need to support the leaves and other parts of the body so that they do not collapse. Supporting structures enable land plants to position the parts of their body that make food so that they are exposed to as much sunlight as possible.

Land plants need to obtain water and minerals. In an aquatic environment, plants are surrounded by water and dissolved minerals. On land, however, water and minerals are located in the soil.

Land plants need to transport food, water, minerals, and other materials from one part of the body to another. In general, water and minerals are taken up by the bottom part of a land plant and food is made in the top part. To supply all the cells of the body with the substances they need, water and minerals must be transported to the top part of the plant and food must be transported to the bottom part.

Land plants need to prevent excess water loss to the environment. Because air contains less water

B ■ 115

Answers

❶ They had to be able to support leaves and other parts without the help of water. They had to be able to obtain water and minerals from soil instead of water. They had to be able to transport food, water, and other materials from one part to another. They had to be able to prevent excess water loss to avoid drying up. In order to reproduce, they had to be able to get their sperm and egg cells together without water's help. (Making inferences)

BACKGROUND INFORMATION
WHO GOT THERE FIRST?

When asked to describe the most important event in the evolution of life on land, many students think of some fishlike creature crawling onto the shore and trying to breathe air. This idea is wrong for several reasons. First of all, this is not how evolution occurs. But equally important, animals could never have moved onto land if plants (their food) hadn't gotten there first.

CONTENT DEVELOPMENT

Point out that land plants met the requirements for living on land in many ways. Be sure students understand the five basic problems plants had to solve. Begin the discussion with these three tasks:

Support—Without water to hold the plant, it needs some kind of strong, stiff structure to support itself. Also, the green parts of the plant that carry on the food-making process must be held up and exposed to the sun.

Water—When the plant was in water, naturally it had plenty of water to take in as it needed. On land, it must get its water from the soil.

Transport—In water, all of a plant's cells took what they needed directly from the water. On land, a plant needs to be able to move water up and food down through all its parts.

• **Are they similar in any ways?** (Both are green plants and require the same materials to survive [oxygen and water] and to make food [carbon dioxide, sunlight, and water].)

GUIDED PRACTICE
▶ *Laboratory Manual*

Skills: Making observations, comparing, analyzing data, applying concepts

At this point you may want to have students complete the Chapter 5 Laboratory Investigation in the *Laboratory Manual* called Observing Mosses and Liverworts. Students will observe and compare two types of nonvascular plants, a moss and a liverwort.

BACKGROUND INFORMATION
SPHAGNUM MOSS

The moss genus *Sphagnum,* also called peat moss, grows especially well in cold freshwater ponds, where it may form floating green mats that cover the surface. Sphagnum moss species make the water around them very acidic, creating special environments called sphagnum bogs, which are inhabited by many rare plants—including orchids and carnivorous plants such as pitcher plants and sundews.

Sometimes the sphagnum mat, together with the roots of other plants, is strong enough to support the weight of a full-grown human. If you jump up and down on these mats, you can see the surface tremble, spreading out from your location like waves. Such places are often called quaking bogs for this reason.

Figure 5–15 *In Washington State's Olympic National Park, mosses thrive in the cool, extremely damp Hoh Rain Forest. There, the mosses cover the trees like shaggy green coats (top). In most other forests, mosses do not cover the trees but may form a soft carpet on the forest floor (bottom).*

Although mosses are not particularly impressive plants, they are quite useful to humans. Certain Japanese-style gardens feature gently curving mounds of soil covered with a plush layer of emerald green moss. At least one modern artist has created living patchwork quilts of mosses. Dried sphagnum (SFAG-nuhm) moss is added to garden soil to enrich the soil and improve its ability to retain water. Sphagnum moss also changes the chemical balance in the soil so that the soil is better for growing certain plants, such as azaleas and rhododendrons. Ground-up sphagnum moss is added to soil and sprinkled around seedlings to help prevent the growth of certain disease-causing bacteria and fungi. In the past, people have used sphagnum moss to treat burns and bruises and to bandage wounds. Can you explain why sphagnum moss may have been a good substance for covering a wound? **1**

Over the course of hundreds of years, layer after layer of sphagnum moss and other plants may accumulate at the bottom of a bog or swamp. Under the right conditions, the deposits of dead plants form a substance called peat. Peat is used as a soil conditioner. It is also dried and used as fuel. Because the conditions in peat bogs greatly slow the process of decay, scientists who study ancient civilizations have found many interesting things within the deposits of peat—including the lifelike body of a man who was buried about 2000 years ago. **1**

5–2 Section Review

1. List five tasks that plants need to perform in order to live on land.
2. How are mosses and their relatives adapted for life on land?
3. What are three ways in which humans use mosses?

Critical Thinking—*Relating Cause and Effect*
4. Explain why mosses and their relatives never grow very large.

5–3 Vascular Plants Develop: Ferns

How would you like to be able to make yourself invisible? Sound like fun? Well, folklore has it that all you have to do is gather fern seed by the light of the moon at midnight on Saint John's Eve. Carrying some fern seed so gathered is guaranteed to make you invisible to human eyes.

But before you go off into the woods on a mid-summer night to look for ferns, you should know there is a catch. Like the algae, mosses, liverworts, and hornworts you read about earlier, ferns do not have seeds.

Unlike the other seedless plants you read about in this chapter, however, ferns do have a number of special adaptations to life on land. For example, they have a waxy covering on their leaves that helps to prevent water loss and roots that enable them to gather water and minerals from the soil. The most important adaptations, however, involve a system of tiny tubes that transport food, water, and other materials throughout the body of the fern. These tiny tubes are known as vascular tissue. **Because they have vascular tissue, ferns are said to be vascular plants.** Plants that lack vascular tissue—such as algae and mosses—are known as **nonvascular plants.**

The first **vascular plants,** which appeared about 400 million years ago, represent a major step in plant evolution. Vascular plants, such as ferns, are much better adapted to life on land than nonvascular plants. Vascular tissue allows materials to be

Guide for Reading

Focus on this question as you read.

▶ How are ferns different from other kinds of plants without seeds?

Figure 5–16 *Ferns do not have seeds. The dots or lines visible on the underside of fern leaves are formed by clusters of spore cases (left). As this 200-million-year-old fossil indicates, ferns are among the most ancient types of vascular plants (right).*

B ■ 119

REINFORCEMENT/RETEACHING

Monitor students' responses to the Section Review questions. If students appear to have difficulty with any of the questions, review the appropriate material in the section.

CLOSURE

▶ *Review and Reinforcement Guide*

At this point have students complete Section 5–2 in the *Review and Reinforcement Guide.*

TEACHING STRATEGY 5–3

FOCUS/MOTIVATION

Display a fern plant (or several of different species, if available) for students to observe. Ask them to examine the leaves, or fronds, closely.

• **What does a fern have that mosses do not have?** (Ferns have leaves.)

• **Look more closely at one frond and one of the smaller leaflets. What do you see running up the middle of each frond**

5–3 Vascular Plants Develop: Ferns

MULTICULTURAL OPPORTUNITY 5–3

If students look carefully, they can find examples of mosses, liverworts, and ferns in most environments throughout the country. Sometimes they are found in a park on a shady tree, in a damp corner of a yard, or near a stream or drainage ditch. Ask students to examine their own environment and find examples of these organisms.

ESL STRATEGY 5–3

Help students develop a "Plants Without Seeds" study chart that includes these headings:

Vascular/Nonvascular	Uses
_____	_____
_____	_____
_____	_____
_____	_____
_____	_____

Have students write a few short sentences or make an illustration to explain why ferns grow larger than mosses and their relatives do.

---

and into each leaflet? (A large vein in the middle of each frond branches out into each leaflet.)

CONTENT DEVELOPMENT

Point out that the veins are part of the fern's vascular tissue, or system of tiny tubes that carry materials throughout the plant. A fern is a vascular plant; algae and mosses are nonvascular plants because they lack this kind of transport system. Ferns are some of the oldest plants on Earth; they are also some of the earliest vascular plants.

● ● ● ● **Integration** ● ● ● ●

Use the discussion of early vascular plants to integrate concepts of evolution into your lesson.

CLUB MOSSES AND HORSETAILS

Club mosses and horsetails are the only living descendants of ancient plants that made up Earth's first forests more than 400 million years ago. But when the climate of the Earth changed, these plants could not compete with new types of plants, and so were replaced by them. Club mosses and horsetails are small plants that live in damp areas in woods or near streams.

Club mosses are small, evergreen, and usually look like miniature pines. Their yellow spores were used as dressings for cuts and coatings for pills. Because the spores explode when burned, they were also used in fireworks. Horsetails are also called scouring rush. Their stems contain silica crystals, which give the stems a rough texture. Because horsetails are very abrasive, they were used to scour pots and pans in Colonial times. People on camping trips still use them for that purpose.

FACTS AND FIGURES

IF YOU CAN'T SEE IT, IT MUST BE INVISIBLE

Fern spores are microscopic in size; the only way they can be seen with the unaided eye is when they cling together in large groups. Because fern spores are so hard to see, people once though that ferns grew from invisible seeds. So they believed that if a person did find a fern seed and carried it around, he or she would become invisible, too!

Figure 5–17 *Vascular plants without seeds include horsetails (left) and club mosses such as the peacock "fern" (right). They also include ferns.*

Figure 5–18 *A close examination of this tree fern leaf reveals that it is divided into smaller leaflike parts, which in turn are divided into still smaller leaflike parts. A few ferns have leaves that are not at all fernlike. For example, the water fern* Marsilea *has rounded leaves that make it look like a four-leaf clover.*

120 ■ B

transported quickly and effectively throughout the body of a plant. In addition, the type of vascular tissue that carries water from the roots in the soil to the leaves in the air is made of cells that have extremely thick, strong cell walls. These sturdy cell walls greatly strengthen the stems. Now can you explain why vascular plants can grow much larger than nonvascular plants? ●

Thanks to their vascular tissue, ferns can grow much taller than mosses and their relatives. In the mainland United States, most ferns range in height from a few centimeters to about a meter. Tree ferns, which grow in rain forests in Hawaii, the Caribbean, and other tropical areas, can be enormous. If laid along the ground, the tallest tree fern would extend from the pitcher to the batter on a baseball diamond, or from the foul line to the pins in a bowling lane.

5–3 (continued)

CONTENT DEVELOPMENT

Emphasize that mosses and ferns are often found growing in the same locations.
• **Why do ferns grow more upright than mosses do?** (Unlike mosses, ferns are vascular plants. Because ferns have transport tubes that carry materials throughout the plants, they can grow away from the soil. Mosses, on the other hand, must grow close to the soil because they do not have transport tubes.)

Point out that as vascular plants, ferns are well adapted for life on land. They can vary greatly in size and form and are found everywhere except in the driest or coldest places. They can be 3 centimeters to 20 meters tall. Most species of ferns grow in damp, shady habitats. Many other species can withstand harsh environments.

● ● ● ● **Integration** ● ● ● ●

Use the discussion on tall tree ferns to integrate concepts of biogeography into your lesson.

REINFORCEMENT/RETEACHING

Have students review the list of adaptations that plants have to make to live on land on pages B115 and B116. Ask them to evaluate how ferns have handled each adaptation. Discuss this topic as a group or have students write their explanations

Like other vascular plants, ferns have true leaves, stems, and roots. As you can see in Figure 5–18, the leaf of a fern is often divided into many smaller parts that look like miniature leaves. In many types of ferns, the developing leaves are curled at the top and look much like the top of a violin. Because of their appearance, these developing leaves are called fiddleheads. As they mature, the fiddleheads uncurl until they reach their full size.

People often mistake the sticklike central portion of a fern leaf for a stem. This is quite understandable when you realize that the stems of most ferns are hidden from view. In some ferns, the stems look like fuzzy brown strands of yarn lying on the ground. The stems are hidden by the feathery fern leaves that emerge from them. In other ferns, the stems run below the surface of the soil.

If you were to pull part of a fern's stem from the ground, you would discover clumps of wiry structures growing down from the stem here and there. These structures are the fern's roots. The roots anchor the fern to the ground. They also absorb water and minerals for the plant.

Although ferns are much better adapted to life on land than other nonvascular plants, they are not fully adapted. Take a moment now to look back at the list of tasks that plants need to perform in order to survive on land (pages 115–116). How do ferns accomplish each task? Can you guess for which task ferns require a watery environment? That's right: Ferns need standing water in order for their sperm cells to swim to their egg cells. Like other plants without seeds, they need an abundant supply of water in order to reproduce.

Ferns are useful to humans in a number of ways. Because ferns have lovely, interestingly shaped leaves and thrive in places that have little sunlight, they are popular houseplants. Products from ferns are often used to grow other kinds of houseplants. For example, orchids may be grown on the tangled masses of fern roots or on fibrous chunks of tree-fern stems. Ferns may also assist in the growth of crops. In southeast Asia, farmers grow a small aquatic fern in rice fields. Tiny pockets in the fern's leaves provide a home for special blue-green bacteria that produce a natural fertilizer. Thus the fern and its microscopic

Figure 5–19 *The developing leaves of most ferns are tightly coiled. This makes them look like the scroll at the top of a violin. What are these developing fern leaves called?* ②

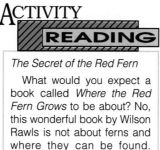

ACTIVITY READING

The Secret of the Red Fern

What would you expect a book called *Where the Red Fern Grows* to be about? No, this wonderful book by Wilson Rawls is not about ferns and where they can be found. However, the red fern plays a small but important part in this story. What is the meaning of the red fern? Why does the author include it in this story? Read, and find out. ②

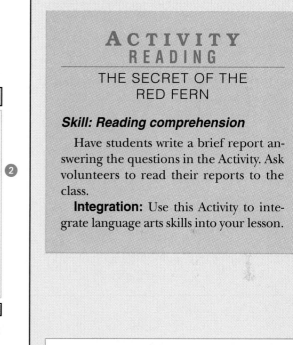

before discussing them as a class. Of the five adaptations that land plants need, ferns have made four. There is still one function for which they require standing water. What is it?

CONTENT DEVELOPMENT

Point out to students that like all vascular plants, ferns have leaves, stems, and roots. But like nonvascular plants, most ferns reproduce using spores. Fern spores are contained in cases called sori on the undersides of the leaves. This is a very

recognizable feature, and ferns are the only plants that have it.

Tell students about three well-known North American ferns. Bracken is the most common; it has large, triangular fronds, and unlike many ferns, it grows in open sunlit areas by roads and fields. It cannot be killed by burning. Its underground stems (rhizomes) simply grow new fronds. It is probably the fern that most people are familiar with. Royal fern grows in swampy places in forests. It has a short, thick stem surrounded by fronds

in a basket shape and is a favorite ornamental fern. Western sword fern grows only in Pacific coast forests.

GUIDED PRACTICE

Skills Development

Skills: Making observations, recording data, comparing

At this point have students complete the in-text Chapter 5 Laboratory Investigation: Comparing Algae, Mosses, and Ferns. Students will have the opportunity to note the similarities and differences between nonvascular plants (a moss and an alga) and vascular plants (a fern).

Figure 5–20 *Ferns are valued ornamental plants. The* Dipteris *fern has paired leaves that look like lacy wings (top right). The leaves of maidenhair ferns are divided into many small, dainty leaflike parts (bottom right). A young leaf from a Sri Lankan fern is a beautiful red color when it first uncoils (bottom left). Ferns may also be valued for reasons other than their beauty. For example, the tiny water fern* Azolla *(top left) helps to fertilize rice fields.*

partner help the farmers to grow food. Some ferns are eaten directly as food. During the spring, edible (fit to be eaten) fern fiddleheads may be sold in specialty food shops, supermarkets, and roadside vegetable stands. When properly cooked, fiddleheads make a delicious vegetable dish. However, unless you are absolutely certain which ferns are edible, you should not gather fiddleheads for food.

5–3 Section Review

1. What is the most important difference between ferns and the other plants without seeds that you studied in previous sections?
2. Describe the structure of a typical fern.
3. How are ferns adapted to life on land?

Critical Thinking—*Designing an Experiment*
4. Design a series of experiments to determine the best conditions for growing ferns.

CONNECTIONS

1 I Scream, You Scream, We All Scream for Ice Cream

You know that *ice cream* is sweet, cold, soft, and creamy—a wonderful treat. But did you know that you can learn a lot about *physical science* by studying ice cream? That's right—ice cream can be the key to understanding many scientific principles.

Consider for a moment the stuff ice cream is made of: milk and cream, sugar, a dash of flavoring, and perhaps a pinch of agar or gelatin. Blended together, these ingredients form a syrupy liquid. This liquid mix is transformed into a solid dessert by removing heat energy—that is, by freezing it. And this illustrates an important science concept: The form, or phase, of matter can be changed by adding or taking away energy.

But heat energy does not simply vanish. It has to go somewhere. (Scientists know this principle as the first law of thermodynamics.) In an old-fashioned ice-cream maker, the heat energy is taken up by a mixture of ice and rock salt. Because nature tends to work to even things out (scientists call this principle the second law of thermodynamics), heat energy is transferred from the ice-cream mix, which has more heat energy

to the mixture of ice and rock salt, which has less heat energy.

Have you ever tried to restore a dish of melted ice cream by putting it into the freezer? If you have, you know that the resulting frozen substance is hard and unappetizing—more like ice than ice cream. Why? One of the most important ingredients of ice cream is air. Turning the crank on an old-fashioned ice-cream maker stirs air into the mix (as well as helps the mix cool evenly). Tiny bubbles of air about 0.1 mm across make ice cream soft. The need for bubbles is exactly the reason agar, carrageenan, or gelatin is often added to ice-cream mixes. These substances help to stabilize the walls of the air bubbles.

The air bubbles are also the reason ice cream is not served on airplanes. You see, the air pressure inside an airplane flying high above the ground is much lower than the air pressure on the ground. At these lower pressures, air escapes from the ice cream. So a quart of soft, delicious ice cream can easily be turned into a pint of hard, not-so-tasty ice cream—hardly a wonderful treat!

ANNOTATION KEY

Integration
1 Physical Science: Heat. See *Heat Energy*, Chapter 1.

CONNECTIONS

I SCREAM, YOU SCREAM, WE ALL SCREAM FOR ICE CREAM

Students will probably be interested in the discussion of how ice cream is made. Before they read the article, ask them what they know about the ingredients and the process involved in making ice cream. They are not likely to have connected laws of thermodynamics with the production of ice cream, so be sure to review the scientific concepts discussed in the article carefully. You may want to obtain an old-fashioned ice-cream maker and the necessary ingredients and perform a practical (and delicious) demonstration of the points covered in the article.

If you are teaching thematically, you may wish to use the Connections feature to reinforce the themes of energy, patterns of change, and systems and interactions.

Integration: Use the Connections feature to integrate physical science concepts of heat into your lesson.

2. Ferns have leaves, stems, and roots. Their leaves are often divided into smaller parts like miniature leaves. The developing leaves begin as curls called fiddleheads and then uncurl as they mature. Ferns' stems are brown strands that lie on the ground or under the soil. Their roots are wiry structures attached to the stems.
3. Ferns have waxy coverings on their leaves that help to prevent water loss. They have roots that take water and minerals from the soil and anchor the plants to the ground. They have vascular tissue that

transports food and water throughout the plants and provides a strong structure to support the plants. They still require abundant water, however, for reproduction.
4. Students' experimental designs will vary but should take into consideration the variables of temperature, water, soil, and light.

REINFORCEMENT/RETEACHING

Review students' responses to the Section Review questions. Reteach any material that is still unclear, based on students' responses.

CLOSURE

▶ *Review and Reinforcement Guide*
Students may now complete Section 5–3 in the *Review and Reinforcement Guide*.

Laboratory Investigation

COMPARING ALGAE, MOSSES, AND FERNS

BEFORE THE LAB

1. Decide whether students are to work individually or in pairs.
2. Gather all materials at least one day prior to the investigation.
3. Obtain at least one living fern, several specimens of algae, and a large cluster of moss plants for students' use.

PRE-LAB DISCUSSION

Have students read the complete laboratory procedure.
- **What is the purpose of the investigation?** (To observe similarities and differences among algae, mosses, and ferns.)
- **What structures do ferns have that mosses and algae do not?** (Leaves, roots, stems, vascular tissue, and waxy covering.)
- **What is the purpose of examining the plant with a hand lens?** (Some structures can be seen more clearly with a hand lens.)
- **What is the purpose of examining the plant with both the low and high powers of the microscope?** (Again, different things can be seen more clearly using different magnifications.)
- **What is the purpose of sketching what you see at each stage?** (To record the different views of each plant for comparison.)

Laboratory Investigation

Comparing Algae, Mosses, and Ferns

Problem

How are algae, mosses, and ferns similar? How are they different?

Materials *(per group)*

> 3 microscope slides
> 3 coverslips
> medicine dropper
> microscope
> hand lens
> metric ruler
> scissors
> brown alga plant
> moss plant
> fern plant

Procedure 🧪 📇

1. Examine the brown alga carefully. Use the hand lens to get a closer look at the plant.
2. Draw a diagram of the alga on a sheet of unlined paper. Try to be as neat, accurate, and realistic as you can.
3. Next to your diagram, write down your observations about the color, texture, flexibility, and any other characteristics of the plant that you think interesting or important.
4. Using the metric ruler, measure the length and width of the entire plant. You should also measure any parts of the plant that you think appropriate. Write these measurements on or next to the appropriate part of your diagram.
5. Using the scissors, carefully cut a piece about 5 mm long from the tip of a "leaf" on the plant.

6. Place the plant piece in the center of a glass slide. With the medicine dropper, place a drop of water on the plant piece. Cover with a coverslip.
7. Examine the plant piece under the low- and high-powers of the microscope. Draw a diagram of what you observe.
8. Examine the moss plant using the procedure outlined in steps 1 through 7.
9. Examine the fern plant using the procedure outlined in steps 1 through 7.

Observations

1. What are the dimensions of the alga, moss, and fern?
2. Which land plant grows larger—the moss or the fern?
3. Which plant has vascular tissue? How can you tell?
4. How do the top and bottom surfaces of the fern leaf differ from each other?

Analysis and Conclusions

1. Why is the fern able to grow larger than the moss?
2. What alga structures appear to be adaptations to life in water? Explain.
3. Why is one side of the fern leaf shinier than the other? How is this an adaptation to life on land?
4. How are the plants similar?
5. How are the plants different?
6. **On Your Own** Examine a liverwort, green alga, and/or red alga. How do these plants compare to the ones you examined in this investigation? How are they adapted to the places in which they live?

TEACHING STRATEGY

1. It is a good idea to station yourself by the plant material from time to time. Students usually take too large a sample for microscopic observation and may attempt to fit a clump of moss under a coverslip, instead of a single moss plant.
2. Make sure students use both the low and high powers of the microscope.

DISCOVERY STRATEGIES

Discuss how the investigation relates to the chapter ideas by asking open questions similar to the following.
- **How is the alga like the moss?** (They are plants without seeds; they have chlorophyll; they live in or near water; they do not have vascular tissue—observing, comparing, relating.)
- **How is the moss like the fern?** (They are plants without seeds; they have chlorophyll; they live on land; they need

Study Guide

Summarizing Key Concepts

5–1 Plants Appear: Multicellular Algae

▲ Algae are nonvascular plantlike autotrophs that use sunlight to produce food. Algae do not have true roots, stems, or leaves.

▲ The algae that are classified as plants are divided into three phyla: brown algae, red algae, and green algae.

▲ Algae were the first kinds of plants to appear on Earth.

▲ Algae must live in or near a source of water in order to survive.

▲ Chlorophyll captures light energy so that it can be used in the food-making process.

▲ Accessory pigments absorb light energy and then transfer it to chlorophyll.

▲ Algae are useful to humans in many ways.

▲ Land plants evolved from green algae.

5–2 Plants Move Onto Land: Mosses, Liverworts, and Hornworts

▲ Adaptations of plants for living on land include structures for support, roots, vascular tissue, structures for minimizing water loss, and methods of reproduction that do not require standing water.

▲ Mosses, liverworts, and hornworts are small nonvascular plants that live in moist places. They are not fully adapted for life on land.

5–3 Vascular Plants Develop: Ferns

▲ Vascular plants, such as ferns, are much better adapted to life on land than nonvascular plants.

▲ Vascular tissue, a system of tiny tubes within vascular plants, allows food, water, and other materials to be transported quickly and effectively throughout the body of a plant.

▲ Water-conducting vascular tissue helps to strengthen and support stems.

▲ Vascular plants have true roots, stems, and leaves.

▲ Ferns are one of the earliest types of vascular plants. Although ferns have many adaptations to life on land, they are still dependent on standing water for reproduction.

Reviewing Key Terms

Define each term in a complete sentence.

5–1 Plants Appear: Multicellular Algae
alga
brown alga
red alga
green alga
pigment

5–3 Vascular Plants Develop: Ferns
nonvascular plant
vascular plant

ANALYSIS AND CONCLUSIONS

1. Ferns have vascular tissue to help deliver water over greater distances and to support larger stems.

2. Depending on the specimen, students may be able to identify the holdfast that anchors the plant and the air bladders that keep the plant floating near the surface of the water.

3. The top side has a waxy coating to prevent water loss through evaporation. This adaptation keeps the plant from drying up.

4. Answers will vary. Much depends on the particular plants studied. Brown algae and mosses will have certain similarities; mosses and ferns will have certain similarities. On the most basic level, all the plants have cells, and none have seeds.

5. Answers will vary. Again, it depends on the particular plants observed. Most important, ferns are different from mosses and algae because they have leaves, stems, roots, and vascular tissue.

6. Students may wish to set up a chart to record the information about the first and second sets of plants for easier comparison. The green alga and red alga will have much in common with the brown alga, as will the liverwort and the moss. All of them lack vascular tissue and the ability to live in dry conditions on land.

GOING FURTHER: ENRICHMENT
Part 1
Students can extend the investigation by studying the spore-containing structures, or sporangia, of algae, mosses, and ferns. Help students locate these structures on the three plants and examine them under the microscope.

Part 2
Students can examine several different species of mosses. In some mosses, such as *Funaria*, the leaflike structures are only one cell thick in most places. In other species, such as *Polytrichum*, the leaflike structures are several cells thick. Have students compare the thickness of these plants with the thickness of a fern plant.

---

water to reproduce—observing, comparing, relating.)

• **How is the fern like the alga?** (They are plants without seeds; they have chlorophyll; they need water to reproduce—observing, comparing, relating.)

• **How are the three plants different?** (Algae live only in water, mosses live on land near water, and ferns live on land. Algae and mosses do not have vascular tissue, true leaves, stems, or roots. Ferns have vascular tissue, true leaves, stems, and roots; therefore, they can grow much larger than algae or mosses can—observing, comparing, relating.)

OBSERVATIONS

1. Answers will vary. The fern will be larger than either the alga or the moss.

2. The fern grows larger.

3. The fern, because you can see the veins in its leaves.

4. The top surface is shinier than the bottom surface.

Chapter Review

Chapter Review

ALTERNATIVE ASSESSMENT

The *Prentice Hall Science* program includes a variety of testing components and methodologies. Aside from the Chapter Review questions, you may opt to use the Chapter Test or the Computer Test Bank Test in your *Test Book* for assessment of important facts and concepts. In addition, Performance-Based Tests are included in your *Test Book*. These Performance-Based Tests are designed to test science process skills, rather than content recall. Since they are not content dependent, Performance-Based Tests can be distributed after students complete a chapter or after they complete the entire textbook.

CONTENT REVIEW

Multiple Choice

1. d
2. b
3. a
4. c
5. c
6. d
7. d
8. d

True or False

1. F, brown algae
2. T
3. F, multicellular and some unicellular
4. T
5. F, contain
6. F, ferns
7. F, take in carbon dioxide and release oxygen
8. F, vascular

Concept Mapping

Row 1: Nonvascular, Autotrophic
Row 2: Monerans, Protists
Row 3: Brown, Green

CONCEPT MASTERY

1. Pigments are the colored chemicals found in the cells of algae. These pigments are responsible for capturing light energy so that it can be used in the process that makes food for the algae.

2. Answers may vary. Green algae may someday be used to produce oxygen for astronauts on spacecraft to breathe. Chemicals from red algae are used to manufacture certain foods. Brown algae

Content Review

Multiple Choice

Choose the letter of the answer that best completes each statement.

1. Which of these is a vascular plant?
   a. moss c. *Sargassum*
   b. hornwort d. fern
2. The most important characteristic in the classification of multicellular algae and their close relatives is the
   a. structure of the leaves.
   b. type of pigments present.
   c. method of producing or obtaining food.
   d. presence or absence of a nucleus.
3. The first kinds of plants to invade the land were probably
   a. algae. c. mosses.
   b. ferns. d. liverworts.
4. Ferns have
   a. simple seeds.
   b. no special adaptations for life on land.
   c. true roots, stems, and leaves.
   d. all of these.
5. Most agar comes from
   a. mosses. c. red algae.
   b. ferns. d. green algae.
6. If they existed today, the ancestors of land plants would probably be classified as
   a. ferns. c. brown algae.
   b. red algae. d. green algae.
7. All ferns, green algae, and mosses
   a. live only in water.
   b. are multicellular.
   c. lack vascular tissue.
   d. contain chlorophyll.
8. Mosses, liverworts, and hornworts
   a. rely on air bladders for support.
   b. can grow as tall as 60 meters.
   c. have true roots, stems, and leaves.
   d. require abundant water for reproduction.

True or False

If the statement is true, write "true." If it is false, change the underlined word or words to make the statement true.

1. *Sargassum* is a type of <u>red algae</u>.
2. Plants without seeds have methods of reproduction that <u>require</u> standing water.
3. In your textbook, <u>all</u> algae are classified as plants.
4. Mosses and their relatives have <u>few</u> special adaptations for life on land.
5. Red and brown algae <u>do not contain</u> chlorophyll.
6. The young leaves of <u>liverworts</u> are known as fiddleheads.
7. In the food-making process, plants <u>take in oxygen and release carbon dioxide</u>.
8. Ferns are <u>nonvascular</u> plants.

Concept Mapping

Complete the following concept map for Section 5–1. Refer to pages B6–B7 to construct a concept map for the entire chapter.

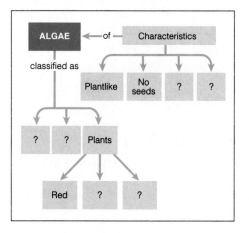

are used as food for people and animals. Moss is used to enrich garden soil and improve its ability to retain water. Ferns are popular houseplants.

3. Mosses and their relatives live only in places where there is plentiful water, so they get their food and water and reproduce just as they would if they lived in water. They do not need a strong external structure or an internal transport system because they are very small.

4. Ferns have roots that obtain water and minerals from the soil. They have vascular tissue that transports food and water throughout the plants and that will support larger stems and therefore larger plants. Ferns' leaves have a waxy coating to prevent water loss. But ferns still need standing water in order to reproduce.

5. Vascular tissue, which carries water from the roots to the leaves, has cells with thick, strong walls. These walls strengthen the stems and will support larger plants. Also, vascular tissue can transport food and water throughout all parts of the plant quickly and efficiently so that

Concept Mastery

Discuss each of the following in a brief paragraph.

1. What are pigments? Why are pigments important to plants?
2. Describe a current or potential use for green algae, red algae, brown algae, mosses, and ferns.
3. How do mosses and their relatives perform the tasks necessary for life on land?
4. How do ferns perform the tasks necessary for life on land?
5. Why can vascular plants grow so much larger than nonvascular plants?
6. What is vascular tissue? Explain why the development of vascular tissue was an important step in the evolution of plants.

Critical Thinking and Problem Solving

Use the skills you have developed in this chapter to answer each of the following.

1. **Making comparisons** How are red, brown, and green algae similar? How are they different?
2. **Applying concepts** A friend tells you that he has seen mosses that were 2 meters tall. Is your friend mistaken? Explain why or why not.
3. **Relating facts** Why do most algae live in shallow water or float on the surface of the water?
4. **Applying concepts** Why can ferns live in drier areas than mosses?
5. **Classifying objects** Examine the accompanying photograph of a plant without seeds. What kind of plant do you think it is? Why? How would you go about confirming your identification?
6. **Using the writing process** Now it's your turn to try being a science teacher. Prepare an outline for a lesson on one of the plants you learned about in this chapter. Your lesson should include an aim, or goal. (For example, the aim of a lesson on classification might be "What are the five kingdoms of living things?") Write a five-question fill-in-the-blank quiz to accompany your lesson and to test whether you have accomplished your aim.

unicellular and multicellular. They live mostly in fresh water or moist land areas. They have some features that are similar to those of land plants, a fact that scientists think means that land plants evolved from green algae.

2. My friend is mistaken. Mosses are nonvascular plants and therefore cannot grow very large.

3. Most algae live in shallow water or float on the surface of water because they need to get as much sunlight as possible to use to make food.

4. Ferns have waxy coatings on their leaves to prevent water loss through evaporation. They have roots to obtain water from the soil and vascular tissue to transport water throughout the body of the plant.

5. Students should look for evidence of vascular tissue to determine if the plant is a fern, a moss, or an alga. Also, does it have leaves, stems, or roots? What about its size and structure? Is its color a clue? Encourage students to share their reasons with their classmates. Students should readily identify the plant as a fern.

6. Make sure students' lessons contain a goal, an outline, and a quiz and that all information is consistent with that presented in the chapter. This would be a good opportunity for students to work in pairs. The partners could present their lessons to each other and then use the quiz to test their knowledge.

KEEPING A PORTFOLIO

You might want to assign some of the Concept Mastery and Critical Thinking and Problem Solving questions as homework and have students include their responses to unassigned questions in their portfolio. Students should be encouraged to include both the question and the answer in their portfolio.

ISSUES IN SCIENCE

The following issue can be used as springboards for discussion or given as a writing assignment:

• Life as we know it would never have evolved without algae. Is this statement true? If so, why is it true?

plants can grow away from the soil.

6. Vascular tissue is a system of tiny tubes that transport food, water, and other materials throughout the body of a plant. Vascular plants are much better adapted to life on land than nonvascular plants are because vascular tissue can transport water from the roots and food from the leaves throughout the plant. Vascular tissue also helps to support the plant.

CRITICAL THINKING AND PROBLEM SOLVING

1. They are all simple plantlike autotrophs that use sunlight to produce their food. They all have chlorophyll, but in brown and red algae, the chlorophyll is masked by other pigments. They all live in or near water. Most brown and red algae are multicellular and live in the ocean. Red algae can live in deeper water than brown or green algae can. Green algae are both

SECTION	HANDS-ON ACTIVITIES
6–1 Structure of Seed Plants pages B130–B142 Multicultural Opportunity, 6–1, p. B130 ESL Strategy 6–1, p. B130	**Student Edition** ACTIVITY (Discovering): Fit to Be Dyed, p. B131 ACTIVITY (Doing): Dead Ringer, p. B137 ACTIVITY (Discovering): Plant-Part Party, p. B140 ACTIVITY BANK: The Ins and Outs of Photosynthesis, p. B181 ACTIVITY BANK: Bubbling Leaves, p. B183 **Laboratory Manual** Identifying Root Structures, p. B45 Structures of Stems, p. B49 Identifying Leaf Structures, p. B55 **Activity Book** CHAPTER DISCOVERY: Discovering Flowers and Seeds, p. B137 ACTIVITY: Observing Root Hairs, p. B145 ACTIVITY: Leaf Prints, p. B153 ACTIVITY BANK: What's the Dirt on These Plants? p. B215 ACTIVITY BANK: A Hairy Situation, p. B219 ACTIVITY BANK: Whence the Water? p. B227 **Teacher Edition** Observing Absorption in Plants, p. B128d
6–2 Reproduction in Seed Plants pages B143–B147 Multicultural Opportunity 6–2, p. B143 ESL Strategy 6–2, p. B143	**Student Edition** ACTIVITY (Discovering): Seed Germination, p. B146 **Activity Book** ACTIVITY BANK: Clone a Plant, p. B229 ACTIVITY BANK: Pet Plants—Or Are They Plant Pets? p. B239
6–3 Gymnosperms and Angiosperms pages B148–B153 Multicultural Opportunity 6–3, p. B148 ESL Strategy 6–3, p. B148	**Laboratory Manual** Observing Reproductive Structures in Angiosperms, p. B59 **Activity Book** ACTIVITY BANK: Flat Flowers, p. B241 **Product Testing Activity** Testing Jeans **Teacher Edition** Observing the Characteristics of Seed Plants, p. B128d
6–4 Patterns of Growth pages B153–B155 Multicultural Opportunity 6–4, p. B153 ESL Strategy 6–4, p. B153	**Student Edition** ACTIVITY (Discovering): Pick a Plant, p. B155 LABORATORY INVESTIGATION: Gravitropism, p. B156 ACTIVITY BANK: Lean to the Light, p. B184 **Laboratory Manual** Investigating Germination Inhibitors, p. B67 **Activity Book** ACTIVITY: Observing a Plant Tropism, p. B157 ACTIVITY BANK: You *Can* Force a Flower to Bloom, p. B243
Chapter Review pages B156–B159	

OTHER ACTIVITIES	MEDIA AND TECHNOLOGY
Student Edition ACTIVITY (Calculating): It Starts to Add Up, p. B138 **Activity Book** ACTIVITY: Looking Inside a Tree, p. B151 **Review and Reinforcement Guide** Section 6–1, p. B43	**Video** Adaptations of Plants (Supplemental) **Courseware** Plant Processes (Supplemental) **English/Spanish Audiotapes** Section 6–1
Activity Book ACTIVITY: Seed Dispersal, p. B147 **Review and Reinforcement Guide** Section 6–2, p. B47	**Interactive Videodisc** On Dry Land: The Desert Biome **English/Spanish Audiotapes** Section 6–2
Student Edition ACTIVITY (Reading): The Birds and the Bees, p. B151 ACTIVITY (Reading): A World of Fun With Plants, p. B152 **Activity Book** ACTIVITY: Observing a Tree, p. B141 ACTIVITY: Plants as Food, p. B149 ACTIVITY: Fruits, Vegetables, and Spices, p. B155 **Review and Reinforcement Guide** Section 6–3, p. B51	**Interactive Videodisc** ScienceVision: EcoVision Paul ParkRanger and the Mystery of the Disappearing Ducks **Video** Mahogany Connections (Supplemental) **Courseware** Food Chains and Webs (Supplemental) **English/Spanish Audiotapes** Section 6–3
Activity Book ACTIVITY: Plant Foods, p. B143 **Review and Reinforcement Guide** Section 6–4, p. B53	**English/Spanish Audiotapes** Section 6–4
Test Book Chapter Test, p. B109 Performance-Based Tests, p. B129	**Test Book** Computer Test Bank Test, p. B115

*All materials in the Chapter Planning Guide Grid are available as part of the Prentice Hall Science Learning System.

CHAPTER OVERVIEW

Plants on the Earth appear in many varieties, each with its own shape, size, and color. Of all the plant life on Earth, seed plants are the most dominant group. Seed plants are divided into two groups, angiosperms and gymnosperms, on the basis of whether or not their seeds are covered by a protective wall.

Seed plants have true roots, stems, and leaves. The roots of a seed plant provide stability and a means by which water and minerals can be absorbed from the soil. The stems of seed plants provide passageways for water, minerals, and food to travel throughout a plant and also provide support for leaves. The leaves of a seed plant possess the machinery and materials from which the plant generates its own food, using light energy. This food-making process is known as photosynthesis.

Reproduction in seed plants is accomplished by a process known as pollination. Pollination produces new seeds, which are dispersed by various means to help ensure future generations of various seed plants.

6-1 STRUCTURE OF SEED PLANTS

THEMATIC FOCUS

The purpose of this section is to introduce the various structures that compose a seed plant and to explain the functions of these structures. Seed plants are vascular plants that produce seeds. Xylem tissue carries water and minerals from the roots up through a plant. Phloem tissue carries food throughout a plant.

The three basic structures of any seed plant include roots, stems, and leaves. Root systems serve to anchor a plant and to absorb water and minerals from the surrounding soil. All stems serve to provide the means by which water, minerals, and food are transported between the leaves and roots of a plant. The function of leaves is to capture energy and use this energy to produce food. This food-making process is known as photosynthesis. Photosynthesis generates a waste product called oxygen.

The themes that can be focused on in this section are scale and structure, energy, and stability.

*Scale and structure: Seed plants have true roots, stems, and leaves. The internal structure of the roots, stems, and leaves reflects and enhances their function. The vascular tissue in stems is composed of xylem and phloem.

Energy: In the leaf of a plant, light energy is used to change water and carbon dioxide into simple sugars called glucose during the process of photosynthesis. Leaves contain the machinery of photosynthesis; the materials are supplied by the plant.

Stability: Photosynthesis helps to maintain the balance of carbon dioxide and oxygen in the environment. This balance is essential for the maintenance of life.

PERFORMANCE OBJECTIVES 6-1

1. List the function of roots, stems, and leaves.
2. Compare xylem and phloem vascular tissue.
3. Compare herbaceous and woody stems.
4. Describe the process of photosynthesis.

SCIENCE TERMS 6-1

xylem p. B130
phloem p. B130
root p. B130
stem p. B130
leaf p. B130
photosynthesis p. B137

6-2 REPRODUCTION IN SEED PLANTS

THEMATIC FOCUS

The purpose of this section is to explain the process of seed-plant reproduction. Seed plants are categorized into two groups on the basis of the kind of seeds they produce: angiosperms and gymnosperms. Angiosperm seed plants produce a seed that is contained within a structure called an ovary. Gymnosperm seed plants produce a seed that is uncovered, or naked.

Seeds of a plant consist of a seed coat, an embryo or young plant, and stored food.

After a plant has produced seeds, the seeds must be dispersed, or distributed, to other places. To increase the likelihood that a seed will germinate, many plants have a mechanism by which its seeds are dispersed.

The themes that can be focused on in this section are scale and structure and systems and interactions.

*Scale and structure: The reproductive structures of seed plants are known as cones and flowers. The seeds that are produced by cones and flowers consist of a seed coat, an embryo or young plant, and stored food.

Systems and interactions: Many angiosperms rely on animals for pollination and seed dispersal. To attract animals, the angiosperm often produces colorful seed coats and sweet fruits.

PERFORMANCE OBJECTIVES 6-2

1. Identify the structure of a seed.
2. Describe the process of seed dispersal by angiosperms and gymnosperms.

SCIENCE TERMS 6-2

cone p. B143
flower p. B143
ovule p. B143
pollen p. B143
pollination p. B143
angiosperm p. B144
gymnosperm p. B144

6–3 GYMNOSPERMS AND ANGIOSPERMS
THEMATIC FOCUS

The purpose of this section is to introduce students to common examples of angiosperms and gymnosperms and to describe the specific structures involved in angiosperm reproduction. Gymnosperms, once the most dominant form of plant life on Earth, exist today in four phyla: conifers, cycads, gingkoes, and gnetophytes. Today, angiosperms represent the most dominant form of plant life on Earth. Flowers are the structures that contain the reproductive organs of angiosperms. Generally speaking, the male reproductive organs in a typical angiosperm include stamens, filaments, and anthers; the female reproductive organs include pistils, stigmas, styles, ovaries, and ovules.

The themes that can be focused on in this section are evolution, patterns of change, and unity and diversity.

*Evolution: Gymnosperms were the first group of seed plants to evolve. They were followed by angiosperms, about 100 million years ago. Through the course of time, both gymnosperms and angiosperms have evolved in ways that have made them better-adapted to life on land than other kinds of plants are.

*Patterns of change: A flower is pollinated when a grain of pollen lands on the stigma. If ovules become fertilized from the pollen, the ovules develop into seeds, and the ovaries develop into fruit.

*Unity and diversity: Plant structures are adapted for a variety of different conditions and special functions and may look quite dissimilar in different plants.

PERFORMANCE OBJECTIVES 6–3
1. Identify the characteristics of gymnosperms and angiosperms.
2. Identify the structures of a flower.
3. Describe the functions of the various structures of a flower.

SCIENCE TERMS 6–3
sepal P. B151
petal p. B151
stamen p. B151
pistil p. B151
fruit p. B152

6–4 PATTERNS OF GROWTH
THEMATIC FOCUS

The purpose of this section is to introduce the terms *annual, biennial,* and *perennial* and to explain how tropisms affect plant growth.

Based on the life span of a plant and how long it takes a plant to produce flowers, plants are classified as annual (one growing season), biennial (two growing seasons), and perennial (more than two growing seasons, or many years).

The growth of a plant toward, or away from, a stimulus is known as a tropism. Common tropisms occur in the presence of light and gravity.

The themes that can be focused on in this section are unity and diversity and patterns of change.

*Unity and diversity: Plants are grouped as annual, biennial, or perennial based on how long it takes them to produce flowers and how long they live.

*Patterns of change: Plants respond to certain stimuli by growing toward or away from the stimulus.

PERFORMANCE OBJECTIVES 6–4
1. Explain the plant terms *annual, biennial,* and *perennial*.
2. Compare a positive tropism to a negative tropism.

SCIENCE TERMS 6–4
annual p. B153
biennial p. B153
perennial p. B153
tropism p. B155

Discovery *Learning*

TEACHER DEMONSTRATIONS MODELING

Observing Absorption in Plants

Obtain two paper cups, a plastic knife, blue and red vegetable coloring, and fresh white carnations. Place some red vegetable coloring into one cup and some blue vegetable coloring in another cup. Carefully slit the entire stem of a carnation into halves. Place one half of the stem into the red vegetable coloring and one half of the stem into the blue vegetable coloring. Have students observe the results.
• **What has happened to the carnation?** (The tips of the flower are now red and blue.)
• **What caused the color change?** (The vegetable coloring.)
• **How did the vegetable coloring get to the petals of the flower?** (It was transported up through the stem.)

Tell students that water and other minerals are transported through the flower in a similar manner.

Observing the Characteristics of Seed Plants

Prepare a display of various seed plants representing those that will be studied in this chapter. If actual specimens are not available, use photographs.
• **Can you identify any of these plants?** (Answers will vary, depending on the display.)
• **Can you identify any plant structures of these plants?** (Responses might include stems, leaves, or roots.)
• **What evidence is there to indicate that these plants carry on photosynthesis, or make their own food?** (Students might suggest that the leaves of plants that perform photosynthesis are green.)
• **Although these plants share many common characteristics, they also have many differences. What are some of these differences?** (Accept all logical answers.)

Point out that despite their differences, these plants have one characteristic in common: They are all seed plants. In this chapter students will learn about the features and diversity of seed plants.

Plants With Seeds

INTEGRATING SCIENCE

This life science chapter provides you with numerous opportunities to integrate other areas of science, as well as other disciplines, into your curriculum. Blue numbered annotations on the student page and integration notes on the teacher wraparound pages alert you to areas of possible integration.

In this chapter you can integrate social studies (p. 132), life science and cells (p. 137), physical science and chemistry (p. 138), physical science and ecology (p. 142), life science and the immune system (p. 143), language arts (pp. 144, 151, 152, 153, 154), and life science and evolution (p. 148).

SCIENCE, TECHNOLOGY, AND SOCIETY/COOPERATIVE LEARNING

Homeowners striving for the "perfect" lawn create several different kinds of environmental problems. They often use tremendous amounts of limited water resources; they apply thousands of gallons of toxic chemicals, which eventually reach the underground water supply; and they create large amounts of yard waste for our already overburdened landfills. Most people do not think of a lawn as an environmental hazard, but it can be.

Various communities are taking steps to help ease the environmental burden of lawn maintenance. Some communities have imposed restrictions on lawn watering so that water reserves and underground pressure are maintained; others have instituted programs encouraging composting of yard clippings. But a major concern of health-care professionals and environmentalists is the chemicals that are routinely used on lawns and gardens.

Herbicides and pesticides sprayed on lawns and gardens are poisons to the weeds and bugs they are designed to destroy, but evidence suggests that they are also harmful to pets, children, adults, and any other living organisms that come into contact with them. Many homeowners fail to read the warning labels or directions of the chemicals they use. Exposure to these chemicals can cause headaches, nausea, dizziness, vision loss, partial paral-

INTRODUCING CHAPTER 6

DISCOVERY LEARNING

▶ *Activity Book*

Begin teaching the chapter by using the Chapter 6 Discovery Activity from the *Activity Book*. Using this activity, students will examine the flowers, fruit, and seeds of some closely related flowering plants.

USING THE TEXTBOOK

Have students observe the picture on page B128.

• **What do you observe in the picture?** (Plants and/or flowers.)

• **What structures of these plants are you able to see?** (Students might suggest stems, leaves, and flowers.)

• **What structures of these plants are you unable to see?** (Responses might include root systems.)

Plants With Seeds

Guide for Reading

After you read the following sections, you will be able to

6–1 Structure of Seed Plants
- Describe the structure of roots, stems, and leaves.
- Give the general equation for photosynthesis.

6–2 Reproduction in Seed Plants
- Discuss the events leading up to the formation of a seed.

6–3 Gymnosperms and Angiosperms
- Describe the four phyla of gymnosperms.
- Describe the structure of a typical flower.

6–4 Patterns of Growth
- Classify plants according to how long it takes them to produce flowers and how long they live.
- Describe some basic ways in which plants grow in response to their environment.

Spotted, striped, or solid-colored; double or single; magenta, pink, scarlet, orange, peach, or white—the brightly colored flowers of the impatiens plant are a familiar sight in suburban gardens and in city window boxes and planters.

Look carefully at the impatiens flowers in the photograph. Can you see a tiny green structure in the center of each flower? After the flower fades and its petals fall off, this structure begins to lengthen, swell, and change color from dark green to a pale yellowish green. Eventually, it is round with a tapered tip and has grooves that run along its length.

Touching the fully grown structure has some startling results. The structure pops off the stem, bursts, and sprays tiny brown objects in all directions. These brown objects are the impatiens's seeds.

Impatiens are just one kind of plant with seeds. What are some other plants with seeds? What do they look like? Where are they found? And why are seeds considered such an important development in plant evolution? Read on to find the answers to these and other questions.

Journal *Activity*

You and Your World Do you remember when you first learned that plants grow from seeds? When you were very young, did you ever plant seeds and watch them grow? In your journal, write about one of your earliest experiences with seeds.

◄ *The tiny green structure in the center of impatiens flowers develops into a seed pod. When ripe, the seed pod bursts open and scatters the seeds within it.*

B ■ 129

---

ysis, and mental disorientation and can possibly result in cancer. Are we creating the "perfect" lawn or an environmental hazard?

Cooperative learning: Using preassigned groups or randomly selected teams, have groups complete one of the following assignments.
- As the environmental task force for your community, prepare a pamphlet educating homeowners about the environmental responsibilities of lawn maintenance. The pamphlet should educate homeowners about lawn maintenance practices that are hazardous to the environment and should include suggestions on how homeowners can contribute to environmental quality and still maintain an attractive lawn.
- According to citizen-advocacy groups, the EPA has been negligent in evaluating herbicides and pesticides used by homeowners for safety. According to their studies, many lawn-care chemicals may be carcinogenic, cause birth defects, damage major organs, or attack the nervous system. Write a letter to the EPA expressing the group's opinion of evaluating and testing lawn-care products. Encourage groups to consider the following questions: Should testing be a priority? Should untested products be available to the public? Who should bear the cost of testing—the government or the chemical manufacturers? Should a pesticide license be required for the purchase and/or application of lawn-care chemicals?

See Cooperative Learning in the *Teacher's Desk Reference.*

JOURNAL ACTIVITY

You may want to use the Journal Activity as the basis for a class discussion. Encourage students to recall the details about the growth of the seeds and their feelings about what they saw. Point out that they will learn about plants with seeds as they complete the chapter. Students should be instructed to keep their Journal Activity in their portfolio.

- **How do you think these plants, or other plants, get food?** (If students suggest that plants get food from the soil, water, or fertilizers, point out that fertilizer and any other substances are only materials; plants must use materials such as these to make their own food.)

Have students read the chapter introduction on page B129.
- **How does an impatiens plant release, or disperse, its seeds?** (A stimulus causes the seeds to spray in many directions.)

- **Do you think all plants spread their seeds the same way as the impatiens does?** (Lead students to respond no.)
- **Describe some of the ways you think other plants might disperse their seeds.** (Descriptions might include wind, or carried by some other organism.)
- **How long do you think impatiens lives?** (Accept all responses.)

Point out that some plants take only one year to complete a growth cycle, others take two years, and still others live many years.

6-1 Structure of Seed Plants

MUTLICULTURAL OPPORTUNITY 6-1

Ask students what they know about spices. Spices are seeds, buds, dried flowers, bark, or roots of different plants. They originate in different parts of the world but are mainly found in tropical regions. Each ethnic cuisine uses its own assortment of spices to create many of the unique flavors of that cuisine. Cinnamon, for instance, is the bark of a tree brought by Arab traders from China to the Middle East centuries ago. From there, it spread to other parts of the world. Turmeric, the ground rhizome of a plant, comes from India. Nutmeg is from the East Indies, and saffron, a very expensive spice, comes from Spain.

Ask students to name some spices used for cooking in their own homes. Can they name (and perhaps describe) some ethnic dishes in which cardamom, turmeric, ginger, or chili peppers are used? Can they name some typical spices used in these cuisines: Mexican, Chinese, Italian, Scandinavian, Creole, Greek?

ESL STRATEGY 6-1

Provide students with drawings of a herbaceous and a woody seed plant's roots, stems, and leaves. Have them illustrate how each transport system functions by color-coding and labeling the pathway taken by the xylem and phloem and showing where photosynthesis and transpiration occur. Also, have them label the epidermis, root hairs, cortex, root cap, bark, cambium, pith, stalk, blade, simple/compound leaves, mesophyll, and stomata (where appropriate).

Activity Bank

What's the Dirt on These Plants? Activity Book, p. B215. This activity can be used for ESL and/or Cooperative Learning.

Guide for Reading
Focus on these questions as you read.
▶ What are seed plants?
▶ How are seed plants adapted to life on land?

6-1 Structure of Seed Plants

Seed plants are among the most numerous plants on Earth. They are also the plants with which most people are familiar. The tomatoes and watermelons in a garden, the pine and oak trees in a forest, and the cotton and wheat plants in a field are all examples of seed plants. What other seed plants can you name? ①

Seed plants are vascular plants that produce seeds. Recall from Chapter 5 that vascular plants have vascular tissue. Vascular tissue forms a system of tiny tubes that transport water, food, and other materials throughout the body of a plant. There are two types of vascular tissue: **xylem** (ZIGH-luhm) and **phloem** (FLOH-ehm). Xylem carries water and minerals from the roots up through the plant. Because xylem cells have thick cell walls, they also help to support the plant. Phloem carries food throughout the plant. Unlike xylem cells, which carry water and minerals upward only, phloem cells carry materials both upward and downward.

Like all vascular plants, seed plants have true **roots, stems,** and **leaves.** It might be helpful for you to review the adaptations that plants need to survive on land, which are found on pages 115–116. Keep these basic adaptations in mind as you read about roots, stems, and leaves in seed plants.

Figure 6–1 *Plants with seeds come in a wide variety of shapes and sizes. Grasses, orchids, and Douglas firs are all plants with seeds.*

TEACHING STATEGY 6-1

FOCUS/MOTIVATION

Note: *Place several radish seeds in culture dishes between damp paper towels two to three days before you perform this lesson.*

Begin the lesson by arranging students into small groups. Remove several radish seedlings from the culture dishes and place them individually on paper towels. Distribute a seedling to each group so that students can examine its root system.

Provide each group with a hand lens to aid in the observations of the roots and the root hairs.

• **Is the root system of this seedling an example of a taproot or a fibrous root system?** (Taproot.)

• **What did you observe that made you decide this is a taproot system?** (The system has a main root and smaller, branching roots extending from it.)

• **Describe the appearance of the root hairs.** (The root hairs are white, cottonlike fibers.)

Roots

Roots anchor a plant in the ground and absorb water and minerals from the soil. Roots also store food for plants.

The root systems of plants follow two basic plans: fibrous roots and taproots. Fibrous roots consist of several main roots that branch repeatedly to form a tangled mass of thin roots. Grass, corn, and most trees have fibrous root systems. Taproot systems consist of a long, thick main root (the taproot) and thin, branching roots that extend out of the taproot. Carrots, cacti, and dandelions are examples of plants with taproots.

Have you ever pulled up weeds? If so, you might have noticed that weeds with fibrous roots, such as crabgrass, tend to take a big chunk of soil with them when they are pulled up. Weeds with taproots, such as dandelions, tend to be difficult to pull out of the ground. How does the structure of root systems cause these two problems in weeding? ②

Refer to Figure 6–2 on page 132 as you read about the structure of a typical root. The outermost layer of the root is called the epidermis (ehp-ih-DER-mihs). The term epidermis (*epi-* means upon; *dermis* means skin) is used to refer to the outermost cell layer of just about any multicellular living thing, including plants, worms, fishes, and humans.

The outer surfaces of the cells of a root's epidermis have many thin, hairlike extensions. These extensions are known as root hairs. The root hairs greatly increase the surface area through which the plant takes in water and minerals from the soil. The water and minerals that are taken up by the root hairs pass into the next layer of the root, the cortex. In many plants, the cells of the cortex store food. They also carry water and dissolved minerals into the center of the root, which is made of vascular tissue, that is, cells of xylem and phloem.

As you can see in Figure 6–2, the very tip of the root is covered by a structure called the root cap. The root cap protects the tip of the root as it grows through the soil. Just behind the root cap is a region that contains growth tissue. This is where new cells are formed.

ACTIVITY DISCOVERING

Fit to Be Dyed

1. Fill a medium-sized jar one-fourth full of water. Then add a few drops of food coloring and stir. **Note:** *Be careful when using dyes, as they can stain.*

2. Place a stalk of celery in the jar so that its leaves are at the top and its base is at the bottom. Only the base should be submerged in the colored water.

3. Place the jar where it will not be disturbed. After 24 hours, examine your stalk of celery. What do you observe?

4. Remove the stalk of celery from the jar. Using a knife, cut off the portion of the stalk that was under water. Be very careful when using a knife. Discard this portion.

5. Examine the base of the stalk. What do you observe? Why does this occur?

■ Florists sometimes sell red, white, and blue carnations for the Fourth of July and other celebrations. Carnations do not naturally occur in this color combination. How do florists color the carnations without painting them?

■ Obtain a white carnation and test your hypothesis.

• **What do you suppose is the function of the root hairs?** (Students should infer that the root hairs increase the working surface area of the root system, or the amount of root surface exposed to the surrounding soil, allowing the plant to gather additional water and nutrients from the soil.)

CONTENT DEVELOPMENT

Have students recall that vascular tissue forms a transportation system throughout the body of a plant. Point out that there are two types of vascular tissue: xylem and phloem. Xylem and phloem function in a way analogous to (or similar to) the function of our streets, roads, and highways. Movement by vehicles on streets can be in one direction—a one-way street, or in two directions—a two-way street. Material moving through the xylem is analogous to vehicle movement on a one-way street—xylem carries materials from the root system in one direction, upward, through a plant. Material moving through the phloem is analogous to vehicle movement on a two-way street—phloem carries materials in two directions, upward and downward, throughout a plant.

Also, point out that the root systems of various vascular plants can differ. These root systems are generally categorized two ways: fibrous roots or taproots. When the primary root of a plant grows longer and thicker than its secondary roots, the plant has a taproot system.

In plants in which taproots do not form, secondary roots grow and branch. These plants, in which no single root grows larger than the rest, have a fibrous root system.

ROOTS ABOVE AND BELOW

Many plants produce roots above ground as well as below. Corn plants, for example, form roots that emerge from the stems and grow toward the ground. Orchids frequently grow perched high on tree trunks and branches. They attach themselves to trees with roots that secrete a kind of cement. Orchids actually grow on other plants. They gain no nourishment from their unintended hosts, obtaining nutrients from leaves that fall and decompose near their roots.

FACTS AND FIGURES

ROOT SYSTEMS

Scientists have calculated that the root system of a rye plant measures more than 600,000 meters in length.

6–1 (continued)

GUIDED PRACTICE

▶ *Laboratory Manual*

Skills Development

Skills: Making observations, identifying relationships, making comparisons

At this point you may want to have students complete the Chapter 6 Laboratory Investigation in the *Laboratory Manual* called Identifying Root Structures. In this investigation students will explore the structures of a root system and identify the function of each structure.

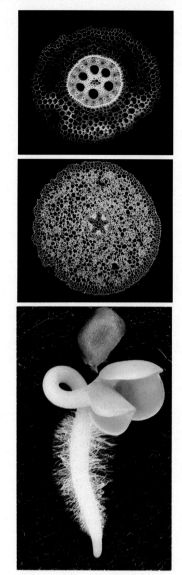

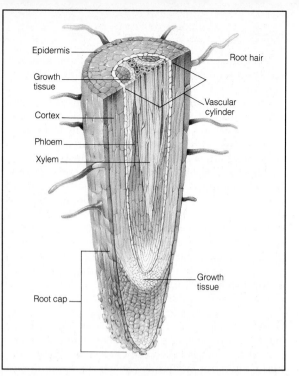

Figure 6–2 *In the root of a buttercup, phloem is located between the arms of a xylem "star" (center left). In the root of a corn plant, yellow circles of xylem alternate with reddish bundles of phloem (top left). Root hairs make this radish sprout look fuzzy (bottom left). What are the functions of the major root structures shown in the diagram?* ①

For thousands of years, people have used roots in many different ways. Some roots are used for food. Carrots, beets, yams, and turnips are among the many roots that are eaten. The root of the cassava plant is used to make tapioca, which may be familiar to you as the small, starchy lumps in certain kinds of puddings and baby food. Marshmallows were originally candies made from the root of the marsh mallow plant. (Modern marshmallows are made out of sugar, cornstarch, and gelatin.) Roasted chicory and dandelion roots are used as substitutes for coffee. Some roots—licorice, horseradish, and sassafras, for example—are used as spices. Roots are also used to make substances other than food, such as medicines, dyes, and insecticides.

132 ■ B

CONTENT DEVELOPMENT

Explain to students that roots may be specialized for functions other than absorption of water and nutrients. Direct their attention to Figure 6–3.

Point out that the swollen roots of some plants, like radishes and carrots, are adapted for food storage. The thick, sturdy roots from some tropical trees spread out on the surface of the ground, forming a stable base. The roots from deciduous trees, like giant tropical trees, are thick and sturdy but extend far beneath the topsoil to anchor the tree and to reach underground water supplies. The roots of parasitic plants, such as mistletoe and dodder, grow into the stem of a host plant and steal water, minerals, and food from the host's vascular tissues. Tiny roots emerge from the stems of various ivy plants to allow them to climb up walls, tree trunks, and other supporting structures. Mention to students that there are many other examples of specialized root systems.

▲ Thick, sturdy roots that spread out on the surface of the ground form a stable base for certain giant tropical trees.

▷ Roots emerge from the branches of banyans and grow down to the ground. These roots eventually thicken into new trunks, and the banyan continues to spread outward.

◁ Tiny roots that emerge from the stems of philodendrons help these plants to climb walls, tree trunks, and other supporting structures.

▲ The mangrove's spreading roots, which look rather like stilts, trap dead leaves and other debris and help create more soil for the plant.

▷ Radishes and carrots—like turnips, yams, beets, and many other plants— have swollen roots adapted for food storage.

Figure 6–3 *Specialized roots*

Stems

Stems provide the means by which water, minerals, and food are transported between the roots and the leaves of a plant. Stems are able to perform this function because they contain xylem and phloem tissue. Stems also hold the leaves of a plant up in the air, thus enabling the leaves to receive sunlight and make food.

Plant stems vary greatly in size and shape. The trunk, branches, and twigs of a tree are all stems. Some plants, such as the strange-looking baobab (BAY-oh-bab) in Figure 6–4 on page 134, have enormous stems that are many meters tall. As you can see, the trunk of the baobab is the most noticeable part of the plant. Other plants, such as the cabbage,

B ■ 133

B ■ 133

BACKGROUND INFORMATION

ANNUAL RINGS

The annual rings of a tree are sometimes used to provide information about the weather conditions of an area. In some trees, the inner part of an annual ring, the springwood, is lighter in color. There is usually more moisture available to the tree in the spring. In response, the tree produces a high concentration of the growth hormone auxin, and new cells in the xylem grow rapidly, becoming quite large before they die. In summer, the cells grow more slowly, due to less water and auxin. Cell walls occupy a greater proportion of the area, so the summer wood appears darker.

A wide band of growth in an annual ring reflects a good growing season (more moisture), while a narrow band forms in a dry year. By comparing the pattern of tree rings in building timber with the patterns found in old trees in the same area, archaeologists and anthropologists have been able to date structures made by humans.

Figure 6–4 *Huge, fat stems and stubby branches make baobab trees look like creatures from another planet. The rectangular patches on the left-hand baobab are places where its bark has been harvested for rope-making.*

Figure 6–5 *Plants bring much color to the world. In spring and summer, flowers such as lupines brighten fields and gardens (bottom). In autumn, the leaves of plants such as the maple turn red and orange (top). Which of these plants is woody? Which is herbaceous?* ❸

134 ■ B

have very short stems. The stem of a cabbage is a tough, cone-shaped core that is hidden beneath tightly closed leaves.

Plants may be classified into two groups based on the structure of their stems: herbaceous (her-BAY-shuhs) and woody. Herbaceous plants have stems that are green and soft. Sunflowers, peas, dandelions, grass, and tomatoes are examples of herbaceous plants. As you might expect, woody plants have stems that contain wood. Wood, as defined by plant biologists, is a hard substance made of the layers of xylem that form when a stem grows thicker. Unlike herbaceous stems, woody stems are rigid and quite strong. Roses, maples, and firs are woody plants. What other woody plants can you name? ❶

The structure of a woody stem is shown in Figure 6–6. The outermost layer of the stem is the bark. The outer bark, which is tough and waterproof, helps to protect the fragile tissues beneath it. The innermost part of the bark is the phloem. What is the function of phloem? ❷

The next layer of the stem is called the vascular cambium. The vascular cambium is a growth region of a stem, for it is here that xylem and phloem are produced. The center of the stem is called the pith. It contains large, thin-walled cells that store water and food.

If you were to cut through the stem of certain woody plants, a pattern of rings-within-rings that looks somewhat like a target might be visible. See Figure 6–7. These tree rings are made of xylem.

6–1 (continued)

GUIDED PRACTICE

Skills Development

Skills: Making computations, calculator

After dividing students into small groups, pose the following situation to the groups.

• **If water moves up the xylem of a 30-meter-tall tree at a rate of 0.15 millimeters per second, how many days will it take the water to reach the top of the tree?** (Several methods exist for determining the correct answer of approximately 2.3 days. Have volunteers from groups with correct answers explain to the class the dif-

ferent methods by which they obtained their answer.)

Have each group create a similar problem to be exchanged with another group. Remind groups to use numbers that represent sensible heights of plants and sensible speeds by which water travels through the plants.

INDEPENDENT PRACTICE

▶ *Activity Book*

Students who need practice on the concept of tree structure should complete

the chapter activity Looking Inside a Tree. In this activity students will examine a cross section of a tree trunk.

GUIDED PRACTICE

▶ *Laboratory Manual*

Skills Development

Skills: Making observations, identifying relationships, making comparisons

At this point you may want to have students complete the Chapter 6 Laboratory

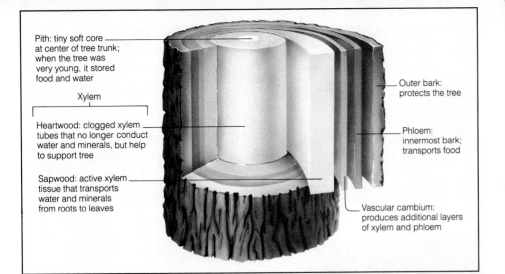

Pith: tiny soft core at center of tree trunk; when the tree was very young, it stored food and water

Xylem

Heartwood: clogged xylem tubes that no longer conduct water and minerals, but help to support tree

Sapwood: active xylem tissue that transports water and minerals from roots to leaves

Outer bark: protects the tree

Phloem: innermost bark; transports food

Vascular cambium: produces additional layers of xylem and phloem

A tree develops rings only if it grows at different rates during the different seasons of the year. For example, many trees in the northern part of the United States grow very little (if at all) in autumn and winter, when the weather is quite cold. They grow rapidly in spring, when it is warm and rainy. And they grow slowly in summer, when it is hot and dry. As a tree grows, a new layer of xylem forms around the old layers. What layer in the stem produces the new xylem cells? ❹

The xylem cells formed in the spring are large and have thin walls. They produce a light brown ring. The xylem cells formed in the summer are small and have thick walls. They produce a dark brown ring. Each pair of light and dark rings represents a year in the life of a tree. How old is a tree with 14 light rings and 14 dark rings? ❻

One of the most important and widely used products of plant stems is wood. Wood is used to

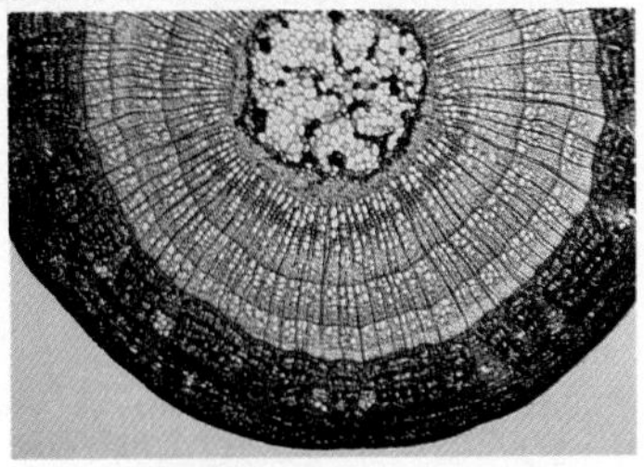

Figure 6–6 *The diagram shows the layers in the stem of a typical tree. The young basswood stem clearly shows its structure. Where is the pith? The vascular cambium? How old is the basswood stem?* ❺

Figure 6–7 *Tree rings can tell you more than just the age of a tree. The spacing and thickness of the rings provide information about weather conditions in the area over time. Thick rings that are far apart indicate years in which conditions were favorable for tree growth—years of much rain. What was the weather like for the first seven years of this larch tree's life? For the last seven years?* ❼

Investigation in the *Laboratory Manual* called Structures of Stems. In this investigation students will identify the structures of a plant stem and determine how materials are transported through the stem.

CONTENT DEVELOPMENT

Emphasize the idea that a plant stem contains specialized tissues suited for particular tasks. Stems vary greatly in size and shape from one plant species to another. Some are tiny; others are 100 meters tall. Some grow entirely underground; others

reach far into the air. Regardless of size and shape, all stems have two important functions: They hold leaves up in the sunlight, and they conduct various substances between the roots and the leaves.

Plants may be classified into two groups on the struture of their stems: herbaceous and woody. Herbaceous plants have stems that are green and soft. Woody plants have stems that contain wood.

SPECIALIZED STEMS

You can see several kinds of specialized stems simply by making a trip to the produce department of a grocery store. The corkscrew-shaped tendril that you might find on a bunch of grapes is a specialized stem that enables grape vines to climb. A potato is an example of a tuber. Tubers are fat underground stems that store food. The "eyes" of a potato are buds. Each bud is capable of developing into a new potato plant. An onion is an example of a short, thick, food-storing underground stem known as a bulb. Bulbs have thick leaves that contain food for the plant. When planted, the bulbs develop into new plants. Fresh ginger is an example of a food-storing underground stem known as a rhizome. Rhizomes grow horizontally beneath or along the ground. As they grow, rhizomes produce buds that grow into new plants.

▲ Long stems that run along the surface of the ground allow morning glories to spread into new areas.

▲ The thick green stem of a cactus stores water.

▼ Tubers, bulbs, and rhizomes are different types of underground food-storing stems. A potato is a tuber. The "eyes" of the potato are buds. Each bud is capable of developing into a new potato plant. Onions and garlic are bulbs. Bulbs have thick, short leaves that contain food for the plant. Rhizomes such as those of irises grow horizontally along the ground. As they grow, the rhizomes produce buds that grow into new plants.

Figure 6–8 *Specialized stems*

make a variety of objects ranging from toothpicks to buildings. Ground-up wood is used to make paper. Can you think of some other uses of wood? ❶

Some stems—such as potatoes, onions, ginger, and sugar cane—are a source of food. Other stems are used to make dyes and medicines. Still others have more uncommon uses. For example, the long, flexible stems of certain trees and vines are woven together to make wicker furniture and baskets. And fibers from the stems of flax plants are used to make linen fabric.

6–1 (continued)

ENRICHMENT

Most students may be familiar with the idea of counting the annual rings of a tree to determine its age. To be accurate, however, the count must be taken on a cut surface as near the soil level as possible. Counts taken higher up the stem will indicate only the age from the time that part of the tree developed.

CONTENT DEVELOPMENT

Explain that the leaves of green plants are the world's oldest solar-energy collectors. Leaves are also the world's most important manufacturers of food. The various sugars, starches, and oils manufactured by plants are sources of virtually all food for land animals. Even animals that do not eat plants must eat other animals that do.

To collect sunlight, most leaves have large, thin, flattened sections called blades. Leaf blades occur in an incredible variety of sizes and shapes. Leaves are classified as either simple or compound, depending on the structure of the blade. A simple leaf has a blade that is in one piece. A compound leaf has a blade that is divided into a number of separate, leaflike parts.

GUIDED PRACTICE

Skills Development
Skill: Making observations

Remind students that the flat part of a leaf is called the blade. Also, remind students that simple leaves have blades that are in one piece and that compound leaves have blades that are divided into a number of separate, leaflike parts. Set up several stations around the classroom where various examples of locally collected simple and compound leaves are

Leaves

Plant leaves vary greatly in shape and size. For example, birch trees have oval leaves with jagged edges. Pines, firs, and balsams have needle-shaped leaves. And maples and oaks have flat, wide leaves. Most leaves have a stalk and a blade. The stalk connects the leaf to the stem; the blade is the thin, flat part of the leaf. The thin, flat shape of most leaves exposes a large amount of surface area to sunlight.

Leaves are classified as either simple or compound, depending on the structure of the blade. A simple leaf has a blade that is in one piece. Maple, oak, and apple trees have simple leaves. A compound leaf has a blade that is divided into a number of separate, leaflike parts. Roses, clover, and palms have compound leaves. Can you name some plants that have simple leaves? Compound leaves? ②

No matter what their shape, leaves are important structures. For it is in the leaves that the sun's energy is captured and used to produce food.

Up to this point, we have been referring to the process in which light energy is used to make food simply as the food-making process. But this important process has its own special name: **photosynthesis** (foht-oh-SIHN-thuh-sihs). The word photosynthesis comes from the root words *photo*, which means light,

ACTIVITY DOING

Dead Ringer

Find a sawed-off tree stump. (Alternatively, find a large circular knot on a piece of wood paneling or piece of lumber.) How old was the tree (or branch) when it was cut? How can you tell? What do the rings tell you about the weather conditions during the life of the tree (or branch)?

① **A**ctivity Bank

The Ins and Outs of Photosynthesis, p.181

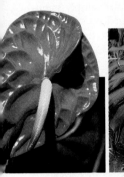

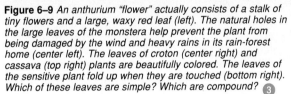

Figure 6–9 *An anthurium "flower" actually consists of a stalk of tiny flowers and a large, waxy red leaf (left). The natural holes in the large leaves of the monstera help prevent the plant from being damaged by the wind and heavy rains in its rain-forest home (center left). The leaves of croton (center right) and cassava (top right) plants are beautifully colored. The leaves of the sensitive plant fold up when they are touched (bottom right). Which of these leaves are simple? Which are compound?* ③

ACTIVITY DOING

DEAD RINGER

Skills: Making observations, making comparisons

Materials: tree stump or branch

The growth rings of most severed tree trunks should be clearly visible. You might caution students that counting rings represents only a good approximation of the age of a tree. It is not exact because in some years more than one growth layer can be added.

---

displayed. If the identity of the leaves is known, write their names on a card beside each leaf. Then have students visit each station and classify the leaves as simple or compound.

As an alternative approach to this activity, you might consider having students collect examples of simple and compound leaves to be brought to class for identification.

ENRICHMENT

▶ *Activity Book*

Students will be challenged by the Chapter 6 activity called Leaf Prints. In this activity students will explore a way to make copies of leaves using ink.

CONTENT DEVELOPMENT

Explain to students that photosynthesis is the process in which food is formed, or synthesized, by plants using light energy. Our understanding of photosynthesis is

the result of the same question being asked for thousands of years: When a seedling with a mass of only a few grams grows into a tall tree with a mass of several tons, where does the tree's increase in mass come from?

● ● ● ● **Integration** ● ● ● ●

Use the discussion of food-making and photosynthesis in plants to integrate the concept of cell processes in your lesson.

Figure 6–10 *Plants use the energy of sunlight to make food. The process that uses sunlight, water, and carbon dioxide to make food and oxygen is called photosynthesis.*

ACTIVITY
CALCULATING

It Starts to Add Up

Use a calculator to answer the following questions about an apple tree that has 200,000 leaves, each of which has a surface area of 20 cm².

1. What is the total leaf surface area of the tree?

2. The lower surface of each leaf contains 25,000 stomata per cm². How many stomata are on the lower surface of a single leaf? Assuming that the stomata are found only on the lower surface of the leaves, how many stomata are on the tree?

3. In a single summer, each leaf loses about 86 mL of water through transpiration. How much water is transpired by the entire tree?

and *synthesis*, which means to put together. Can you see why this name is appropriate? ➊

Photosynthesis is the process in which food is synthesized using light energy. Photosynthesis is the largest and most important manufacturing process in the world. It is also one of the most complex. Let's take a closer look at this amazing process.

In photosynthesis, the sun's light energy is captured by chlorophyll, which is the green pigment you read about in Chapter 5. Through a complex series of chemical reactions (which will not be discussed here), the light energy is used to combine water from the soil with carbon dioxide from the air. One of the products of the chemical reactions is food, which is generally in the form of a sugar called glucose.

Glucose can be broken down to release energy. Cells need energy to carry out their life functions. Glucose can also be changed into other chemicals. Some of these chemicals are used by a plant for growth and for repair of its parts. Other chemicals are stored in special areas in the roots and stems.

The other product of photosynthesis is oxygen. Oxygen is important to you, and to almost every other living thing on Earth. Do you know why? ➋

An equation can be used to sum up what occurs during photosynthesis. (Do not be alarmed by the appearance of equations! Equations are simply scientific shorthand for describing chemical reactions.) Here is the equation for photosynthesis in words and in chemical notation:

$$\text{carbon dioxide} + \text{water} \xrightarrow[\text{chlorophyll}]{\text{sunlight}} \text{glucose} + \text{oxygen}$$

$$6\ CO_2 + 6\ H_2O \xrightarrow[\text{chlorophyll}]{\text{sunlight}} C_6H_{12}O_6 + 6\ O_2 \quad ➋$$

Why does most photosynthesis occur in leaves? To answer this question, you need to know something about the internal structure of a leaf. Refer to Figure 6–11 as you read about the structure and function of a typical leaf.

The outermost layer of a leaf is called the epidermis. The cells of the epidermis are covered with a

---

6–1 (continued)

CONTENT DEVELOPMENT

Make sure students understand the basic concepts of photosynthesis. Remind them that photosynthesis is the process in which food is synthesized, or formed, using light energy. Then explain that specifically, photosynthesis is the process by which green plants use carbon dioxide, water, and energy to produce glucose and oxygen. Glucose is a simple sugar that provides organisms with energy. To plants, oxygen is a waste product.

• **What value does the waste product oxygen have to other living things?** (Most organisms need oxygen to maintain life.)

Also, explain that most photosynthesis occurs in the leaves of green plants. Some photosynthesis, however, may also occur in the stems of green plants.

Because photosynthesis is a complicated process, scientists use an equation to describe what occurs.

● ● ● ● **Integration** ● ● ● ●

Use the equation for photosynthesis in chemical notation to integrate the physical science concepts of chemistry into your life science lesson.

GUIDED PRACTICE

▶ *Laboratory Manual*

Skills Development

Skills: Making observations, identifying relationships, making comparisons

At this point you may want to have students complete the Chapter 6 Laboratory Investigation in the *Laboratory Manual* called Identifying Leaf Structures. In this investigation students will observe the var-

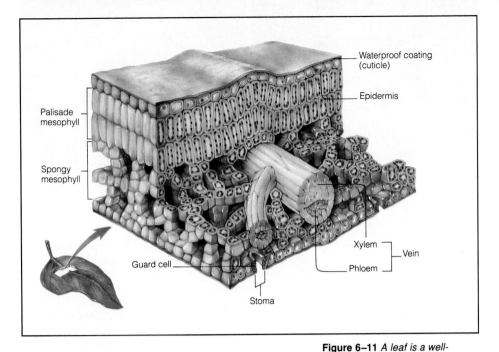

Palisade mesophyll

Spongy mesophyll

Waterproof coating (cuticle)

Epidermis

Xylem

Phloem

Vein

Guard cell

Stoma

Figure 6–11 *A leaf is a well-designed factory for photosynthesis. How does the structure of each leaf part help the leaf to perform its function?* ③

waxy, waterproof coating that helps to prevent excess water loss. This coating makes some real plants look as if they are made of shiny plastic.

Light passes through the epidermis to reach inner cells known collectively as mesophyll. (*Meso-* means middle; *-phyll* means leaf.) The cells of the mesophyll are where almost all photosynthesis occurs. The shape and arrangement of the upper cells maximize the amount of photosynthesis that takes place. The many air spaces between the cells of the lower layer allow carbon dioxide, oxygen, and water vapor (water in the form of a gas) to flow freely.

Carbon dioxide enters the air spaces within a leaf through microscopic openings in the epidermis. These openings are called stomata (STOH-mah-tah; singular: stoma). The Greek word *stoma* means mouth—and stomata do indeed look like tiny mouths!

In addition to permitting carbon dioxide to enter a leaf, stomata allow oxygen and water to exit. The process in which water is lost through a plant's

Ⓐctivity Bank

Bubbling Leaves, p. 183

ious structures of a leaf and identify their functions.

ENRICHMENT

Students may be interested in a more specific description of photosynthesis. Photosynthesis takes place in two major stages. In the first stage, light energy is used to split water molecules into their component parts—hydrogen and oxygen. The oxygen is discarded as a waste product at this point. In the second stage,

hydrogen combines with carbon dioxide to form a simple sugar called glucose.

In addition to the simple sugar glucose that is formed during the process of photosynthesis, plants can also make other compounds. These include complex sugars, starches, fats, and proteins. Complex sugars and starches provide energy for plants and the animals that eat plants. Fats also provide organisms with energy. Proteins are used in the building and repair of plant cells.

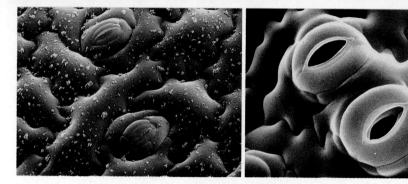

Figure 6–12 *Stomata open to allow carbon dioxide, a raw material of photosynthesis, into the leaf and allow oxygen, a waste product of photosynthesis, out of the leaf. Why do stomata close?* ❶

ACTIVITY

DISCOVERING

Plant-Part Party

Working with a friend or two, make a list of all the foods in your house that come from plants. Remember to include things like ketchup, canned foods, snack foods, and spices. Next to each entry on your list, write the part of the plant from which the food is obtained. If you and your friends cannot identify the source of a food, try to discover its identity using an encyclopedia, cookbook, or botany (plant science) textbook. Compare your list with your classmates'.

▪ What foods come from plants?

▪ Which plant part was the source of the most foods?

▪ One common vegetable looks like stems but is actually the stalks of leaves. What is the name of this vegetable? Were you fooled by these disguised leaves?

140 ▪ B

leaves is known as transpiration. Plants lose huge amounts of water through transpiration. For example, a full-grown birch tree releases about 17,000 liters of water through transpiration during a single summer season. And the grass on a football field loses almost 3000 liters of water on a summer day.

Although transpiration is necessary for the movement of water through most plants, too much water loss can cause cells to shrivel up and die. One way that plants avoid losing too much water through transpiration is by closing the stomata. Each stoma is formed by two slightly curved epidermal cells called guard cells. When the guard cells swell up with water, they curve away from each other. This opens the stoma. When the water pressure in the guard cells decreases, the guard cells straighten out and come together. This closes the stoma. In general, stomata are open during the day and closed at night. Can you explain why this is so? (*Hint:* When does photosynthesis take place?) ❷

As the cartoon character Popeye tells it, eating spinach is good for you. Popeye is right. In fact, several kinds of leaves are eaten by humans. Spinach, parsley, sage, rosemary, and thyme are just a few examples. What other kinds of leaves are used as food or to season food? ❸

Leaves are also the source of drugs such as digitalis, atropine, and cocaine, of deadly poisons such as strychnine and nicotine, and of dyes such as indigo and henna.

▼ The leaves of a kalanchoe are thick and fleshy, allowing the plant to store lots of water. Kalanchoes also produce tiny offspring plants along the edges of their leaves. Eventually, the leaves fall off and the young plants take root.

▲ The spines of a cactus are actually leaves, as are the threadlike tendrils of peas.

Figure 6–13 *Specialized leaves*

6–1 Section Review

1. What are seed plants?
2. Describe the structure and importance of roots, stems, and leaves.
3. What is photosynthesis? What is the chemical equation for photosynthesis?

Critical Thinking—*Applying Concepts*

4. Stomata are usually located on the lower surface of leaves. Why do plants with floating leaves, such as water lilies, have stomata on the top surface of their leaves?

B ■ 141

ECOLOGY NOTE

GREEN PLANTS AND OXYGEN

The process of photosynthesis in green plants generates a waste product called oxygen. Have interested students research the impact that the destruction of tropical rain forests may be having on the oxygen content of our atmosphere. Students should share their findings with the class.

---

which green plants use light energy to combine water and carbon dioxide to produce glucose and oxygen. The chemical equation for photosynthesis is

$$6CO_2 + 6H_2O \xrightarrow[\text{chlorophyll}]{\text{light}} C_6H_{12}O_6 + 6O_2$$

4. If stomata were located on the bottom surfaces of floating leaves, the leaves would have difficulty obtaining carbon dioxide, may have difficulty completing transpiration, and may allow excessive amounts of water to enter the plants. Stomata on the top surface of floating leaves allow carbon dioxide to enter the leaf, oxygen to exit, and transpiration to occur.

REINFORCEMENT/RETEACHING

Monitor students' responses to the Section Review questions. If students appear to have difficulty with any of the questions, review the appropriate material in the section.

CLOSURE

▶ *Review and Reinforcement Guide*

At this point have students complete Section 6–1 in the *Review and Reinforcement Guide.*

Integration
1 Physical Science: Energy Resources. See *Ecology: Earth's Natural Resources*, Chapter 1.
2 Life Science: Immune System. See *Human Biology and Health*, Chapter 8.

CONNECTIONS
PLANT POWER FOR POWER PLANTS

This feature presents to students the idea that the process of photosynthesis in plants could be modified to provide power for electricity-generating power plants. Explain to students that all power plants, whether they are powered by fossil fuels or nuclear energy, generate electricity but face two serious ramifications. The first is that all fuel used today to produce electricity is found in the Earth in limited quantities. Second, the use of these fuels damages the environment and contributes to global warming by releasing pollutants. Point out that the process of modifying photosynthesis in plants to produce alternative fuel supplies may sound like science fiction to some but that these efforts, and others, must be explored and developed if we expect to have energy sources for the future.

Integration: Use the Connections feature to integrate ecology into your science lesson.

CONNECTIONS

Plant Power for Power Plants 1

Imagine a busy power plant: Workers hurry about, checking gauges and adjusting machinery. The air vibrates with the low roar of the giant wheels that spin to produce electricity. Now imagine a peaceful countryside: A tree lifts its branches to the sun. The wind whispers through the leaves. A bird sings. What do these two scenes have in common?

Not that much—right now. However, power plants and green plants may soon use the same chemical reactions to obtain the *fuel* and food they need to function. Scientists recently reported that they are on the brink of producing artificial versions of the "machinery" of photosynthesis.

The reactions of photosynthesis in plants take place in complex, precisely arranged groups of molecules. Special molecules absorb the energy in light and pass it from one set of molecules to the next. The energy runs the chemical reactions that produce food for the plants—that is, glucose.

By copying the machinery of photosynthesis, scientists will be able to produce fuel using some of the least expensive and most abundant raw materials—sunlight, water, and carbon dioxide. With some slight adjustments, hydrogen gas and methane (natural gas), rather than glucose, will be the end products.

Creating artificial forms of photosynthesis may solve the *energy crisis*. And it may also solve two other serious problems facing us—*global warming* and *air pollution*. Here is why. Global warming is caused by excess carbon dioxide in the air. Photosynthesis—and artificial photosynthesis—uses carbon dioxide. So these processes remove excess carbon dioxide from the air, thus reducing global warming. Much air pollution is caused by burning fuels such as oil and gasoline. Hydrogen, on the other hand, does not pollute when it burns. It simply produces water. As you can see, it makes a lot of sense to use plant power in power plants!

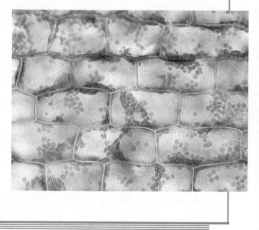

TEACHING STRATEGY 6–2

FOCUS/MOTIVATION

Have students examine the internal parts of seeds. Soak various seeds overnight; then distribute them for dissection. Using hand lenses, have students observe the inside of each seed and identify as many structures as possible. Discuss with students the similarities and differences among the various types of seeds.

CONTENT DEVELOPMENT

Remind students that the Earth's environments did not remain constant through the course of time. Changing environments created new species of plants with new adaptations. These new species of plants either survived in a previously empty niche or their adaptations made them better suited to their environments than existing species that did not possess the new adaptations. Over time, the better-adapted species survived, and the older species became extinct. In short, the species of plants that exist today are different from the species of plants that existed in the past.

Stress that seed plants have made two evolutionary leaps that make them completely adapted for life on land. One adaptation is that seed plants do not require water to reproduce. The other adaptation is that young seed plants are encased in a structure that provides food and protection.

6–2 Reproduction in Seed Plants

Seed plants have made two evolutionary leaps that make them fully adapted for life on land. **Seed plants do not require water to reproduce.** Sperm cells are carried straight to the waiting egg; the sperm cells do not have to swim to the egg. **Young seed plants are encased in a structure that provides food and protection.**

Fertilization in Seed Plants

The reproductive structures of seed plants are known as **cones** and **flowers.** Female cones and flower parts contain structures called **ovules** (AH-vyoolz). Each mature ovule contains an egg cell. Male cones and flower parts produce tiny grains of **pollen** (PAHL-uhn), which you can think of as containing sperm cells. The contents of a pollen grain are enclosed in a tough cell wall. This cell wall may be covered with strange and beautiful patterns of spikes and ridges.

If you have allergies, you may know pollen as an irritating form of dust. However, pollen is nothing to sneeze at! Carried by the wind or by animals, each grain of pollen is capable of delivering a sperm cell to an egg cell. The process by which pollen is carried from male reproductive structures to female reproductive structures is called **pollination.**

If everything goes right, pollination is followed by fertilization. The contents of the pollen grain break

Figure 6–14 *A breeze causes pollen from a cluster of male pine cones to drift away in a dusty yellowish cloud. A few of these pollen grains may eventually reach a female pine cone and fertilize its ovules. Some seed plants, such as the African tulip tree, have flowers rather than cones.*

B ■ 143

6–2 Reproduction in Seed Plants

MULTICULTURAL OPPORTUNITY 6–2

Have students research the origins of different plants used for food that is part of their everyday diet. For instance, barley and wheat were grown in Iraq some 9000 years ago. Corn was domesticated in Mexico, and millet and rice in China. Okra comes from Africa, where it has been prized from prehistoric times. An important ingredient in Cajun cooking, okra is used to thicken stews by means of a gummy substance in its pod.

ESL STRATEGY 6–2

Remind students that words such as *like, just as, too,* and *as well* may be used to express comparison and that words such as *however, but, whereas,* and *on the other hand* may be used to express contrast. Ask students to work in small cooperative groups as they do the following activities.
1. Write a paragraph comparing similarities and contrasting differences in angiosperm and gymnosperm seeds.
2. Create a cartoon character that illustrates the parts of a seed.

Activity Bank

Clone a Plant, Activity Book, p. B229. This activity can be used for ESL and/or Cooperative Learning.

Also, stress to students that, in a sense, some seed plants can be thought of as being male or female. The reproductive structures of seed plants are known as cones and flowers. Cones and flower parts from a female seed plant contain structures called ovules. Each mature ovule contains an egg cell. Cones and flower parts from a male seed plant contain tiny structures known as pollen. Pollen can be thought of as containing sperm cells that are enclosed in a tough, protective cell wall.

● ● ● ● **Integration** ● ● ● ●

Use the discussion of pollen's relationship to allergies to integrate concepts about the immune system into your life science lesson.

ENRICHMENT

Ask students to describe a situation they have seen or heard about in which there was some type of overreaction. Then inform students that allergies are caused by an overreaction of the body's immune system to plant pollen, dust, molds, and so forth. Scientists do not thoroughly understand the reasons why some individuals are oversensitive to certain particles like pollen, but allergies of some kind affect about 20 percent of the population. In an allergic reaction, certain cells in the body activate and release chemicals known as histamines. Histamines increase the flow of blood and fluids to affected areas—especially the linings of the nasal passages. Histamines produce the sneezing, runny eyes and nose, and other irritations that make a person with allergies so uncomfortable.

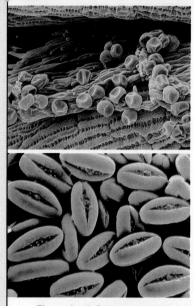

Figure 6–15 *Pollen grains may be spherical or oval, textured or smooth. The pollen grains from a flowering horse chestnut resemble loaves of French bread.*

Figure 6–16 *Fertilized ovules develop into seeds. The seeds of gymnosperms, such as pines, are not covered by an ovary. Some of the seeds in this pine cone have been removed from their normal position on the top of the woody scales that make up the cone so that they can be clearly seen (top left). The seeds of angiosperms— such as cantaloupes (top right), pomegranates (bottom left), and avocados (bottom right)— are enclosed by an ovary.*

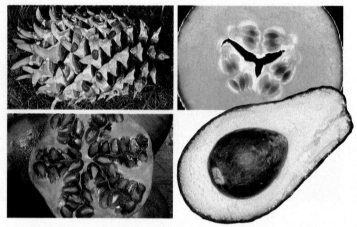

144 ■ B

out of their hard cell wall. They then grow a long tube that delivers the sperm cell to the egg cell in the ovule. The sperm cell joins with the egg cell, fertilizing it. The fertilized egg and the ovule that surrounds it develop into a seed.

In some seed plants, the ovules and the seeds that develop from them are contained within a structure called the ovary (OH-vah-ree). These plants are known as **angiosperms** (AN-jee-oh-sperms). The Greek word *angion* means vessel; *sperma* means seed. Thus angiosperms are plants whose seeds are contained in a vessel (the ovary).

In other seed plants, the ovules and seeds are not surrounded by an ovary. These plants are known as **gymnosperms** (JIHM-noh-sperms). The Greek word *gymnos* means naked. Thus, gymnosperms are plants whose seeds are naked; that is, not covered by an ovary.

Seeds

Although seeds look quite different from one another, they all have basically the same structure. **A seed consists of a seed coat, a young plant, and stored food.**

The seed coat is a tough, protective covering that develops from the ovule wall. Some familiar seed coats include the "skins" on lima beans, peanuts, and corn kernels. The brown, winglike covering on pine seeds is also a seed coat. So is the fleshy red covering of a yew seed.

6–2 (continued)

CONTENT DEVELOPMENT

Have students note that plants known as angiosperms have seeds that are contained in a vessel, or ovary. Plants known as gymnosperms have seeds that are naked.

• **A cantaloupe plant is an example of an angiosperm plant because the seeds of a cantaloupe are contained in a vessel, or ovary. Name several other examples of angiosperm plants.** (Students might suggest apples, watermelons, and tomatoes.)

• **A pine tree is an example of a gymnosperm plant because the seeds of a pine tree are naked. Name several other examples of gymnosperm plants.** (Students might suggest various conifer trees.)

● ● ● ● **Integration** ● ● ● ●

Use the discussion of the derivation of the words *angiosperm* and *gymnosperm* to integrate language arts concepts into your life science lesson.

GUIDED PRACTICE

Skills Development

Skill: Making comparisons

Have students collect and bring to class seed cones of various conifers. In groups, have them compare and contrast the sizes, textures, and shapes of the male and female cones of different conifers. You might wish to have each group create a display of the collected cones and identify the tree that produced each cone.

CONTENT DEVELOPMENT

The seeds of different types of plants vary greatly in size. For example, one kind of coconut tree can produce a seed that has a mass of almost 23 kilograms. On the other hand, orchid seeds are so tiny that 30,000 of them have a mass of about 1 gram. But regardless of its size, every seed contains all that is necessary for the development of a new plant. A seed consists of a seed coat, a young plant, and stored food.

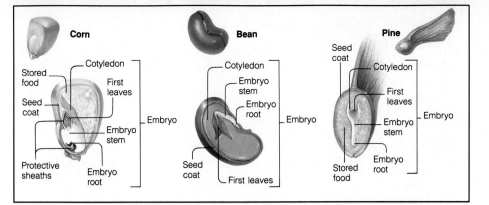

Corn   Bean   Pine

Corn: Cotyledon · Stored food · First leaves · Seed coat · Embryo · Embryo stem · Protective sheaths · Embryo root

Bean: Cotyledon · Embryo stem · Embryo root · Embryo · Seed coat · First leaves

Pine: Seed coat · Cotyledon · First leaves · Embryo · Embryo stem · Embryo root · Stored food

Enclosed within the seed coat is a tiny young plant, or embryo. The embryo develops from the fertilized egg. As you can see in Figure 6–17, the embryo is basically a miniature plant. The top of the embryo eventually gives rise to the leaves and upper stem of the plant. The middle stemlike portion of the embryo becomes the bottom part of the plant's stem. The bottom portion of the embryo becomes the plant's roots.

You probably have noticed that the embryo makes up only a small portion of the seed. In the seeds of most plants, the embryo stops growing when it is quite small and enters a state of "suspended animation." While "sleeping" inside the seed, the embryo can survive long periods of cold, heat, or dryness. When conditions are favorable for growth, the embryo becomes active and the young plant begins to grow once more. Once the young plant begins to grow, it uses the food stored in the seed until it can make its own.

Figure 6–17 *In some seeds, such as those of pines, the embryo is surrounded by stored food. In other seeds, such as corn kernels, some of the food is stored inside a seed leaf, or cotyledon. In still other seeds, such as beans, all of the food is stored inside the cotyledons. Why do most seeds consist mostly of stored food?* ①

Figure 6–18 *These tropical plants have seeds and fruits that are quite exotic in appearance. Can you identify their seeds and fruits?* ②

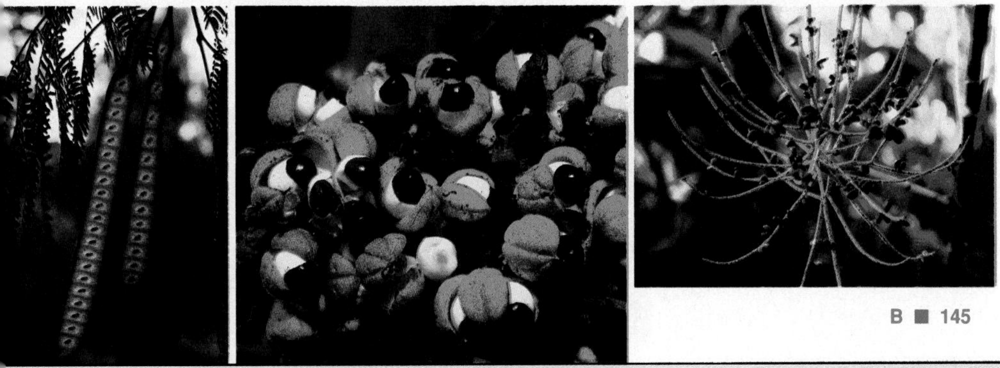

B ■ 145

Direct students' attention to Figure 6–17. Point out that a tiny young plant, or embryo, is enclosed within the seed coat. The seed coat is a tough covering for the embryo. It provides protection for the embryo and its food supply and also prevents them both from drying out. The embryo is essentially a miniature plant. The top portion of the embryo develops into the leaves and upper stem of a new plant. The middle portion of the embryo develops into the lower stem of the new plant, and the bottom portion of the embryo develops into the root system of the new plant.

The embryo, when it is still in the seed, stops growing while it is quite small. It can remain in this dormant state for weeks, months, or even years. Seeds, and the embryos inside them, can survive long periods of bitter cold, extreme heat, or drought and begin to grow only when conditions are once again correct. Once the embryo or young plant begins to grow, it uses the food stored in the seed until it can make its own through photosynthesis.

ACTIVITY
DISCOVERING
SEED GERMINATION

Discovery Learning

Skills: Predicting, making observations

Materials: 20 dried beans or unpopped popcorn kernels, 20 test tubes, paper towels

In this activity students should observe that light and warmth contribute to a greater seed germination rate. If students predicted that light and warmth would positively influence the germination rate of the seeds, their results should generally match their predictions. A possible source of error in the experiment might include the temperature ranges: If the range of temperatures between groups of seeds is too great or too small, the results of the experiment may be skewed.

INTEGRATION
HEALTH

Students will recall that a portion of the reproductive processes of seed plants involves pollen. Have interested (or allergic) students volunteer to examine newspapers or television broadcasts for a piece of weather-related data known as the pollen count. Have them discover what the pollen count is, how it is determined, and what types of pollen are included. Findings should be shared with the class.

ACTIVITY
DISCOVERING

Seed Germination

1. Obtain 20 dried beans or unpopped popcorn kernels, 20 test tubes, and some paper towels.

2. Design a two-part experiment that uses these materials to determine whether light and warmth are needed for the seeds to germinate.

3. Write down what you plan to do in your experiment and what results you expect.

4. Soak the seeds in water overnight. Then put together the apparatus for your experiment.

■ What were the results of your experiment? Did the results match your predictions? What were some possible sources of error?

Seed Dispersal

After seeds have finished forming, they are usually scattered far from where they were produced. The scattering of seeds is called seed dispersal (dih-SPER-suhl). Seeds are dispersed in many ways. Have you ever picked a dandelion puff and blown away all its tiny fluffy seeds? If so, you have helped to scatter the seeds. Usually, dandelion seeds are scattered by the wind. Other plants whose seeds are scattered by the wind include maples and certain pines. The winglike structures on these seeds cause them to spin through the air like tiny propellers.

Humans and animals play a part in seed dispersal. For example, burdock seeds have spines that stick to people's clothing or to animal fur. People or animals may pick up the seeds on a walk through a field or forest. At some other place, the seeds may fall off and eventually start a new plant.

The seeds of most water plants are scattered by floating in oceans, rivers, and streams. The coconut, which is the seed of the coconut palm, floats in water. This seed is carried from one piece of land to another by ocean currents.

Other seeds are scattered by a kind of natural explosion, which sends the seeds flying into the air. This is how the impatiens you read about at the beginning of this chapter disperses its seeds.

Figure 6–19 *The seeds of the coconut palm are dispersed by water (center). A tumbleweed is said to disperse its seeds mechanically. The main body of the plant serves as a device for scattering seeds (top right). The seeds of the unicorn plant hitch a ride to new places when spines hook around the hooves (or hiking shoes) of animals that walk by (bottom right). How do you think milkweed seeds are dispersed (left)?* ❶

6–2 (continued)

CONTENT DEVELOPMENT

The process of distributing seeds away from parent plants is called seed dispersal.
• **Seed dispersal is very important to plants. Why?** (Lead students to understand that if the seeds of a plant are not dispersed and instead fall to the ground beneath the parent plant, the seedlings will compete with one another and with the parent plant for sunlight, water, and nutrients. This competition will reduce the chances of survival for the growing seeds.)

Point out that seed dispersal also enables plants to colonize new environments. Although adult plants cannot move around, their seeds can be carried to new environments.

ENRICHMENT

• **Why are many unripe fruits green and have a bitter taste?** (Accept all reasonable answers.)

Point out that many fruit plants manufacture bitter-tasting chemicals that are pumped into fruits as fruits develop.

These bitter-tasting compounds discourage animals from eating fruits that are not ripe. If animals would eat unripened fruit, the seeds of the fruit would be dispersed with the animal, but the seeds would be too immature to develop. Consequently, the plant would not spread. When the seeds of fruit are mature enough for dispersal, the plant either removes the bitter-tasting compound or chemically breaks them down, replacing them with sugars. Sugars make the fruit sweet and cause them to change color.

Most seeds remain dormant, or inactive, for a time after being scattered. If there is enough moisture and oxygen, and if the temperature is just right, most seeds go through a process called germination (jer-mih-NAY-shuhn). Germination is the early growth stage of an embryo plant. Some people call this stage "sprouting."

6–2 Section Review

1. Explain how reproduction in seed plants is adapted to life on land.
2. What is a seed? What are the main parts of a seed?
3. How do angiosperms differ from gymnosperms?

Critical Thinking—*Sequencing Events*
4. Put these events in the correct order: germination, fertilization, seed dispersal, release of pollen, seed formation, pollination. Briefly describe each event.

PROBLEM ??? Solving

They Went Thataway!

Seeds are dispersed in many different ways. Seed dispersal is important because it helps plants spread to new areas. It also improves the chances that some of a plant's seeds will grow and survive to produce seeds of their own. Examine the accompanying photographs carefully. Then answer the following questions for each photograph.

Making Inferences
1. How is the seed dispersed? How can you tell?
2. How is the seed or fruit adapted for this method of dispersal?

The fruits are then easily visible and tasty to animals—who unknowingly disperse the mature seeds.

INDEPENDENT PRACTICE

▶ *Activity Book*

Students who need practice on the concept of the scattering of seeds should complete the chapter activity Seed Dispersal. In this activity students will identify the methods by which various seeds are generally dispersed.

INDEPENDENT PRACTICE
Section Review 6–2

1. Seed plants do not require water to reproduce, and young seed plants are encased in a structure that provides food and protection.

2. A seed is a structure that protects the fertilized egg cell of a plant. A seed consists of a seed coat, a young plant, and stored food.

3. An angiosperm is a flowering plant whose seeds develop within ovaries. A gym-

ANNOTATION KEY

Answers

① Students might suggest wind. (Interpreting illustrations)

PROBLEM SOLVING
THEY WENT THATAWAY!

This feature reinforces the concept of seed dispersal. Student answers will vary.

Poppy seeds are dispersed mechanically. The pod acts rather like a salt shaker to scatter the seeds. Burr seeds are carried on the coats of animals such as sheep. The seeds of many weeds have fluffy, feathery structures that allow them to be blown about by the wind.

BACKGROUND INFORMATION
MEDICINAL PLANTS

Seed plants have been an important source of medicine since the dawn of civilization. For example, the ancient Greeks discovered that chewing the leaves of a certain willow tree (genus *Salix*) cured aches and pains. Later, doctors found that those willows contain a compound called salicylic acid. Many centuries later, we still use this compound, commonly called aspirin.

nosperm is a plant whose seeds develop naked—they do not develop within ovaries.
4. The correct order of events is (1) release of pollen, (2) pollination, (3) fertilization, (4) seed formation, (5) seed dispersal, and (6) germination. Check students' descriptions of these events for scientific accuracy.

REINFORCEMENT/RETEACHING

Review students' responses to the Section Review questions. Reteach any material that is still unclear, based on students' responses.

CLOSURE

▶ *Review and Reinforcement Guide*
Students may now complete Section 6–2 in the *Review and Reinforcement Guide*.

6-3 Gymnosperms and Angiosperms

MULTICULTURAL OPPORTUNITY 6-3

Suggest that students conduct a survey of their schoolyard (if there are no trees in the schoolyard, use a local park). Divide the area into small, relatively equal-sized sections and have pairs of students classify the trees in their assigned section. When all the data have been collected, create a profile of trees in the area.

ESL STRATEGY 6-3

Have students write a short paragraph identifying the four phyla of gymnosperms. Also, ask small groups to create research reports that include a brief historical account of gymnosperms' existence on Earth, their typical method of pollination, an illustration of a pine cone and its winged seeds (showing how they grow upon its scales), and an explanation of why, in students' opinion, there are more angiosperms than gymnosperms on Earth today.

Guide for Reading

Focus on these questions as you read.

▶ What are the different phyla of gymnosperms?

▶ How do flowering plants reproduce?

Figure 6–20 *Cycads (top) and ginkgoes (bottom) are gymnosperms that were quite common during the age of the dinosaurs. Unlike a conifer, a cycad or a ginkgo produces either male cones or female cones, not both.*

6-3 Gymnosperms and Angiosperms

Gymnosperms are the most ancient group of seed plants. They first appeared about 360 million years ago—about the same time as the first land animals. Throughout the age of the dinosaurs, approximately 65 to 245 million years ago, gymnosperms were the dominant form of plant life on Earth.

Near the end of the age of dinosaurs, about 100 million years ago, a new phylum of plants appeared on the scene. The plants in this new phylum—the angiosperms—soon replaced the gymnosperm phyla as the Earth's dominant form of plant life.

Gymnosperms

Although the reign of gymnosperms has ended, four phyla of gymnosperms have survived to the present day. **The four phyla of gymnosperms are commonly known as cycads, ginkgoes, conifers, and gnetophytes.**

Cycads (SIGH-kadz) are tropical plants that look like palm trees. Some cycads grow up to 15 meters tall, but most grow no taller than a human. The trunk of a cycad is topped by a cluster of feathery leaves. In the center of the leaves of a mature cycad are its cones.

Although ginkgoes (GING-kohz) were fairly common during the age of the dinosaurs, only one species exists today. The ginkgo is sometimes known as the maidenhair tree. Interestingly, the ginkgo does not seem to exist in the wild. The ginkgoes that grow along many city streets in the United States and elsewhere are the descendants of plants from gardens in China. Ginkgo seeds have an incredibly bad odor.

With about 550 species, conifers (KAHN-ih-ferz) make up the largest group of gymnosperms. Most conifers are large trees with needlelike or scalelike leaves. See Figure 6–21. Conifers are found throughout the world, and are the dominant plants in many forests in the Northern Hemisphere. Conifers include the world's tallest trees (coast redwoods) and longest-lived trees (bristlecone pines).

TEACHING STRATEGY 6-3

FOCUS/MOTIVATION

Obtain some pictures of various types of flowering plants. Display the pictures. Ask these questions.

• **Can anyone identify the flowers in these pictures?** (Answers will vary.)

• **In what ways are all these plants similar?** (All have flowers, roots, stems, leaves, and seeds.)

• **In what ways are all these plants different?** (All have various shapes, various colors, leaves of different shapes, different flowers, and are native to different geographical regions.)

Write the word *angiosperm* on the chalkboard. Point out to students that all the flowers in the pictures are examples of angiosperms.

Write the word *gymnosperm* on the chalkboard. Direct students' attention to the giant redwood tree in Figure 6–21.

• **Describe what you see.** (A car moving through a hole in the trunk of a living tree.)

Point out to students that the giant redwood tree is an example of a gymnosperm. They will learn more about angiosperms and gymnosperms in this section.

CONTENT DEVELOPMENT

Have students recall from previous studies that gymnosperms are seed plants that produce uncovered, or naked, seeds. Point out that, at one time, gymnosperms were the most dominant form of plant life on Earth. Four phyla of gymnosperms have survived until today. These four phyla are commonly known as conifers, cycads, ginkgoes, and gnetophytes.

Direct students' attention to Figure 6–21. Point out that conifers represent

The word conifer literally means cone-bearer. This name is quite appropriate. Like cycads and ginkgoes, most conifers bear cones that produce pollen or ovules.

A few conifers, such as larches and bald cypresses, shed their leaves in autumn. However, most conifers—such as pines, firs, spruces, cedars, and hemlocks—are evergreen. Evergreen plants have leaves year round. Old leaves drop off gradually and are replaced by new ones throughout the life of the plant. The needles or leaves may remain on an evergreen tree for 2 to 12 years.

Conifers are of great importance to people. They are a major source of wood for building and for manufacturing paper. Useful substances such as turpentine, pitch, and rosin are made from the sap of conifers. The seeds of certain pines are rich in protein and are used for cooking and snacking. Juniper seeds are used to flavor food. Recently, scientists discovered that taxol, a substance in the bark of the Pacific yew tree, is showing promise as an effective treatment for certain kinds of cancer.

Gnetophytes (NEE-toh-fights) are a diverse group of plants that share characteristics with both gymnosperms and angiosperms. Gnetophytes include climbing tropical vines with oval leaves, bushes with jointed branches and tiny scalelike leaves, and a peculiar-looking desert plant whose two straplike leaves grow throughout its lifetime.

Figure 6–21 *Some conifers, such as the Port Oxford cedar, have tiny scalelike leaves (left). Others, such as the Douglas fir, have needlelike leaves (center). A few conifers are amazingly large. Why is this giant redwood tree able to live and grow in spite of having a car-sized tunnel cut through it?* ❶

Figure 6–22 Welwitschia mirabilis *is one of the few organisms that can survive in the extremely hot, dry Namib desert of Africa. Although* Welwitschia *has only two leaves, the leaves soon become tattered and torn so that they look like many leaves. To which phylum of gymnosperms does* Welwitschia *belong?* ❷

B ■ 149

BACKGROUND INFORMATION
LIVING GINKGOES

The ginkgo trees, with their unusual fan-shaped leaves, existed long before humans. For some reason, they are highly resistant to damage from polluted air. Because of this trait, many have been planted in cities in recent years. The ginkgoes produce male and female cones on separate plants. The female ginkgo seeds are surrounded with a fleshy seed coat that produces butyric acid, which has the strong odor of rancid butter. For this reason, when ginkgoes are planted as ornamentals, it is best to plant only male trees to avoid a very smelly fruit fall.

the largest group of gymnosperms (more than 550 species) and probably are the most familiar phylum to students. Conifer means "cone-bearing."

Direct students' attention to Figure 6–20. Inform them that cycads are tropical trees and that their trunks are topped by a cluster of feathery leaves. Ginkgoes, once a relatively common tree (there is only one species remaining), have fan-shaped leaves.

Stress that along with the gnetophyte phylum (Figure 6–22), conifers, cycads, and ginkgoes all are gymnosperms because they produce naked seeds.

● ● ● ● **Integration** ● ● ● ●

Use the discussion of plant life during the time of the dinosaurs to integrate concepts of evolution into your life science lesson.

INDEPENDENT PRACTICE

🎧 Media and Technology

To reinforce students' understanding of the relationships between living organisms in a biome, you can now use the Interactive Videodisc called EcoVision. Using this videodisc, students will become eco-agents. They will investigate the interrelationships between plants and other organisms in a biome and explore the effects of pollution on plants.

REINFORCEMENT/RETEACHING

▶ *Activity Book*

Students who need practice on the concept of changes in living things should complete the chapter activity Observing a Tree. In this activity students will observe and record the changes in a tree during the course of time.

CARNIVOROUS PLANTS

Some of the more interesting angiosperm plants are carnivorous. There are three basic types of carnivorous plants. Some, like the Venus' flytrap, are called active traps. Others, such as the varieties of pitcher plants, are pitfall traps. The third type is the flypaper trap, such as the sundew. All three have certain characteristics in common. The trap mechanism is always formed from modified leaves and has nectar glands that exude substances attractive to insects. The leaves of these plants also have glands that produce digestive enzymes. Another feature of carnivorous plants is the modification of leaf hairs in a way that aids in capturing prey. Most of these plants grow in nutrient-deficient soil; they absorb nitrogen and minerals from trapped and digested insects.

Flat Flowers, Activity Book, p. B241. This activity can be used for ESL and/or Cooperative Learning.

6–3 (continued)

ENRICHMENT

▶ *Activity Book*

Students will be challenged by the Chapter 6 activity called Fruits, Vegetables, and Spices. In this activity students will classify various foods, vegetables, and spices according to the plant part they represent.

CONTENT DEVELOPMENT

Introduce angiosperms by explaining that they are similar to gymnosperms in that both groups of plants form seeds and have leaves, stems, and roots. But angiosperms differ from gymnosperms in several important ways—the most noticeable is that only angiosperms produce flowers.

Figure 6–23 *Some flowers, such as sweet alyssum, impatiens, geraniums, and marigolds, are common garden flowers (top left). Others, such as the strange silversword plant of Hawaii, are delightfully uncommon (right). The silversword looks quite different after it develops its enormous stalk of flowers (bottom left). After it produces its seeds, the silversword dies.*

Angiosperms

Angiosperms make up the largest group of plants in the world—there are more than 230,000 known species, with perhaps hundreds of thousands more as yet undiscovered.

Angiosperms vary greatly in size and shape. They can be found in just about all of Earth's environments, including frozen wastelands near the North Pole, steamy tropical jungles, and practically waterless deserts.

Because they produce flowers, angiosperms are also called flowering plants. The flowers of angiosperms fill the Earth with beautiful colors and pleasant smells. But flowers serve a more important purpose. **Flowers are the structures that contain the reproductive organs of angiosperms.**

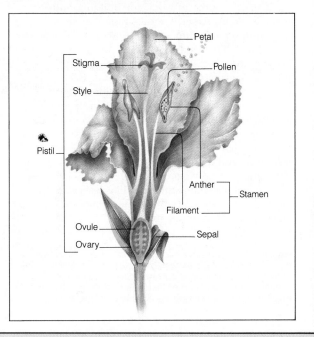

Figure 6–24 *The parts of a flower work together to accomplish the task of reproduction. Which are the male parts of a flower? Which are the female parts?* ①

Today angiosperms are the most widespread of all land plants. Angiosperms live everywhere from frigid mountains to blazing deserts. Some angiosperms even live underwater.

INDEPENDENT PRACTICE

▶ *Product Testing Activity*

Have students perform the product test on jeans from the Product Testing Activity worksheets. In this activity students will test various natural and synthetic fibers in fabrics.

REINFORCEMENT/RETEACHING

Ask the following questions to reinforce the concepts of flowering plants.

• **Have you ever leaned over to sniff a brightly colored flower and been startled to find a bee inside the flower?** (Answers will vary.)

• **What is the name given to the beautifully shaped and colored leaflike structures of a flower?** (Petals.)

Figure 6–25 *Flowers and the animals that pollinate them have evolved in response to one another. How are the hummingbird, bee, and bat adapted to feeding from flowers? Can you tell what kind of animal pollinates the bee orchid, which looks and smells like a female bumblebee (top right)? Plants have also evolved ways that help them to disperse their seeds. How are blackberries adapted for seed dispersal by animals?* ②

As you can see in Figure 6–23, flowers come in all sorts of sizes, colors, and forms. As you read about the parts of a typical flower, keep in mind that the descriptions do not apply to all flowers. For example, some flowers have only male reproductive parts, and some flowers lack petals.

When a flower is still a bud, it is enclosed by leaflike structures called **sepals** (SEE-puhls). Sepals protect the developing flower. Once the sepals fold back and the flower opens, colorful leaflike structures called **petals** are revealed. The colors, shapes, and odors of the petals attract insects and other animals. These creatures play a vital role in the reproduction of flowering plants.

Within the petals are the flower's reproductive organs. The thin stalks topped by small knobs are the male reproductive organs, or **stamens** (STAY-muhns). The stalklike part of the stamen is called the filament, and the knoblike part is called the anther. The anther produces pollen.

The female reproductive organs, or **pistils** (PIHS-tihls), are found in the center of the flower. Some

The Birds and the Bees

The process of evolution has resulted in some amazing relationships between angiosperms and the animals that pollinate them. Discover some of these fascinating relationships in *The Clover and the Bee: A Book of Pollination* by Anne Dowden.

①

ACTIVITY
READING
THE BIRDS AND THE BEES

Skills: Reading comprehension

You may want to have students describe the evolutionary relationship they found most fascinating in their reading.

Integration: Use this Activity to integrate language arts into your lesson.

mystery of the waterfowls' failure to return to their seasonal home. Challenge students to find out how angiosperms might have contributed to the ducks' disappearance. (*Hint:* What are some common crop plants? To which phylum do these plants belong?)

GUIDED PRACTICE

▶ *Laboratory Manual*
Skills Development

Skills: Making observations, recording data, applying concepts

At this point you may want to have students complete the Chapter 6 Laboratory Investigation in the *Laboratory Manual* called Observing Reproductive Structures in Angiosperms. In this investigation students will observe and identify the reproductive structures of flowering plants and observe various fruits and seeds.

• **What is the function of the brightly colored petals of a flower?** (The petals help to attract insects and other animals.)
• **Why is it important that a flower attract insects and other animals?** (Insects and other animals play a vital role in the reproduction of flowering plants.)
• **What is the name given to the male organs of a flowering plant?** (Stamens.)
• **A stamen consists of two parts. Name the two parts of a stamen.** (Filament and anther.)
• **What is the name given to the female or-**gans of a flowering plant (Pistils.)
• **In most flowers, a pistil consists of three parts. Name the three parts of a pistil.** (Ovary, style, and stigma.)
• **Does every flowering plant have all these structures?** (No.)

INDEPENDENT PRACTICE

🎯 **Media and Technology**

In the Interactive Videodisc called Paul ParkRanger and the Mystery of the Disappearing Ducks, students help solve the

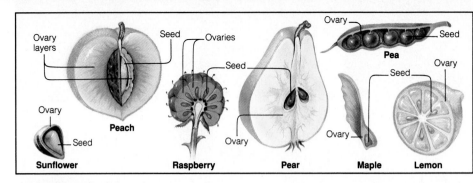

Answers

❶ Maple seeds have a shape that lends itself to flight; various fruits are sweet and colorful to attract the attention of animals. (Applying concepts)

❷ A stigma is sticky so that it can collect, or retain, pollen from itself and other flowers. (Applying concepts)

Integration

❶ Language Arts

❷ Language Arts

ACTIVITY
READING

A WORLD OF FUN WITH PLANTS

Skill: Reading comprehension

You may wish to have volunteers demonstrate some of the interesting things they learned in the book they read.
Integration: Use this Activity to integrate language arts into your lesson.

6–3 (continued)

INDEPENDENT PRACTICE

▶ *Activity Book*

Students who need practice on the concept of plant structures should complete the chapter activity Plants as Food. In this activity students will identify the various parts of a plant from which different vegetables originate.

INDEPENDENT PRACTICE

Section Review 6–3
1. Conifers, cycads, ginkgoes, and gnetophytes.
2. A structures that contains the reproductive organs of angiosperms. Accept all logical responses.
3. A flower is pollinated when a grain of pollen lands on a stigma. The pollen grain then breaks open, and its contents produce a tube that grows down through the

Figure 6–26 *Fruits have evolved in ways that help to disperse seeds. How are maple seeds specialized for being dispersed by the wind? How are peaches, pears, lemons, and raspberries adapted for seed dispersal by animals?* ❶

ACTIVITY
READING

A World of Fun With Plants

❶ Do you know how to make a shrill whistle with a blade of grass? Learn how to do this and other neat things by reading *Hidden Stories in Plants* by Anne Pellowski.

flowers have two or more pistils; others have only one. The sticky tip of the pistil is called the stigma. A slender tube, called the style, connects the stigma to a hollow structure at the base of the flower. This hollow structure is the ovary, which contains one or more ovules.

A flower is pollinated when a grain of pollen lands on the stigma. (Can you explain why the stigma is sticky?) If the pollen is from the right kind of ❷ plant, the wall of the pollen grain breaks open. The contents of the pollen grain then produce a tube that grows down through the style and into an ovule. When the tube has finished growing, a sperm cell emerges from the tube and fertilizes the egg cell in the ovule.

Like gymnosperms, some angiosperms are pollinated by the wind. The flowers of these plants are usually small, unscented, and produce huge amounts of pollen. Most angiosperms, however, are pollinated by insects, birds, and other animals. Flowers that rely on animals for pollination are colorful, scented, and/or full of food. This helps them to attract the animals that pollinate them.

As you have learned, after the egg cell is fertilized, the ovule develops into a seed. As the seed develops, the ovary also undergoes some changes and becomes a fruit. A **fruit** is a ripened ovary that encloses and protects the seed or seeds. Apples and cherries are fruits. So are many of the plant foods you usually think of as vegetables, such as cucumbers and tomatoes. What other kinds of fruits do you eat? The next time you are enjoying a favorite fruit, try to identify the seed, seed coat, and ripened ovary.

style and into an ovule. A sperm cell emerges from the tube and fertilizes the egg cell in the ovule.
4. Female ginkgoes produce seeds that have an unpleasant smell.

REINFORCEMENT/RETEACHING

Monitor students' responses to the Section Review questions. If students appear to have difficulty with any of the questions, review the appropriate material in the section.

CLOSURE

▶ *Review and Reinforcement Guide*

At this point have students complete Section 6–3 in the *Review and Reinforcement Guide.*

TEACHING STRATEGY 6–4

FOCUS/MOTIVATION

Write the terms *stimulus* and *response* on the chalkboard. Ask these questions.

1. What are the four phyla of gymnosperms?
2. What is a flower? How are flowers involved in reproduction in angiosperms?
3. Describe pollination and fertilization in angiosperms.

Connection—*You and Your World*
4. Why are almost all the ginkgoes that are planted male trees?

6–4 Patterns of Growth

Why do some garden plants die and have to be replaced each year? Why do you have to turn a houseplant that is growing on the windowsill every now and then? To answer questions such as these, you have to know something about the patterns of growth in plants.

Annuals, Biennials, and Perennials

Plants are placed into three groups according to how long it takes them to produce flowers and how long they live. The three groups of plants are **annuals, biennials,** and **perennials.**

Some plants grow from a seed, flower, produce seeds, and die all in the course of one growing season. Such plants include marigolds, petunias, and many other common garden flowers. **Plants that complete their life cycle within one growing season are called annuals.** The word annual comes from the Latin word *annus*, which means year. Most annuals have herbaceous (nonwoody) stems. Wheat, rye, and tobacco plants are other examples of annuals.

Plants that complete their life cycle in two years are called biennials. (The Latin word *bi* means two.) ❷ Biennials sprout and grow roots, stems, and leaves during their first growing season. Although the stems and leaves may die during the winter, the roots survive. During the second growing season, a

Guide for Reading

Focus on these questions as you read.

▶ How do annuals, biennials, and perennials differ from one another?

▶ How do plants grow in response to their external environment?

B ■ 153

6-4 Patterns of Growth

From ancient times, forests have sheltered, fed, and healed people. Have students make a list of trees and shrubs with which they are familiar and research the role those plants have in different cultures. The tough bark of the white birch, for example, has been used by Native Americans to make canoes. The leaves of palm trees have been used to make thatched roofs in Indonesia. Stems of bamboo plants that grow in Burma have been used to make musical instruments. Students should find other examples both from home and abroad. Ask volunteers to share their findings with the class.

ESL STRATEGY 6-4

Refer students to their textbook, which explains that the prefixes *photo-* and *gravi-* indicate a particular form of tropism. Ask for the definitions of these prefixes and then explain that the word *tropism* comes from a Greek word meaning "to turn."

● ● ● ● **Integration** ● ● ● ●

Use the derivation of the term *biennials* to integrate language arts concepts into your life science lesson.

GUIDED PRACTICE
▶ *Laboratory Manual*

Skills Development

Skills: Making observations, recording data, analyzing concepts

At this point you may want to have students complete the Chapter 6 Laboratory Investigation in the *Laboratory Manual* called Investigating Germination Inhibitors. In this investigation students will explore the control mechanisms inside a tomato seed that regulate metabolism and growth processes.

REINFORCEMENT/RETEACHING
▶ *Activity Book*

Students will be challenged by the Chapter 6 activity called Plant Foods. In this activity students will identify the parts of plants they eat.

• **Has anyone ever seen a "bug zapper"?** (Yes.)
• **Describe a "bug zapper."** (It is a type of light used to attract and eliminate bugs such as mosquitoes.)
• **How is the light considered to be a stimulus for the insects?** (The light attracts, or causes a reaction in, insects.)
• **Describe a time when you turned on a light at night and saw a bug run across the floor.** (Descriptions will vary.)
• **Did the bug have a reaction to the light?** (Yes.)

• **Can the light be considered a stimulus in this situation?** (Yes.)
• **What was the reaction, or response, of the bug to the stimulus of the light?** (Answers might include it was startled and ran away to hide.)

CONTENT DEVELOPMENT

Stress to students that plants are classified into three groups based on how long it takes them to produce flowers and how long they live. The three groups of flowers are annuals, biennials, and perennials.

BACKGROUND
INFORMATION
ROOT GROWTH

Biologists do not thoroughly understand what causes the downward growth of roots. Somehow, a root senses its orientation with respect to gravity.

It is known that small amounts of auxin (plant growth hormone) inhibit root-cell elongation. Thus, downward growth is thought by some to be due to the accumulation of auxin on the lower side of a horizontally growing root.

Other biologists believe that auxin is not involved at all in normal root curvature. Rather, they think another hormone, abscissic acid, moves up from the root tip to control curvature.

While scientists continue to speculate on the causes of the downward growth of roots, the precise chemicals and mechanisms that influence roots to grow downward remain an unsolved puzzle.

Activity Bank

You *Can* Force a Flower to Bloom, Activity Book, p. B245. This activity can be used for ESL and/or Cooperative Learning.

Figure 6–27 *Petunias bloom cheerfully throughout the warm months of the year, but die in the winter (top). During its second and final growing season, the foxglove produces a spike of many flowers. The spotted "trail" on a foxglove flower directs bees to the nectar and pollen inside (right). Begonias live for many years, producing flowers each summer (bottom). Which of these plants is a perennial? Which is an annual? Which is a biennial?* ❶

biennial grows new stems and leaves and then produces flowers and seeds. Once the flowers produce seeds, the plant dies. Examples of biennials include sugar beets, carrots, celery, and certain kinds of foxgloves.

Still other plants live for more than two growing seasons. **Plants that live for many years are called perennials.** (The Latin word *per* means through; a perennial lives through the years.) Some perennials, such as most garden peonies, are herbaceous. Their leaves and stems die each winter and are renewed each spring. Most perennials, however, are woody. The long lives of perennials permit them to accumulate lots of layers of woody xylem in their stems. Pine trees and rhododendron bushes are examples of woody perennials. What other plants are woody perennials? ❷

Tropisms

When studying behavior in plants and other living things, it is helpful to be familiar with two terms: stimulus (STIHM-yoo-luhs; plural: stimuli, STIHM-yoo-ligh) and response. A stimulus is something in a

Activity Bank

Lean to the Light, p. 184

6–4 (continued)

REINFORCEMENT/RETEACHING

Remind students of the plant terms *annual*, *biennial*, and *perennial* by asking the following questions.
• **How many of you or your family members have ever planted a garden?** (Answers will vary.)

• **What varieties of plants are usually planted in a garden?** (Answers might include tomatoes, carrots, and onions.)
• **Generally speaking, are the majority of plants in any garden annual, biennial, or perennial plants?** (Annual.)
• **What is the name of a plant that completes its life cycle in two years?** (Biennial.)
• **What does the prefix *bi*- mean?** (Two.)
• **What is the definition of a perennial plant?** (A plant that lives for many years.)
• **Many people choose to plant tulips around their homes or in flower gardens.**

Every spring, the tulip bulbs in the ground grow to produce very beautiful flowers. Is a tulip considered an annual, biennial, or perennial flowering plant? (A perennial flowering plant.)

GUIDED PRACTICE

Skills Development

Skills: Making observations, making comparisons

At this point have students complete the in-text Chapter 6 Laboratory Investiga-

living thing's environment (both its internal and external environment) that causes a reaction, or response.

Plants respond to stimuli in a variety of ways. One way is by adjusting the way they grow. The growth of a plant toward or away from a stimulus is called a **tropism** (TROH-pihz-uhm). If a plant grows toward a stimulus, it is said to have a positive tropism. If it grows away from a stimulus, it is said to have a negative tropism.

There are several kinds of tropisms. The two most important tropisms involve responses to light and to gravity.

All plants exhibit a response to light called phototropism. (Recall that the Greek word *photos* means light.) Think about the houseplant mentioned at the start of this section. If a houseplant is not turned regularly, it soon begins to bend toward the window, the source of its light. Do leaves and stems show positive or negative phototropism? ③

Plants also show a response to gravity, or gravitropism. Roots show positive gravitropism—they grow downward. Stems, on the other hand, show negative gravitropism—they grow upward, against the pull of gravity.

Figure 6–28 *Because this houseplant was not turned regularly, it started to grow in a crooked way. Why did this happen?* ④

6–4 Section Review

1. What are annuals, biennials, and perennials?
2. What are tropisms? Describe two tropisms.

Critical Thinking—*Making Predictions*
3. Suppose a tree is knocked over by a storm, but it survives the experience. What will this tree look like a few years after the storm?

Pick a Plant

In the late winter or early spring, purchase a packet of seeds for an annual plant you would like to grow. You may wish to get together with a classmate or two and split the cost of the seeds. (A packet usually costs $1 to $2.) Following the directions on the seed packet, plant 3 seeds in a clean pint-sized milk carton. Water the seeds and resulting plants every day or two.

■ How long does it take your seeds to sprout? How do your plants change over the course of the year? Make a poster showing the more interesting stages in the life cycle of your plant. The stages shown in your poster should include the seed, the plant when it first begins to sprout, the plant when it has two leaves, the plant when it develops flowers, and the plant when it develops fruits or seeds.

ACTIVITY
DISCOVERING
PICK A PLANT

Skills: Making observations, relating concepts

Materials: seeds, milk cartons, soil, fertilizer

This activity allows students to observe and record the growth and development of a plant that germinates from a seed. The data collected by each student and the posters that are created will vary.

tion: Gravitropism. In this investigation students will investigate the effect of gravity on germinating seeds.

ENRICHMENT

▶ *Activity Book*

Students will be challenged by the Chapter 6 activity called Observing a Plant Tropism. In this activity students will explore the effect of light on a germinating seed.

INDEPENDENT PRACTICE

Section Review 6–4

1. An annual is a plant that completes its life cycle within one growing season; a biennial is a plant that completes its life cycle in two growing seasons; a perennial is a plant that lives for many years.

2. Tropisms are the growth of plants toward, or away from, a stimulus. Examples of tropisms might include a plant that bends toward the light entering a room from a window, and the roots of a plant growing toward a constant source of water.

3. The fallen tree should display growth upward, or perpendicular to its trunk.

REINFORCEMENT/RETEACHING

Review students' responses to the Section Review questions. Reteach any material that is still unclear, based on students' responses.

CLOSURE

▶ *Review and Reinforcement Guide*

Students may now complete Section 6–4 in the *Review and Reinforcement Guide.*

Laboratory Investigation

GRAVITROPISM

BEFORE THE LAB

1. Gather all materials at least one day before the activity. Be sure to have enough corn seeds to give four seeds to each group of students.

2. At least 24 hours before the activity, soak all seeds to be used in water.

PRE-LAB DISCUSSION

Have students read the complete laboratory procedure.

• **What is a tropism?** (A tropism is the growth of a plant toward, or away from, a stimulus.)

• **This investigation is titled Gravitropism. What do you suppose the prefix *gravi-* means?** (Lead students to suggest that it is an abbreviated form of the word *gravity*.)

• **What do you think the term *gravitropism* means?** (The growth of a plant toward, or away from, the pull of gravity.)

Laboratory Investigation

Gravitropism

Problem

How does gravity affect the growth of a seed?

Materials *(per group)*

4 corn seeds soaked in water for 24 hours	paper towels
	masking tape
petri dish	glass-marking pencil
scissors	clay

Procedure 🧪 📼

1. Arrange the four seeds in a petri dish. The pointed ends of all the seeds should face the center of the dish. One of the seeds should be at the 12 o'clock position of the circle, and the other seeds at 3, 6, and 9 o'clock.

2. Place a circle of paper towel over the seeds. Then pack the dish with enough pieces of paper towel so that the seeds will be held firmly in place when the other half of the petri dish is put on.

3. Moisten the paper towels with water. Cover the petri dish and seal the two halves together with a strip of masking tape.

4. With the glass-marking pencil, draw an arrow on the lid pointing toward 12 o'clock. Label the lid with your name and the date.

5. With pieces of clay, prop the dish up so that the arrow is pointing up. Place the dish in a completely dark place.

6. Predict what will happen to the seeds. Then observe the seeds each day for about one week. Make a sketch of them each day. Be sure to return the dish and seeds to their original position when you have finished.

Observations

What happened to the corn seeds? In which direction did the roots and the stems grow?

Analysis and Conclusions

1. Which part of the germinating seeds showed positive gravitropism? Which part showed negative gravitropism?

2. What would happen to the corn seeds if the dish were turned so that the arrow was pointed toward the bottom of the dish? If it were turned to the right or the left?

3. Why is it important that the petri dish remain in a stable position throughout the investigation?

4. Explain why the seeds were placed in the dark rather than near a sunny window.

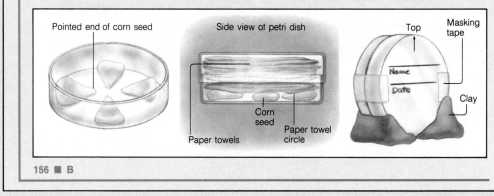

Pointed end of corn seed — Side view of petri dish — Top — Masking tape — Name — Date — Clay — Corn seed — Paper towel circle — Paper towels

---

TEACHING STRATEGY

1. Before arranging seeds in the petri dishes, make sure students are familiar with the concept of 3, 6, 9, and 12 o'clock positions.

2. Remind students that the clay will be used to hold the covered petri dishes in a vertical rather than a horizontal position.

DISCOVERY STRATEGIES

Discuss how the investigation relates to the chapter by asking open questions similar to the following.

• **How do you think gravity influences the root system, stem, and leaves of a plant?** (Accept all answers—analyzing, generalizing.)

• **Describe what will have occurred if some of the structures of the young corn plants in this activity display positive tropism.** (They will have displayed growth

Study Guide

Summarizing Key Concepts

6-1 Structure of Seed Plants

▲ There are two types of vascular tissue. Xylem carries water and dissolved minerals. Phloem carries food.

▲ Roots anchor the plant in the ground and absorb water and minerals from the soil.

▲ Stems transport materials between the roots and the leaves and support the leaves.

▲ In most plants, leaves are the structures in which photosynthesis occurs.

▲ Photosynthesis is the process that uses light energy to change carbon dioxide and water into glucose and oxygen.

6-2 Reproduction in Seed Plants

▲ Seed plants do not require standing water to reproduce.

▲ The process in which pollen is carried from male reproductive structures to female reproductive structures is called pollination.

▲ After pollination, the contents of the pollen grain grow a tube into an ovule and release a sperm cell. The sperm cell fuses with the ovule's egg cell to produce a fertilized egg in a process called fertilization.

▲ The ovules and seeds of angiosperms are covered by an ovary; those of gymnosperms are uncovered.

▲ Seeds consist of a seed coat, an embryo, and stored food.

6-3 Gymnosperms and Angiosperms

▲ There are four living phyla of gymnosperms: cycads, ginkgoes, conifers, and gnetophytes.

▲ Because their reproductive structures are contained in flowers, angiosperms are known as flowering plants.

6-4 Patterns of Growth

▲ Plants are classified as annuals, biennials, and perennials according to how long it takes them to produce flowers and how long they live.

▲ The growth of a plant in response to an environmental stimulus is called a tropism.

Reviewing Key Terms

Define each term in a complete sentence.

6-1 Structure of Seed Plants
xylem
phloem
root
stem
leaf
photosynthesis

6-2 Reproduction in Seed Plants
cone

flower
ovule
pollen
pollination
angiosperm
gymnosperm

6-3 Gymnosperms and Angiosperms
sepal
petal
stamen

pistil
fruit

6-4 Patterns of Growth
annual
biennial
perennial
tropism

ANALYSIS AND CONCLUSIONS

1. The roots showed positive gravitropism. The stems showed negative gravitropism.
2. The roots would curl around and grow downward, whereas the stems would curl around and grow upward.
3. Changing the position of the dish would alter the variable being studied.
4. To make sure that there was only one variable in the experiment; to rule out the possibility that the seeds were responding to a factor other than gravity.

GOING FURTHER: ENRICHMENT
Part 1
Students can investigate the effect of light on a plant. Position a seedling in such a way that the source of light is almost horizontal to the plant. Ask students to predict what might occur. Then have them design and perform an experiment that tests their prediction.
Part 2
Have students continue their study of tropisms by selecting a stimulus other than light or gravity that may generate a response in the root or stem system of a plant. An experiment should be designed so that it contains a prediction and a method to test for the chosen stimulus.

---

toward the pull, or influence, of gravity—generalizing, concluding.)
• **Describe what will have occurred if some of the structures of the young corn plants in this activity display negative tropism.** (They will have displayed growth away from the influence, or pull, of gravity—generalizing, concluding.)
• **You have read the laboratory procedure for this investigation. What do you predict will happen?** (Predictions will vary—predicting, inferring.)

OBSERVATIONS

The corn seeds germinated; the roots grew toward the Earth and the stems grew away from the Earth.

Chapter Review

ALTERNATIVE ASSESSMENT

The *Prentice Hall Science* program includes a variety of testing components and methodologies. Aside from the Chapter Review questions, you may opt to use the Chapter Test or the Computer Test Bank Test in your *Test Book* for assessment of important facts and concepts. In addition, Performance-Based Tests are included in your *Test Book*. These Performance-Based Tests are designed to test science process skills, rather than factual content recall. Since they are not content dependent, Performance-Based Tests can be distributed after students complete a chapter or after they complete the entire textbook.

CONTENT REVIEW

Multiple Choice

1. c
2. b
3. a
4. a
5. d
6. b
7. a
8. d
9. b
10. c

True or False

1. T
2. F, cambium
3. F, compound
4. F, taproot
5. F, seed coat
6. F, cones
7. F, sepals
8. T

Concept Mapping

Row 1: Stems, Leaves
Row 2: Phloem, Absorption and
Stability, Transportation and
Strength

CONCEPT MASTERY

1. Seed coat: tough, protective covering; embryo: a young plant; stored food: used until plant is able to complete photosynthesis on its own.
2. Answers will vary but might include decorative purposes, protection from wind, and habitat for wildlife.

Content Review

Multiple Choice

Choose the letter of the answer that best completes each statement.

1. Roots grow away from light. This is an example of
   a. negative gravitropism.
   b. positive gravitropism.
   c. negative phototropism.
   d. positive phototropism.
2. Which phylum has seeds that are enclosed by an ovary?
   a. gnetophytes   c. conifers
   b. angiosperms   d. cycads
3. The movement of gases in and out of the leaf is regulated by the
   a. stomata.   c. cortex.
   b. mesophyll.   d. cambium.
4. Many flowering plants rely on animal partners for
   a. pollination.   c. fertilization.
   b. germination.   d. transpiration.
5. Photosynthesis produces
   a. chlorophyll.   c. water.
   b. carbon dioxide. d. oxygen.

6. The main water-conducting tissue in a plant is
   a. phloem.   c. cambium.
   b. xylem.   d. pith.
7. The functions of anchoring the plant and absorbing water are performed primarily by a plant's
   a. roots.   c. leaves.
   b. stems.   d. anthers.
8. Which structure is not part of a flower's pistil?
   a. style   c. ovary
   b. stigma   d. anther
9. A pine tree is best described as
   a. annual.   c. biennial.
   b. perennial. d. herbaceous.
10. The union of the sperm cell and the egg cell is known as
   a. pollination.   c. fertilization.
   b. germination.   d. transpiration.

True or False

If the statement is true, write "true." If it is false, change the underlined word or words to make the statement true.

1. Pollen is produced in the <u>anther</u>.
2. The <u>cortex</u> produces new xylem and phloem cells.
3. A four-leafed clover is an example of a <u>simple</u> leaf.
4. Carrots and dandelions have <u>fibrous root</u> systems.
5. The protective outer covering of a seed is called the <u>epidermis</u>.
6. The ovules of a conifer are located in its <u>flowers</u>.
7. The structures that protect a flower bud are called <u>stamens</u>.
8. A <u>fruit</u> is a ripened ovary.

Concept Mapping

Complete the following concept map for Section 6–1. Refer to pages B6–B7 to construct a concept map for the entire chapter.

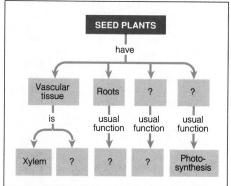

---

3. Dispersal is the scattering of seeds away from parent plants; humans and animals help to disperse seeds; dispersal allows seeds a greater likelihood of germination and growth.

4. Herbaceous plants have green and soft stems, whereas woody plants have woody stems. No; explanations will vary but might suggest that annuals do not live long enough to develop the multiple layers of xylem that form wood.

5. The shape of a stem lends itself to accommodate structural tubes for the effi-

cient transportation of materials from the root system to the leaves. The structure of a stem also lends strength to the plant and support for the leaves.

6. Photosynthesis is the process by which food is synthesized using light energy.

7. Both are similar in that they are plants that produce seeds. They are different in that gymnosperms produce uncovered seeds and do not produce flowers and fruits, whereas angiosperms produce flowers or fruits that contain seeds. Examples of gymnosperms include conifers, cycads,

Concept Mastery

Discuss each of the following in a brief paragraph.

1. List the three main parts of a seed and describe their function.
2. Give three examples of ways in which people use conifers.
3. What is seed dispersal? How does it take place? Why is it important?
4. Distinguish between herbaceous and woody plants. Would you expect an annual plant to be woody? Explain.
5. How does the structure of a stem help it to carry out its functions?

6. What is photosynthesis?
7. How are angiosperms and gymnosperms similar? How are they different? Give three examples of each kind of plant.
8. Describe the two most important kinds of tropisms in terms of stimulus and response.
9. Unlike the first plants to appear on Earth, seed plants do not depend on water for reproduction. Explain why.

Critical Thinking and Problem Solving

Use the skills you have developed in this chapter to answer each of the following.

1. **Summarizing information** List the major parts of a typical flower. Briefly describe the function of each part.
2. **Relating concepts** Describe the adaptations in seed plants that help them to avoid excess water loss.
3. **Developing a hypothesis** You observe that hungry deer have eaten the bark on an apple tree as far up as they can reach. At first, the tree seems to remain healthy. However, it eventually dies. Develop a hypothesis to explain your observations. How might you test your hypothesis?
4. **Identifying relationships** When life first appeared on Earth more than 3.5 billion years ago, there was no oxygen gas in the air. But for the past 1.8 billion years or so, the air has been about 20 percent oxygen. How did photosynthesis cause this change in the composition of Earth's air and maintain it? Why is it important to most of Earth's organisms to have so much oxygen in the air?
5. **Making inferences** Pesticides are designed to kill harmful insects. Sometimes, however, they kill helpful insects. What effect could this have on angiosperms?

6. **Designing an experiment** Design an experiment to determine whether or not water is needed for seed germination. Describe your experimental setup. Be sure to include a control.
7. **Using the writing process** Imagine that you are the seed plant of your choice. Describe your life from your earliest memories as a forming seed to the time you produced seeds of your own. (*Hints:* What were seed dispersal and germination like? What changes did you undergo as you grew?)

and ginkgoes; examples of angiosperms include watermelons, apples, and peaches.

8. In phototropism, the stimulus is light. The response is growth toward the light (positive) or growth away from the light (negative). In gravitropism, the stimulus is the pull of gravity. The response is growth downward, toward the perceived source of the stimulus (positive), or growth upward (negative).

9. The first plants on Earth existed in a moist or wet environment. In time, as areas of the Earth became drier, plants either adapted to the changing environment or became extinct. The seed plants of today reflect the successful adaptation from a wet to a drier environment.

CRITICAL THINKING AND PROBLEM SOLVING

1. Sepals: protect flower bud. Petals: attract pollinators. Pistil: female reproductive parts. Stigma: traps pollen. Style: connects stigma to ovary. Ovary: contains ovules, which develop into seeds. Stamen: male reproductive parts. Filament: bears anther. Anther: produces pollen, which contain sperm that fertilize ovules.

2. The coatings on leaves and the opening and closing of stomata are adaptations that help seed plants control excess water loss.

3. The hypotheses might suggest that the deer have destroyed the vascular tubes needed to carry food, water, and minerals from the soil, and the tree slowly starved to death. Tests for each hypothesis will vary.

4. Photosynthesis changed the content of Earth's atmosphere by constantly contributing oxygen released by the food-making processes of plants over time. Without oxygen, most organisms on Earth would die.

5. Angiosperms that depend on insects for pollination may not be able to reproduce if the number and variety of insects in a particular ecosystem are altered.

6. Individual designs will vary, but each should include a description and a control.

7. Written reports will vary but should include descriptions of seed dispersal, germination, and growth changes.

KEEPING A PORTFOLIO

You might want to assign some of the Concept Mastery and Critical Thinking and Problem Solving questions as homework and have students include their responses to unassigned questions in their portfolio. Students should be encouraged to include both the question and the answer in their portfolio.

ISSUES IN SCIENCE

The following issues can be used as springboards for discussion or given as writing assignments.

Scientists are constantly trying to make plants bigger, more beautiful, or more generally useful to humans. This is achieved by interfering in the natural reproduction processes of plants.

1. Identify several hybrid species of plants and describe how these plants differ from earlier varieties of the same plant.

2. Sometimes plants are chemically treated or artificially pollinated to achieve a desired effect. Should people interfere with the natural processes of plants?

COLLEEN CAVANAUGH EXPLORES THE UNDERWATER WORLD OF TUBE WORMS

Background Information

A worm is any of many kinds of boneless animals with a soft slender body and no legs. There are thousands of different kinds of worms. They usually live in soil or water. The size of worms ranges from microscopic to several meters long. There are many phyla of worms. The three most important phyla are flatworms, roundworms, and segmented worms.

Flatworms are the simplest worms. Some appear to be oval leaves, whereas others look like ribbons. Roundworms are the largest group of worms and look like a piece of thread. Segmented worms have a ringed appearance. Some segmented worms live in the sea and along the shore. Many segmented worms have tentacles on their head and leglike projections on each segment. Some of them live in tubes on the ocean floor.

The tube worms described here are so unlike other worms that their classification is a subject of controversy. They have been grouped with segmented worms and with beard worms. They have even been assigned a phylum of their own!

Dr. Colleen Cavanaugh is studying the tube worm's ability to survive and reproduce under the harsh conditions of the Pacific Ocean floor.

GAZETTE

Colleen Cavanaugh Explores the Underwater World of

TUBE WORMS

160 ■ GAZETTE

The seabed lies 2500 meters below the ocean's surface. Here, the pressure is nearly 260 times that at the Earth's surface, and the temperature is close to the freezing point. The region is always dark, as sunlight cannot penetrate these ocean depths.

Almost all organisms depend on light as their source of energy in manufacturing food. So scientists had expected to find only the simplest creatures inhabiting this dark area of the Pacific Ocean floor near the Galápagos Islands. To the surprise of a group of scientists from the Woods Hole Oceanographic Institution, however, communities of strange sea animals were found in this forbidding environment.

Among the animals observed by the Woods Hole team are giant clams that measure one-third meter in diameter, oversized mussels, and crabs. But perhaps the most striking organisms are giant tube worms. These worms are so different from anything seen before that scientists have placed them in a new family of the animal kingdom.

Although most of the worms are only 2.5 to 5 centimeters in diameter, they can be as long as 2 meters! Like a sausage in a tube, each worm lives inside a tough rigid casing. One end of the tube worm is anchored to the seabed. At the other end, a red plumelike structure made of blood-filled filaments waves in the ocean water.

Internally, the body of the worm is most unusual. Approximately half of its body is made of a colony of densely packed bacteria. These bacteria are the key to a worm's ability to survive in such a harsh environment. But exactly what is the relationship between the bacteria and the tube worms?

Enter Dr. Colleen Cavanaugh, a marine biologist and microbiologist who studies these bacteria and their relationship to the larger life forms they support. She has found

---

TEACHING STRATEGY: ADVENTURE

FOCUS/MOTIVATION

Show students a piece of garden hose about 2 meters long and an earthworm.
• **How are these two things different?** (Accept all logical answers.)
• **How are they alike?** (Accept all logical answers.)

• **How might a worm as large as the piece of hose be different from a common earthworm?** (Accept all logical answers.)
• **Where do you think a worm as large as a hose might live?** (Accept all logical answers.)

CONTENT DEVELOPMENT

After students have read the article, point out that the Galapagos Islands are near the equator in the Pacific Ocean. Point out that they lie just off the west coast of South America.

Explain that many different or unusual kinds of animals have been found around the Galapagos Islands. Point out that perhaps this is why studies were being done in that area of the world.
• **What were some of the unusual kinds of animals that this expedition found?** (Oversized mussels, crabs, and tube worms.)
• **Which of these animals did the scientists observe to be the most unusual? Why?** (The tube worms—because the body was packed with bacteria.)

that the bacteria break down hydrogen sulfide, a common sulfur compound, to produce energy. This energy is then used by the tube worms. In this way, the seemingly worthless hydrogen sulfide—which smells like rotten eggs and is poisonous to most organisms—provides the energy for a fascinating aquatic ecosystem.

Hydrogen sulfide is not ordinarily found in sea water. But plenty of hydrogen sulfide is found near hydrothermal vents such as those in the Pacific Ocean areas where the tube worms were first discovered. Hydrothermal vents are openings in the ocean floor that allow ocean water to come into contact with the earth's hot, molten interior. In the vents, ocean water is heated to temperatures as high as 350 degrees Celsius. The hot water is ejected back into the ocean, laced with many different chemical compounds, including hydrogen sulfide. Tube worms and other animals that are able to use sulfur abound in these regions.

Research on tube worms requires interest, dedication, and a broad knowledge of science. So it is not surprising that Colleen Cavanaugh is part of the team at Woods Hole exploring this fascinating underwater world. As a high-school student growing up in Michigan, Colleen Cavanaugh explored many fields of science. She then went on to study general ecology and biology at the University of Michigan. As a graduate student at Harvard University, she developed an interest in microbial ecology. Now she combines these three fields, as well as others such as oceanography, in her work. The work Dr. Cavanaugh has already done on tube worms will be of help in her more general study of the nature of symbiosis, or cooperation, between animals and bacteria. As she points out, the sulfides and sulfide-based ecosystems such as that of the tube worms are not limited just to the ocean depths. They are also found in salt marshes, mud flats, and many other marine environments closer to home. Thus, scientific work in the dark reaches of the sea may shed new light on the nature of life on the surface as well.

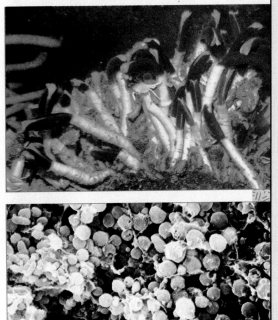

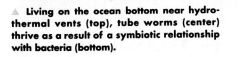

▲ **Living on the ocean bottom near hydrothermal vents (top), tube worms (center) thrive as a result of a symbiotic relationship with bacteria (bottom).**

Additional Questions and Topic Suggestions

1. What made scientists look in this area of the Pacific Ocean? (Darwin first found and studied many unusual animals in this area of the world. Later, from his studies of the many unusual animals of the Galapagos Islands, he developed the theory of evolution.)

2. Why do you predict there are so many unusual animals in that area of the world? (Accept all logical answers.)

3. What kind of energy does hydrogen sulfide produce for the tube worm? (Accept all logical answers. The energy is stored in hydrogen sulfide gas as chemical energy that bonds the hydrogen and sulfur together to form the hydrogen sulfide molecules.)

Critical Thinking Questions

1. Why could the information Colleen Cavanaugh discovers about how the bacteria convert hydrogen sulfide to energy have an effect on life itself? (Accept all logical answers. Because her study might prove that animal life can be sustained by an element other than oxygen, or that bacteria with the help of some element can produce oxygen from other substances.)

2. Why does the environment around the Galapagos Islands allow unusual animals to evolve? (Accept all logical answers.)

Point out that Colleen Cavanaugh is studying these bacteria because they are capable of breaking down hydrogen sulfide into a source of energy. Explain that most animal life needs oxygen to utilize food energy.

Point out that the tube worm uses the energy released when bacteria break down hydrogen sulfide.

GUIDED PRACTICE

Skills Development

Skill: Making predictions

• **What might happen to tube-worm evolution if the amount of hydrogen sulfide gas in the Galapagos water increases?** (Accept all logical answers. The tube worms might evolve into even larger worms or might become more prevalent.)

• **What might happen to tube-worm evolution if the amount of hydrogen sulfide gas in the Galapagos water decreases?** Accept all logical answers. The tube worms might evolve into smaller worms, use a different energy source, become less prevalent, or even become extinct.)

INDEPENDENT PRACTICE

▶ *Activity Book*

After students have read the Science Gazette article, you may want to hand out the reading skills worksheet based on the article in the *Activity Book.*

ISSUES IN SCIENCE

PESTS OR PESTICIDES: WHICH WILL IT BE?

Background Information

Because of the use of pesticides, approximately one-third less food is lost due to pests in the United States than is lost in the rest of the world. Because of the use of pesticides, food prices in the United States are 30 percent to 50 percent lower than they would be otherwise. Pesticides have increased the amount of food production per hectare of land, making it possible to grow more food in less space.

So much for the good news. The bad news is that pesticides can and do damage the environment. Because it is difficult to make a pesticide that is totally specific, organisms that are helpful to human beings and food production may be accidentally killed. Pesticides can upset the ecological balance in an area by accidentally destroying the natural enemies of pests. If a pesticide is released into the environment carelessly or in excessive amounts, it can pollute land, air, and water and cause illness or death to animals, fish, plants, birds, and humans. In addition, pesticides can actually cause an increase in the population of certain pests by forcing the natural selection of the hardiest members of the species. These pests ultimately produce a generation that is resistant to the pesticide and is stronger than ever.

Another problem with some pesticides is that they break down slowly and, therefore, remain in the environment for many years. Most hazardous in this way are the chlorinated hydrocarbons, which include such insecticides as DDT, aldrin, dieldrin, and heptachlor. In the 1970s, the Environmental Protection Agency (EPA) banned the widespread use of these insecticides, although their use is permitted in an emergency.

Since the banning of the chlorinated hydrocarbons, major changes have taken place in pest-control research. Today, nearly 70 percent of the Department of Agriculture's budget for pest-control research is devoted to finding alternatives to pesticides. These alternatives include biological control, which introduces the natural enemies of pests into the environment;

PESTS OR PESTICIDES:
WHICH WILL IT BE?

▲ This greatly enlarged photo of a Mediterranean fruit fly was made with a scanning electron microscope. The fruit fly feeds on food crops such as oranges, tomatoes, walnuts, and peaches. But attacking it with chemical insecticides may do even greater damage.

here is a war being fought right now that has been going on for thousands of years. It is a fight for survival against armies that far outnumber the Earth's entire human population. It is the war people wage against pests, such as certain insects, rats, mice, weeds, and fungi.

Some insects, rats, and mice carry deadly diseases such as malaria, bubonic plague, and typhoid fever. Throughout history, diseases carried by pests have killed hundreds

of millions of people. These animals also threaten food supplies around the world.

Other threats to world food supplies are weeds and fungi. Weeds compete with crops for water and nutrients. Some fungi cause plant diseases that result in huge crop losses. If people lose the war against these organisms, human disease, suffering, and starvation will occur worldwide.

Today people have many weapons to use in the fight against pests. During the past few decades, scientists have made or discovered many chemicals that kill pests. These chemicals, which are called pesticides, have

---

TEACHING STRATEGY: ISSUE

FOCUS/MOTIVATION

Display items such as the following: a mousetrap, a can of aerosol insecticide, flypaper, weed killer, and mothballs.
• **What do all these items have in common?** (Accept all answers.)

CONTENT DEVELOPMENT

Point out that all the displayed items represent methods of pest control. Not all the items are pesticides, however. Pesticides are, by definition, chemicals used to kill pests.

Pesticides can be classified according to the type of pest they seek to kill. Insecticides are directed at insects; rodenticides, at rodents such as mice or rats; fungicides, at fungi; and herbicides, at weeds.

helped to increase food production by killing the pests that destroy crops and eat stored food. Pesticides have also saved many human lives by killing disease-carrying pests.

Unfortunately, people have not always used pesticides wisely. Farmers and others who use pesticides have accidentally killed useful organisms such as honeybees, spiders, songbirds, fish, falcons, dogs, and cats. In addition, the overuse of pesticides has made some pests resistant to the pesticides. In other words, the pests have become able to withstand assaults of deadly chemicals and are now much harder to kill.

To make matters worse, pesticides are sometimes washed into rivers and streams, where they kill fish and other animals. Winds can spread pesticides over hundreds or thousands of kilometers. This pollutes areas far from where the chemical was used to kill pests. Pesticides have injured and killed people in all parts of the world.

So now we face a difficult problem. How can we save crops and kill disease-carrying pests without harming the environment?

THE OTHER SIDE OF PESTICIDES

There are now about 35,000 different pesticides on the market. Each year in the Unit-ed States alone, about one-half billion kilograms of these chemicals are used by farmers, homeowners, and industry.

Specific pesticides called herbicides and fungicides have helped farmers to reduce crop losses due to weeds and harmful fungi. Insect-killing pesticides, or insecticides, have not been as successful in reducing losses. In fact, the amount of crops destroyed by harmful insects has nearly doubled during the past 30 years—even though farmers have been using more powerful insecticides in greater quantities than ever before.

Insecticides are also failing to protect people from insect-borne diseases. For example, malaria, a sometimes fatal disease that is carried by certain mosquitoes, has made an alarming comeback in countries where it had been practically wiped out.

Why are the insecticides failing to protect people and crops? One reason is that the use of insecticides has caused harmful insects to become resistant to them. When an area is sprayed with pesticides, most of the pests are killed. A few individuals, however, may be naturally resistant to the pesticide and may survive. These survivors produce offspring that, like themselves, are resistant. In a short time, all of the pests are resistant.

Another reason is that insecticides can also kill the natural enemies of insect pests. For example, insecticides have killed ladybugs

the use of natural pesticides, which can be extracted from plants or animals that produce substances to ward off their natural enemies; the genetic engineering of pest-resistant crops; and the use of cultivation methods that discourage the buildup of certain pests. Also under consideration is the use of integrated pest management. Integrated pest management seeks to combine many techniques of pest control (including limited use of pesticides when necessary) in as environmentally compatible a manner as possible.

Additional Questions and Topic Suggestions

1. Have you had any personal experiences with pest control in which there were unwanted side effects? Describe what happened and your feelings about the incident. (Accept all answers. Students may have had experiences such as using an insect repellent, only to find that the fumes made them feel sick or caused illness in a pet cat or dog.)

2. Go to the supermarket, hardware store, or local garden center and find out what kinds of pesticides are being sold. Read the labels and note the contents of each product, how the product is to be used, and the kind of pest it is intended to kill. Also note any cautions or warnings that might be included about using the pesticide.

3. Get together with several classmates and research the kinds of pests most often found in your state. Then make a map to show where these pests are concentrated.

▼ **One way to combat insect pests is to spray them from the air with chemical insecticides. But this method, called crop dusting, also affects living things other than pests.**

GAZETTE ■ 163

CONTENT DEVELOPMENT

Emphasize to students the need for some form of pest control. Point out that approximately half of the food produced in the world each year is lost as a result of the damage caused by pests. The variety of pests is amazing—in the United States alone, there are approximately 19,000 different species of agricultural pests.

Another problem with pests is that they can bring disease to humans. For example, malaria is a serious, sometimes fatal disease that is transmitted from person to person by a type of mosquito.

REINFORCEMENT/RETEACHING

Review with students the alternatives to pesticides discussed in the textbook. One of these alternatives is understanding how crops and pests interact with their environment. Another alternative involves cultivation techniques, such as crop rotation. A third alternative is the use of natural chemicals, such as the oil found in oranges.

Class Debate

Divide the class into teams of four to six students. Challenge each team to imagine that they represent members of the Environmental Protection Agency (EPA). Ask students to dramatize a meeting of the agency in which they debate the pros and cons of new laws governing the use of pesticides. You may wish to have students base their discussion on actual laws that have been recently enacted by the EPA, or you may wish to have students write their own laws based on the issues discussed in this article.

▲ Farmers use pesticides against such destructive insects as the tobacco hornworm. The hornworm attacks many kinds of crops, including these tomato plants. However, pesticides can also harm helpful animals and even people.

in some apple orchards. Now there are no more ladybugs in these orchards to keep apple-eating insects, such as aphids, under control.

Because of the harmful effects of pesticides, some people believe that these chemicals are too dangerous to use. But those in favor of using pesticides disagree. They say that pesticides would not be so dangerous if people used them properly.

NEW WEAPONS

One way to lessen our dependence on pesticides is to understand how crops and insects interact with their environments. The study of how living things interact with their environments is called ecology. Ecology provides clues to how to control pests by changing the environment of either the pest or its victim.

For example, a plant called barberry often carries a fungus that causes a disease called black stem rust. When barberry grows near wheat, the rust spreads from the barberry to the wheat. Some farmers have protected their wheat from black stem rust by destroying nearby barberry plants rather than by spraying fungicide.

Other diseases can be controlled by crop rotation. Periodically planting crops other than cabbage in a cabbage field is an example. This method prevents disease-causing organisms that attack only cabbage from building up in the soil.

Pests may also be controlled by finding naturally-occurring things that are harmful only to the pests—preferably things to which the pests cannot become resistant. For example, Japanese beetles are killed by dusting an area with bacteria that causes a fatal beetle disease. Other insects are sprayed with synthetic versions of their own body chemicals. This upsets their normal chemical balance and causes them to grow abnormally and then to die. And because no pest can build up a resistance to being eaten, farmers may introduce helpful insect-eating animals—praying mantises, spiders, and ladybugs, for example—into their fields and orchards.

Not all new types of pest control are so down-to-earth, however. By studying how the crop-eating desert locust interacts with its environment, scientists have been able to use satellites to fight this pest.

The desert locust is a flying insect related to the grasshopper. From time to time, millions of these locusts gather and sweep across Africa and India, eating every crop and blade of grass in their path. No weapons have been able to stop these insects once they take flight. But in recent years, satellites orbiting the Earth have been taking pictures of areas in Africa and India where locusts might breed. Scientists can identify possible breeding areas by the amount of moisture they contain. If satellite photos show that an area is moist enough for locust eggs to survive there, scientists warn the people that may be threatened. The people can then concentrate their pesticide spraying in these areas.

The better we understand how living things interact with their environments, the more clues we will find for controlling pests without using large amounts of chemicals. Perhaps someday we will not have to use chemicals at all!

164 ■ GAZETTE

ISSUE (continued)

INDEPENDENT PRACTICE

▶ *Activity Book*

After students have read the Science Gazette article, you may want to hand out the reading skills worksheet based on the article in the *Activity Book*.

The Corn Is As High As A Satellite's Eye

Farms of the future will move into space, and crops will be far out, too.

The old man held his granddaughter's hand as the high-speed elevator zoomed 180 floors to the top of the Triple Towers. Out on the observation deck, the young girl stared in wonder at the huge city that fanned out to meet the horizon in all directions. Far below, crowds of little dots moved along in neat columns. "Look at all those people!" the girl exclaimed. "There must be millions of them."

The old man nodded. "Twenty-two million, to be exact," he said softly.

GAZETTE ■ 165

FUTURES IN SCIENCE

THE CORN IS AS HIGH AS A SATELLITE'S EYE

Background Information

A major concern of twentieth-century scientists is how to increase and improve food production so as to accommodate a growing world population. A great obstacle to farming is lack of adequate land. The article states that one-quarter of the Earth's land is used for farming; the precise figure is 30 percent. Of the land that is not suitable for farming, 20 percent is too dry; 20 percent is too mountainous; 20 percent is sea- or snow-covered; and 10 percent is land without topsoil.

Clearly, it is to the advantage of agriculturists to maximize the use of available farmland and also to find alternative areas that can be used for farming. The article emphasizes the possibility of farming in outer space. Although it is true that this idea is being developed, its application is aimed primarily at providing food for people in space colonies or for astronauts who spend extended periods (for example, two years) in space. Much more germane to life on Earth are the techniques of genetic engineering that seek to maximize the yields of essential food crops. Also important are the use of alternative growing methods such as hydroponics (growing plants without soil) and aquaculture (farming in the ocean).

Several techniques of future farming not discussed in the article include the use of genetic engineering to improve the nutritional quality of food; the use of plant "vaccines" to immunize plants against certain diseases and other environmental hazards; and the genetic engineering of microorganisms that could be used to target insects, weeds, and other pests that are harmful to crops.

TEACHING STRATEGY: FUTURE

FOCUS/MOTIVATION

• **Which do you think would produce more fruit—a tree 2 1/2 m tall, or a tree 1 m tall?** (Answers may vary, but most students will probably say the tree 2 1/2 m tall.)

Additional Questions and Topic Suggestions

1. Make a graph to show the percentages of the Earth's land that are suitable for farming. For the land that is not suitable for farming, find out what percentages of land are too dry, too mountainous, lacking in topsoil, frozen, and underwater.

2. Farming in the ocean is called aquaculture. Find out where in the world aquaculture is being used and what kind of crops are grown.

3. Write a science-fiction adventure story about farming in space. You may wish to use the idea of a space colony in which people farm and carry out other occupations.

▲ Short food plants, such as lettuce and many other kinds of vegetables, can be grown within revolving drums. On Earth, these devices make it possible for many plants to be grown in a small amount of space. In the future, similar devices may simulate gravity and thereby enable orbiting space stations to grow food crops.

"But Grandpa, how do all these people keep from starving? I read that years ago people in cities all over the world couldn't get enough to eat. And there are even more people now than there were back then."

The old man frowned as he remembered his own youth. "Yes, that's true, Lisa," he said with a sigh. "Twenty percent of the world was going hungry when I was a young man in 1992. The farmers could use only a quarter of the Earth's land for farming."

"Why couldn't they use more of the land, Grandpa?" Lisa asked.

"Well," he answered, "the rest was either sizzling hot deserts, barren mountains, or frozen tundras. Of the available farmlands, some has to be used for nonfood crops like cotton. Less food was available and more and more people were being born."

"Are people still starving, Grandpa?" Lisa asked in a worried voice.

"Well, dear, some are, but because of techniques that were developed during the late

twentieth century, farmers can now produce more food."

"Oh, Grandpa, that sounds interesting. Will you tell me how more food is made?"

"Of course I will, Lisa. Let's sit down on that bench and I'll tell you all about it," the old man said with a smile.

SPACE CROPS

"Look up there. What do you see?" the old man asked.

"Stars, Grandpa. Why?"

"Well, Lisa, much of the world's food is now being grown in space. We have space stations that contain rotating drums filled with plants. The rotation simulates the Earth's gravity. The plants are bathed in fluorescent light, which simulates the sun. Crops that do not grow very tall or that grow in the ground, like lettuce, potatoes, radishes, and carrots, are grown in the drums," the old man stated proudly.

"Wow, that's neat, Grandpa. I never knew that salad came from outer space," Lisa said excitedly.

Laughing, the old man said, "Well, most of it does. We also have space stations where plants grow out of the sides of walls or are moved along conveyor belts through humid air."

GENETICS MAKES THE DIFFERENCE

"Is all our food grown in outer space, Grandpa?" asked Lisa, curious.

"Why, of course not. Methods of growing plants on Earth have been improved too. When I was young, Lisa, 20 percent of the crops that were grown on Earth died off. Now, because of changes scientists have made in the genetic material of plants, very few crops die. The genetic changes have resulted in plants with increased resistance to temperature changes, diseases, herbicides, and harsh environments. Plants have been

FUTURE (continued)

CONTENT DEVELOPMENT

Use the previous question to introduce the idea of "dwarf" tress that can produce as much or more fruit than trees that are much taller.

• **Assuming that the quality of the fruit is the same, what would be the advantage of the smaller trees?** (Answers may vary. Guide students to understand that smaller trees take up less space. Also, though less important, smaller trees might be easier and more efficient to harvest.)

Point out that the development of dwarf fruit trees is just one application of a technique called protoplast fusion, which is a form of genetic engineering. In protoplast fusion, the walls of cells from two different types of plants are dissolved by an enzyme. The contents of the cells are called protoplasts. Once the cell walls have been dissolved, the protoplasts can

be combined, or fused, to bring together genetic material of species that could not otherwise be crossed. After the two protoplasts are fused, they are placed in a solution that stimulates the growth of a new cell wall.

CONTENT DEVELOPMENT

You may wish to point out to students that a forerunner of protoplast fusion is a process called tissue culture. In this process scientists use tiny pieces of plant tissue to reproduce cells that later develop

into full-sized plants. Plants produced in this way have the advantage of being genetically uniform and of taking less time to reach maturity than plants grown from seeds. In addition, chemicals can be added to the cell solution in order to produce plants that are stronger and more resistant to disease.

ENRICHMENT

Another way in which scientists hope to improve crops is by genetically engineering new, environmentally safe microor-

made resistant by several methods. Some plants were altered by a process called protoplast fusion," the old man said.

"That sounds complicated."

"It's not really very complicated," he replied. "The plant cell walls are dissolved by enzymes. The part remaining is called a protoplast. This protoplast is then fused to another protoplast from a different strain of the same plant or from an entirely different plant. The cell wall is regrown around the fused protoplasts, and a new plant develops. We now have many varieties of such plants. And you might think it strange, but many are smaller than those that used to be grown."

"Smaller? But that doesn't make sense," Lisa protested.

"Well, the smaller plants can produce as much as or more than the big ones because they have been genetically changed to be highly productive," explained the old man. "Also, more of them can be grown in the same amount of space. Two examples of the smaller plants are dwarf rice plants and dwarf peach trees."

"Wow, that's neat. Tell me more," Lisa said.

"All right. We also have plants that can get the nitrogen they need directly from the air. Plants need nitrogen to build amino acids, proteins, and other cell chemicals. In the old days, most plants got their nitrogen from the soil. But this process was using up all the nitrogen in the soil, and the plants didn't grow well. So fertilizer had to be added to the soil. Soybeans, legumes, alfalfa, and clover have always been able to use nitrogen directly from the air with the help of nitrogen-fixing bacteria. Other plants that can use nitrogen directly from the air were developed by combining genes from these nitrogen-fixing bacteria with genes from plants that normally took their nitrogen from the soil. Now almost all the plants take their

▲ Dwarf trees being grown right now may one day supply enough fruit to fill most of our needs. What is the advantage of growing dwarf plants?

nitrogen directly from the endless supply in the air."

"I didn't know that. Are any other plants different from the plants you saw when you were young?" Lisa asked.

"Oh, yes," the old man declared. "Many of the crops are now considered to be high-yield. That means they can be harvested faster than they could before, so that more can be grown. By redesigning the plants genetically, scientists have shortened the plants' growing time."

"It seems that genetic changes have kept people from starving," Lisa said thoughtfully. "That and the growing of food in space or on land."

"That's true," said her grandfather. "But there's still a third place. Some of the crops are grown in the ocean. This leaves more room for crops that can be grown only on land."

"That's a neat story, Grandpa!"

"It's not a story, Lisa. It's the truth. But scientists still have to keep working to improve crop production because the population continues to grow," the old man observed.

GAZETTE ■ 167

Critical Thinking Questions

1. How might farms in space extend and improve the science of space exploration? (Instead of having to carry with them the food needed for a space mission, astronauts would be able to grow their own food. This would certainly increase the amount of time that an astronaut could stay away from Earth. In addition, the presence of farms in space might eventually make it possible to develop "space colonies"—groups of people who actually live in space.)

2. How do you think the occupation of farming might change as a result of the innovations discussed in this article? What type of person might be attracted to farming in the future? (Accept all answers. It is likely that farming will require knowledge of genetically engineered plants and new growing techniques; some farmers may want to actually carry out processes such as protoplast fusion; still other farmers may travel to outer space! It is likely that a person with scientific curiosity and a taste for adventure and experimentation may be attracted to farming in the future.)

3. According to the article, can you expect to see those innovations in farming during your lifetime? How can you tell? (Yes. The story is being told by an older man who was a young man in 1992.)

ganisms that can attack and destroy specific weeds, insects, and other pests. Have students find out how, in 1981, the microorganism *Bacillus thuringiensis* was used to destroy gypsy moths in the northeastern United States.

ENRICHMENT

The growing of plants without soil is called *hydroponics*. Have students find out the techniques of hydroponics, where this method is being used, and how successful

the method has been. Students may be able to gain information on this topic by contacting the Kraft Foods company, which is experimenting with hydroponics, or by contacting the EPCOT in Orlando, Florida.

REINFORCEMENT/RETEACHING

Review with students the four ways discussed in this article that scientists hope to improve food crops through the use of genetic engineering.

1. By developing smaller plants that can produce as much or more food in less space.

2. By producing plants that are stronger so that space is not wasted by plants that die.

3. By producing plants that can use nitrogen directly from the air.

4. By producing plants that require shorter growing time.

INDEPENDENT PRACTICE

▶ *Activity Book*

After students have read the Science Gazette article, you may want to hand out the reading skills worksheet based on the article in the *Activity Book*.

For Further Reading

> If you have been intrigued by the concepts examined in this textbook, you may also be interested in the ways fellow thinkers—novelists, poets, essayists, as well as scientists—have imaginatively explored the same ideas.

Chapter 1: Classification of Living Things

Jarrell, Randall. *The Animal Family.* New York: Pantheon.

Paterson, Katherine. *Jacob Have I Loved.* New York: Avon.

Chapter 2: Viruses and Monerans

Christopher, John. *No Blade of Grass.* New York: Simon & Schuster.

De Kruif, Paul. *Microbe Hunters.* New York: Harcourt, Brace & World, Inc.

Chapter 3: Protists

Foster, Alan Dean. *The Thing.* New York: Bantam.

Perez, Norah A. *The Passage.* Philadelphia: Lippincott.

Chapter 4: Fungi

Cameron, Eleanor. *The Wonderful Flight to the Mushroom Planet.* Boston: Little Brown and Co.

Hughes, William Howard. *Alexander Fleming and Penicillin.* Hove, U.K.: Wayland. LTD.

Jones, Judith, and Evan Jones. *Knead It, Punch It, Bake It! Make Your Own Bread.* New York: Thomas Y. Crowell.

Chapter 5: Plants Without Seeds

Abels, Harriette E. *Future Food.* Mankato, MN: Crestwood House.

Kavaler, Lucy. *The Wonders of Algae.* New York: The John Day Co.

Sterling, Dorothy. *The Story of Mosses, Ferns, and Mushrooms.* Garden City, NY: Doubleday & Co.

Chapter 6: Plants With Seeds

Busch, Phyllis B. *Wildflowers and the Stories Behind Their Names.* New York: Charles Scribner's Sons.

Elbert, Virginie Fowler. *Grow a Plant Pet.* Garden City, NY: Doubleday & Co.

Activity Bank

Welcome to the Activity Bank! This is an exciting and enjoyable part of your science textbook. By using the Activity Bank you will have the chance to make a variety of interesting and different observations about science. The best thing about the Activity Bank is that you and your classmates will become the detectives, and as with any investigation you will have to sort through information to find the truth. There will be many twists and turns along the way, some surprises and disappointments too. So always remember to keep an open mind, ask lots of questions, and have fun learning about science.

Chapter 1	**CLASSIFICATION OF LIVING THINGS**	
	A KEY TO THE PUZZLE	**170**
Chapter 2	**VIRUSES AND MONERANS**	
	A GERM THAT INFECTS GERMS	**171**
	YUCK! WHAT ARE THOSE BACTERIA DOING IN MY YOGURT?	**172**
Chapter 3	**PROTISTS**	
	PUTTING THE SQUEEZE ON	**174**
	SHEDDING A LITTLE LIGHT ON EUGLENA	**176**
Chapter 4	**FUNGI**	
	SPREADING SPORES	**177**
	YEAST MEETS BEST	**178**
Chapter 5	**PLANTS WITHOUT SEEDS**	
	SEAWEED SWEETS	**180**
Chapter 6	**PLANTS WITH SEEDS**	
	THE INS AND OUTS OF PHOTOSYNTHESIS	**181**
	BUBBLING LEAVES	**183**
	LEAN TO THE LIGHT	**184**

B ■ 169

Activity Bank

COOPERATIVE LEARNING

Hands-on science activities, such as the ones in the Activity Bank, lend themselves well to cooperative learning techniques. The first step in setting up activities for cooperative learning is to divide the class into small groups of about 4 to 6 students. Next, assign roles to each member of the group. Possible roles include Principal Investigator, Materials Manager, Recorder/Reporter, Maintenance Director. The Principal Investigator directs all operations associated with the group activity, including checking the assignment, giving instructions to the group, making sure that the proper procedure is being followed, performing or delegating the steps of the activity, and asking questions of the teacher on behalf of the group. The Materials Manager obtains and dispenses all materials and equipment and is the only member of the group allowed to move around the classroom without special permission during the activity. The Recorder, or Reporter, collects information, certifies and records results, and reports results to the class. The Maintenance Director is responsible for cleanup and has the authority to assign other members of the group to assist. The Maintenance Director is also in charge of group safety.

For more information about specific roles and cooperative learning in general, refer to the article "Cooperative Learning and Science—The Perfect Match" on pages 70–75 in the *Teacher's Desk Reference*.

ESL/LEP STRATEGY

Activities such as the ones in the Activity Bank can be extremely helpful in teaching science concepts to LEP students—the direct observation of scientific phenomena and the deliberate manipulation of variables can transcend language barriers.

Some strategies for helping LEP students as they develop their English-language skills are listed below. Your school's English-to-Speakers-of-Other-Languages (ESOL) teacher will probably be able to make other concrete suggestions to fit the specific needs of the LEP students in your classroom.
• Assign a "buddy" who is proficient in English to each LEP student. The buddy need not be able to speak the LEP student's native language, but such ability can be helpful. (**Note:** *Instruct multilingual buddies to use the native language only when necessary, such as defining difficult terms or concepts. Students learn English, as all other languages, by using it.*) The buddy's job is to provide encouragement and assistance to the LEP student. Select buddies on the basis of personality as well as proficiency in science and English. If possible, match buddies and LEP students so that the LEP students can help their buddies in another academic area, such as math.
• If possible, do not put LEP students of the same nationality in a cooperative learning group.
• Have artistic students draw diagrams of each step of an activity for the LEP students.

You can read more about teaching science to LEP students in the article "Creating a Positive Learning Environment for Students with Limited English Proficiency," which is found on pages 86–87 in the *Teacher's Desk Reference*.

Activity Bank

BEFORE THE ACTIVITY

1. Prior to this activity, you may want students to perform the activity called Sorting It Out on page 181 of the Activity Book so that they will have experience working with classification diagrams.
2. Divide the class into groups of three to six students.

PRE-ACTIVITY DISCUSSION

Introduce this activity by pointing out that classification diagrams, such as the one in the Laboratory Investigation on p. B30 and in Sorting It Out, are handy for showing how a large group is broken into smaller groups—provided, of course, that you don't have too many groups!
• **What happens if there are a lot of groups in a classification diagram?** (If there are a lot of groups, the diagrams quickly become large, complicated, and difficult to use.)

Tell students that they will explore a better way of showing how things are classified in this activity.

TEACHING STRATEGY

1. You may wish students to work individually on questions 1 and 2 so that you will be able to determine if any individuals are having trouble using taxonomic keys.
2. Encourage students to modify their taxonomic keys in response to the feedback they get from their classmates. Have the groups discuss how and why the keys need to be changed. You may want to review the concept of constructive criticism before the groups begin their discussions.

DISCOVERY STRATEGIES

Discuss how the activity relates to the chapter ideas by asking questions similar to the following.
• **Why are classification systems useful?** (They organize and show the relationships among large numbers of objects.)
• **How might the discovery of new organisms affect a classification tree? A taxonomic key?** (Answers will vary. Students may suggest that new categories may need to be made if the organisms do not clearly fit into existing categories in a classifi-

One way of showing how objects are classified is a taxonomic key. A taxonomic key consists of many pairs of opposing descriptions. Only one of the descriptions in a pair is correct for a given object. Following the correct description is an instruction that directs you to another pair of descriptions. By following each successive description and instruction in a taxonomic key, you will eventually arrive at an object's correct classification group.

Don't worry—this sounds more complicated than it actually is. Let's take a look at a simple taxonomic key for the five kingdoms.

> 1a Made up of cells that contain a nucleus. Go to 2.
> 1b Made up of one cell that does not contain a nucleus. *Monera.*
>
> 2a Is unicellular. Go to 3.
> 2b Is multicellular. Go to 4.
>
> 3a Belongs to the green, red, or brown algae phylum. *Plantae.*
> 3b Does not belong to the green, red, or brown algae phylum. Go to 7.
>
> 4a Is an autotroph. *Plantae.*
> 4b Is a heterotroph. Go to 5.
>
> 5a Evolved from autotrophs. *Plantae.*
> 5b Did not evolve from autotrophs. Go to 6.
>
> 6a Has a cell wall. *Fungi.*
> 6b Does not have a cell wall. *Animalia.*
>
> 7a Is a yeast. *Fungi.*
> 7b Is not a yeast. *Protista.*

1. Using the taxonomic key for the five kingdoms, determine the kingdom to which each of the organisms described in the following stories belongs.
a. While looking at a sample of pond water through a microscope, you notice a rod-shaped unicellular heterotroph that has a cell wall and no nucleus.
b. In the same sample, you see a unicellular heterotroph that has a nucleus and moves by means of whiplike "tails." Because yeasts don't have "tails," you know it is not a yeast.
c. While walking through the woods one day, you notice some white, candy-cane-shaped multicellular heterotrophs growing on a dead log. The cells contain a nucleus and have a cell wall. Chemical study of the hereditary material reveals that they are descended from autotrophs.
d. In a tide pool, you see an autotroph that looks like sheets of green cellophane. It is made up of many cells that contain a nucleus. It looks exactly like the photograph in your field guide of the green alga known as sea lettuce.
e. Further examination of the organisms in the tide pool reveals a small, nonmoving, multicellular blob. At first, you think it's some kind of seaweed, but you later find out that it is a heterotroph whose cells lack cell walls.

2. According to the taxonomic key, how are animals and fungi similar? Different?

3. Invent a taxonomic key for pets. Test your key by using it to determine the kind of pet a friend or classmate is thinking of.

cation tree. New questions may need to be added to a taxonomic key.)

ANSWERS

1. a. Monera **b.** Protista **c.** Plantae **d.** Plantae **e.** Animalia
2. They are similar in that both are multicellular heterotrophs. They are different in that fungi have cell walls.
3. Student keys will vary, but should be complex enough to classify most typical pets.

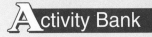

A GERM THAT INFECTS GERMS

In everyday speech, the term germ refers to any microorganism that causes disease. Because bacteriophages infect bacteria, you can say that they are germs that infect germs! In this activity you will build a model of a bacteriophage so that you can better understand how it works.

1. Use the diagram on page B40 and any of the following materials to build your own model bacteriophage: pipe cleaners, construction paper, screws, nuts, bolts, scissors, tape, crayons, glue, screwdriver.

2. Make a drawing of your model and label the parts. Which part contains the hereditary material? Which part would the bacteriophage use to attach itself to a bacterial cell?

nanometers, or billionths of a meter, and range in size from about 10 to 300 nm across.)

• **What is a bacteriophage?** (A virus that infects bacteria.)

• **Describe the life cycle of a typical bacteriophage virus.** (First, a bacteriophage attaches to a bacterium and injects its hereditary material. The bacteriophage's hereditary material takes control of the bacterium's functions, causing bacteriophage parts to be made. The bacteriophage parts assemble into complete bacteriophages. Finally, the bacterium bursts, releasing the bacteriophages.)

• **Why are viruses considered to be parasites?** (They live inside another living thing and do it harm.)

TEACHING STRATEGY

1. Encourage students to be creative within the limits set by the assignment. Point out that there are many different, equally correct solutions to the problem of making a helpful, reasonably accurate model. Note that they can use as many or as few of the materials as they wish—they need not incorporate all the materials into the model.

2. Circulate around the classroom and make sure that all the members of each group are participating in designing and building the group's model.

DISCOVERY STRATEGIES

Discuss how the activity relates to the chapter ideas by asking questions similar to the following.

• **How are viruses similar to one another?** (Student responses should accurately reflect the material in the textbook. Viruses are parasites that require a host cell to carry out their life functions. All viruses consist of a core of hereditary material surrounded by a protein coat.)

• **How are viruses different from one another?** (Student responses should accurately reflect the material in the textbook. Viruses differ in shape, size, the kind of cells they infect, and the effects they have on their host.)

ANSWERS

Check student models and drawings for accuracy. In a bacteriophage, the head contains the hereditary material. A bacteriophage uses its tail to attach itself to a bacterial cell.

Activity Bank

A GERM THAT INFECTS GERMS

BEFORE THE ACTIVITY

1. Divide the class into groups of three to six students. Students may also perform this activity individually.

2. Gather all materials at least one day prior to the activity.

3. You may wish to provide illustrations of different kinds of viruses so that each group makes a different model. If you select viruses that are not bacteriophages, be sure that students know which viruses are bacteriophages and which are not.

PRE-ACTIVITY DISCUSSION

Review basic concepts about viruses by asking questions such as the following.

• **What is a virus?** (Tiny noncellular particles, typically disease-causing, that can invade living cells.)

• **How large are viruses?** (Viruses are extremely small. They are measured in

Activity Bank

BEFORE THE ACTIVITY

1. Divide the class into groups of three to six students. You may wish to assign roles to the members of the group based on individual student strengths and needs.
2. Gather all materials at least one day prior to the activity.

BACKGROUND INFORMATION

Casein, a protein in milk, forms spherical globules that are covered with negatively charged amino acids. Because like charges repel each other, casein molecules stay away from each other and remain suspended in milk. When acid is added to milk, the negative charges are neutralized. This makes it possible for the casein molecules to clump together and form insoluble lumps.

Yogurt-making bacteria obtain the energy they need by digesting lactose (milk sugar) to produce lactic acid, carbon dioxide, and water. When enough lactic acid accumulates to lower the pH to about 4.5, casein clumps, thickening milk into yogurt.

PRE-ACTIVITY DISCUSSION

Discuss the concept of pH by asking questions such as the following. If student responses indicate that they are not familiar with the concepts, give a simplified overview and define terms.

• **What is pH?** (A solution's pH indicates how acidic or basic it is. The pH of a solution is generally defined as a measure of its concentration of hydrogen ions [H^+] in a solution.)

• **What is an acid? What are some examples of acids?** (An acid is a substance that dissolves in water to produce hydrogen ions [H^+]. Because hydrogen ions are protons, acids can be defined as proton donors. Examples include aspirin, citric acid [vitamin C], carbonic acid [soda or seltzer water], and sulfuric acid.)

• **What is a base? What are some examples of bases?** (A base is a substance that dissolves in water to produce hydroxide ions [OH^-], which can react with H^+ ions to form water. Thus bases can be defined as

---

Introduction

Even newly opened, fresh-from-the-refrigerator yogurt is loaded with bacteria. But don't throw your yogurt away in disgust. Bacteria are supposed to be in yogurt. Why? Find out in this activity.

Materials

50 mL milk	10 mL water
10 mL diluted chocolate syrup	pH paper
10 mL lemon juice	5 paper cups
10 mL tea	graduated cylinder
10 mL vinegar	5 spoons

If you don't have a graduated cylinder, use a measuring spoon. One teaspoon equals 5 mL.

Procedure

1. Assign roles to each member of the group. Possible roles include: Recorder (the person who records observations and coordinates the group's presentation of results), Materials Manager (the person who makes sure that the group has all the materials it needs), Maintenance Director (the person who coordinates cleanup), Principal Investigator (the person who reads instructions to the group, makes sure that the proper procedure is being followed, and asks questions of the teacher on behalf of the group), and Specialists (people who perform specific tasks such as preparing the cups of liquids to be tested or testing the pH of the liquids). Your group may divide up the tasks differently, and individuals may have more than one role, depending on the size of your group.

2. On a separate sheet of paper, prepare a data table similar to the one shown in the Observations section.

3. Label the paper cups Chocolate, Lemon, Tea, Vinegar, and Water. Pour the appropriate liquid into each cup.

4. Dip a pH strip into each liquid. Compare the color of the strip to the key on the pH paper box. Record the pH in the appropriate place in your data table.

5. Pour 10 mL of milk into each of the five cups. Use a different spoon to mix the milk with each of the other liquids. Record your observations in your data table.

6. Measure the pH of the mixture in each cup. Record the pH in the appropriate place in your data table.

---

proton acceptors. Examples include lye and milk of magnesia.)

• **What is the pH scale? What is the significance of the number 7 on the pH scale?** (The pH scale is shown as extending from 0 to 14, because almost all solutions fall within that range of values. A substance of pH 7 is neutral—neither basic or acidic. Substances with a pH greater than 7 are said to be basic. The higher the number, the more basic the substance. Substances with a pH less than 7 are said to be acidic. The lower the number, the

more acidic the substance.)

Tell students that edible acidic substances—tomatoes, grapefruits, yogurt, and sour cream, for example—taste sour. Edible basic substances are often bitter. Stress that not all acids and bases are edible—some are very poisonous.

TEACHING STRATEGY

You may wish to have students prepare and wear name-tag stickers to indicate their role on their team. If you often use cooperative learning groups, you may

Observations

DATA TABLE

Liquid	pH Before	Observations	pH After
Chocolate			
Lemon			
Tea			
Vinegar			
Water			

Analysis and Conclusions

1. Describe what happened to the milk when it was added to the vinegar. Look at the substances formed when the milk reacted with the vinegar. How might these substances relate to the production of yogurt?

2. Did any other liquids cause the milk to change in the same way as the vinegar?

3. Relate the pH of the liquid to the way it reacts with milk.

4. Using what you have observed in this activity, explain why yogurt is thick and almost solid whereas milk is a thin liquid.

5. The essential ingredients of yogurt are milk and bacteria. (The bacteria are called "active cultures" on the ingredients label on a container of yogurt.) Without the bacteria, most yogurt would simply be fruit-flavored milk. What is the purpose of the bacteria in yogurt?

arated into lumps, or curdled. The thick, solid lumps of curdled milk, when separated from the watery liquid, have a texture similar to yogurt.

2. Yes. The lemon juice also caused the milk to curdle.

3. Milk curdles when added to liquids that have a pH of 5 or less, but mixes smoothly with liquids that have a pH greater than 5.

4. Yogurt contains acid, which causes milk to thicken into curds.

5. The bacteria produce acid, which causes the milk to thicken into yogurt.

ENRICHMENT

Tell students that yogurt-making bacteria produce lactic acid. Challenge students to find out what lactic acid has to do with exercise. (Muscles may also produce lactic acid when they are not getting enough oxygen, as sometimes happens during vigorous exercise. The buildup of lactic acid in muscles causes cramps and "stitches," or "side pains.")

wish to make reusable tags by laminating paper tags or using plastic pin-on name-tag holders.

DISCOVERY STRATEGIES

Discuss how the activity relates to the chapter ideas by asking questions similar to the following.

• **How do bacteria affect the food that you eat?** (Answers will vary. Some bacteria cause food to spoil. Other kinds of bacteria are necessary to the manufacture of certain foods.)

• **How do you think the milk in this activity will be affected by the other liquids?** (Accept all logical answers.)

OBSERVATIONS

Actual pH readings may vary slightly. Vinegar and lemon juice should, however, have the lowest pH readings—roughly 3 and 2, respectively. Water should have a pH of about 7.

ANALYSIS AND CONCLUSIONS

1. When milk was added to vinegar, it sep-

Activity Bank

BEFORE THE ACTIVITY

1. Divide the class into groups of three to six students.
2. Gather all materials at least one day prior to the activity. Prepare the four different paramecium cultures.

PRE-ACTIVITY DISCUSSION

Have students read the entire laboratory procedure. Review the use of the microscope by asking questions such as the following.

• **What is the proper way of putting a coverslip onto a drop of water on a microscope slide?** (Touch one edge of the coverslip to the drop, then gently lower the coverslip onto the drop.)

• **How do you carry a microscope?** (Using both hands—one hand grasping the arm and the other supporting the base.)

• **Can you use sunlight as the source of light for a microscope?** (No! Doing so can cause permanent damage to the retina.)

• **How do you focus when using the high-power objective?** (Watching the microscope carefully from the side, lower the objective so that it almost touches the coverslip. Then focus upward while looking through the microscope. Never focus downward while looking through the high-power objective. Depending on the microscopes used by the students, you may need to apply this focusing rule to the low- and medium-power objectives as well.)

TEACHING STRATEGY

1. Make sure that students know how to use the microscope and how to prepare wet-mount slides. A useful cooperative-learning strategy is to train one student per group as a microscope specialist. This specialist is then responsible for teaching microscope skills to the other members of the group. Meet with the specialists to train them at a convenient time prior to the first microscope activity.
2. Keep the cold paramecium culture in the refrigerator. Have one person per group take a sample of the culture when their group is ready.

Amebas are not the only protists with contractile vacuoles. In fact, most freshwater protists have contractile vacuoles. In this activity you will observe how this tiny structure works under different conditions.

Materials

3 *Paramecium caudatum* cultures at room temperature (25°C): fresh water, 0.5 percent table salt solution, 1.0 percent table salt solution

refrigerated (2°C) *Paramecium caudatum* culture in fresh water

medicine dropper

4 glass microscope slides

4 coverslips

microscope

cotton ball

Procedure 🧪

1. Using a medicine dropper, put a drop of water containing *Paramecium caudatum* in the center of a clean microscope slide.
2. Pull apart a small piece of cotton and put a few threads in the drop of water.
3. Cover the drop with a coverslip.

4. Locate and focus on a paramecium. Notice the alternating contractions of the two contractile vacuoles. How long does it take for a contractile vacuole to contract, refill, and contract once again? Record this information in the appropriate place in a data table similar to the one shown in Observations.

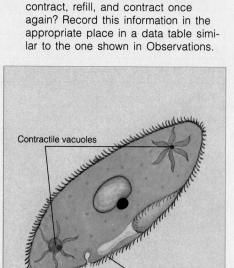

Contractile vacuoles

Paramecium

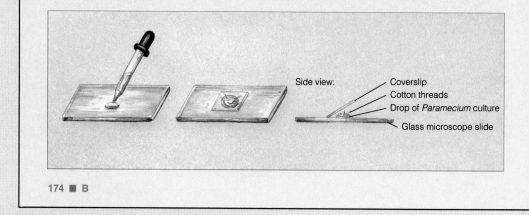

Side view:
Coverslip
Cotton threads
Drop of *Paramecium* culture
Glass microscope slide

DISCOVERY STRATEGIES

Discuss how the activity relates to the chapter ideas by asking questions similar to the following.

• **How are animallike protists similar to animals? How are they different?** (Animallike protists, as animals, have a nucleus in their cells and are heterotrophs. Protists are unicellular, whereas animals are multicellular.)

• **Why is it useful to study protists?** (Accept all logical answers. Elicit the response that

findings about protist cells may be generalized to other cells that contain nuclei, including those that make up humans.)

• **How did the cold temperatures affect the functioning of the paramecium?** (The cold slowed the life functions of the paramecium.)

• **Why are patients about to undergo heart surgery, organ transplants, or other long, difficult operations sometimes literally put on ice?** (The ice cools the patient's

5. Repeat steps 1 through 4 for a paramecium in a 0.5 percent table salt solution, a 1.0 percent table salt solution, and very cold water. Record your data.

6. Observe a paramecium in the very cold water. What happens as the water warms up?

Observations

DATA TABLE

Percent Table Salt	Temperature	Time Between Contractions
0	25°C	
0.5	25°C	
1	25°C	
0	2°C	

Analysis and Conclusions

1. What is the purpose of contractile vacuoles?

2. How does an increase in the concentration of salt in its environment affect a paramecium's contractile vacuoles?

3. What can you infer about the rate at which water enters the paramecium in salt solutions? Explain.

4. How is the rate at which a paramecium performs life functions, such as pumping out excess water, affected by very cold temperatures? What evidence do you have for this?

5. Compare your results with those obtained by your classmates. Did you obtain the same results? Why or why not?

ANALYSIS AND CONCLUSIONS

1. To pump excess water from inside a protist.

2. As the concentration of salt increases, the rate at which the contractile vacuoles contract decreases.

3. As the concentration of salt increases, the rate at which water enters the paramecium decreases. This is because contractile vacuoles do not have to work as fast to remove excess water from the cell.

4. The rate is slowed down until it is almost stopped. No contractile vacuole activity was observed in the protists in the very cold water, but activity resumed as the protists warmed up.

5. Accept all logical answers. Stress that the honest reporting of data and the thoughtful consideration of possible sources of error is far more important than getting the "right" answer.

cells, slowing their life functions and reducing the amount of oxygen needed by the cells and the amount of wastes produced by the cells.)

PROCEDURE

4. Roughly once every 6 seconds.

6. The contractile vacuoles begin to work again. Eventually, they are "back up to speed."

OBSERVATIONS

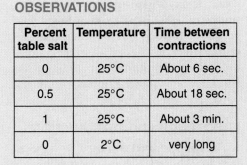

Percent table salt	Temperature	Time between contractions
0	25°C	About 6 sec.
0.5	25°C	About 18 sec.
1	25°C	About 3 min.
0	2°C	very long

Activity Bank

SHEDDING A LITTLE LIGHT ON EUGLENA

BEFORE THE ACTIVITY

1. Divide the class into groups of three to six students. This activity is also appropriate for students to do individually.

2. Gather all materials at least one day prior to the activity.

3. Prepare the concentrated euglena culture no more than a day before the activity. This can be done by centrifuging 20 mL batches of the dilute culture. Alternatively, obtain a vial, a one-hole rubber stopper that fits the vial, a short (7–10 cm) length of glass tubing, and some carbon paper or black construction paper. Insert the tubing into the stopper so that it extends 2–3 cm from the bottom of the stopper. Fill the vial with dilute culture and put the stopper in place. Then wrap the vial with the carbon paper. Do not cover the glass tubing protruding from the stopper. The euglenas, which are positively phototactic and negatively geotactic, will migrate into the glass tubing. When the euglenas are concentrated in the glass tubing, loosen the stopper. Cover the opening of the tubing with your finger. Keeping your finger in place, lift the stopper-tubing assembly out of the vial. Place the bottom end of the tubing into a small beaker or vial, then lift your finger off the top of the tubing, releasing the concentrated euglenas.

PRE-ACTIVITY DISCUSSION

Have students read the entire activity procedure.

• **What materials do you need?** (Concentrated culture of *Euglena*, petri dish, aluminum foil or cardboard.)

• **What do you do?** (Pour some of the culture into the dish, then cover half of the dish.)

DISCOVERY STRATEGIES

Discuss how the activity relates to the chapter ideas by asking questions similar to the following.

• **What other organisms require light to make food?** (Organisms include blue-green bacteria, green plants, and other plantlike protists.)

SHEDDING A LITTLE LIGHT ON EUGLENA

The plantlike protists known as euglenas are quite popular in scientific research—they are small, relatively easy to take care of, and fairly simple in structure. In Section 3-3, you read about the structure of euglenas. In this activity you will discover how the euglenas' structure enables them to function in a particular situation.

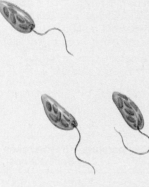

Procedure and Observations

1. Pour a concentrated culture of *Euglena* into a petri dish. What color is the concentrated culture? Why is the culture this color?

2. Cover half the dish with aluminum foil or a small piece of cardboard.

3. After 10 minutes, uncover the dish. What do you observe?

Analysis and Conclusions

1. How do euglenas respond to light?

2. Why do you think this occurs?

3. What structures in euglenas make this possible?

• **Can these organisms move toward a source of light? Explain. How might you go about finding out whether you are right?** (Accept all logical answers. You may want to point out that students will know for sure whether plants move toward light when they study Chapter 6.)

PROCEDURE AND OBSERVATIONS

1. The concentrated culture is light green in color. There are so many green euglenas in it that they make the water look green.

3. The covered side of the dish is very pale green or colorless, whereas the uncovered side is green.

ANALYSIS AND CONCLUSIONS

1. They move toward it.

2. The euglena move to the lighted side of the dish because they need light for photosynthesis.

3. Light-sensitive eyespots for detecting light and flagella for moving toward light.

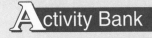
SPREADING SPORES

Have you ever opened a bag of bread and found fuzzy black mold growing on the slices you had planned to use? Or discovered spots of mildew that seemed to appear overnight in a clean bathroom? You may wonder how fungi manage to get just about everyplace. In this activity you will discover the secret of fungi's success.

What Do I Do?

1. Obtain a round balloon, cotton balls, tape, a stick or ruler about 30 cm long, modeling clay, and a pin.
2. Stretch the balloon so that it inflates easily. Do not tie off the end of the balloon!
3. Pull a cotton ball into five pieces about the same size. Roll the pieces into little balls no more than 1 cm in diameter.
4. Insert the little cotton balls through the opening in the neck of the balloon.

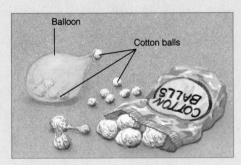

5. Continue making little cotton balls and putting them into the balloon until the balloon is almost full.
6. Inflate the balloon and tie a knot in its neck to keep the air inside.
7. Tape the knotted end of the balloon to the top of the stick.

8. Put the bottom of the stick into the modeling clay. Shape the modeling clay so that the stick stands upright.

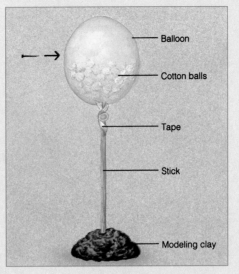

9. You have now made a model of the fruiting body of a bread mold. The balloon represents the spore case, the cotton balls represent spores, and the stick represents the stalk (a hypha) that supports the spore case. Make a drawing of your model. Label the spores, spore case, and stalk.
10. After you have finished admiring your model, jab the balloon with the pin.

What Did I Learn?

1. What happened when you jabbed the balloon with the pin?
2. Relate what you observed to reproduction in fungi. Why are fungi found just about everywhere?

food to obtain the energy they need.)
• **How do fungi obtain their food?** (They release chemicals that digest the substance on which they are growing and then absorb the digested food.)
• **Are fungi multicellular, unicellular, or both?** (Both. Some fungi, such as yeasts, consist of a single cell. Others, such as mushrooms, are made up of many cells. If a student does not bring up the point first, you might wish to remind students that fungi are not as strictly cellular as plants and animals.)

TEACHING STRATEGY

1. You may want to have all the groups pop their balloons at the same time—this makes for an impressive display of flying "spores." Alternatively, pop the balloons separately to see how the height of the sporecase (on the floor, on a table, on top of a cabinet, and so on) and wind (from an electric fan) affect the dispersal of the spores.
2. Allow sufficient time for cleanup.

DISCOVERY STRATEGIES

Discuss how the activity relates to the chapter ideas by asking questions similar to the following.
• **Why is important to fungi to scatter their spores?** (It enables the fungi to colonize new areas. It also may help to minimize competition among a fungus's offspring and between the offspring and the parent fungus.)
• **What are some variable that might affect the spread of a fungus's spores? What effects might these variables have?** (Accept all logical answers.)

WHAT DID I LEARN?

1. The balloon popped and the cotton balls went flying all over the place.
2. The fruiting body of a fungus is structured in a way that helps to scatter its spores. Being small and lightweight, spores can be carried great distances by the wind.

Activity Bank

SPREADING SPORES

BEFORE THE ACTIVITY

1. Divide the class into groups of three to six students. This activity is also appropriate for students to do individually, especially as a demonstration done at home for their family.
2. Gather all materials at least one day prior to the activity.

3. Make sure you have brooms, dustpans, brushes, and wastepaper baskets at hand for cleaning up after the activity.

PRE-ACTIVITY DISCUSSION

Have students read the entire activity procedure. Review basic concepts about fungi by asking questions such as the following.
• **Are fungi autotrophs, heterotrophs, or both? What does this mean?** (Heterotrophs. This means that they cannot make their own food; they must take in

Activity Bank

BEFORE THE ACTIVITY

1. Divide the class into groups of three to six students.

2. Gather all materials at least one day prior to the activity.

3. Have students read the paragraph introducing the forms of fungi on page 90 and the section on yeasts on pages 92–93.

PRE-ACTIVITY DISCUSSION

Ask questions to check whether students understand basic concepts about yeasts and know what they are supposed to do in this activity.

• **What are yeasts?** (Single-celled fungi.)

• **Describe what happens as yeast grows.** (The cells multiply through the process of budding and produce carbon dioxide gas.)

• **What do you predict will happen to the ballons in this activity? Why?** (If the yeast grows in the bottle, then the balloons will inflate as they fill up with carbon dioxide gas.)

DISCOVERY STRATEGIES

Discuss how the activity relates to the chapter ideas by asking questions similar to the following.

• **What do you think is the purpose of the sugar in the bottles?** (The sugar is the source of food for the yeast.)

• **What do you think is the purpose of the salt in the bottles?** (Accept all logical answers. Although the salt affects the osmotic potential of the system, it acts primarily as a distractor in this activity.)

• **Predict what would happen to the balloon on a bottle that contained warm water, 2 mL yeast, and 100 mL sugar.** (Accept all logical answers. The concentration of sugar in this bottle would probably be high enough to kill the yeast cells by causing them to lose water via osmosis.)

WHAT DID YOU OBSERVE?

The balloons on bottles A, B, and C will inflate. The balloon on bottle A will not inflate as rapidly as those on bottles B and C.

What do yeasts need to grow? What are the best conditions for growing yeast? Find out by doing this activity.

What You Need

5 small narrow-necked bottles (the 355-mL or 473-mL plastic or glass ones used for soda and juice work well)

china marker or marking pen

5 round balloons, stretched so that they inflate easily

5 plastic straws

sugar

salt

warm (40-45°C) water

activated yeast

250 mL beaker or small glass

graduated cylinder or measuring spoons (1 teaspoon equals 5 mL)

What To Do

1. Using the china marker, label the bottles A, B, C, D, and E.

2. Fill the bottles about two-thirds full with warm water.

3. Put 5 mL of sugar into bottle A.

4. Put 30 mL of sugar into bottles B, C, and E.

5. Put 5 mL of salt into bottles C, D, and E.

6. Put 2 mL of dry powdered yeast into bottle A and stir with a clean straw. Remove the straw.

7. Quickly place a balloon over the opening of bottle A. Make sure that the balloon fits tightly around the neck of the bottle.

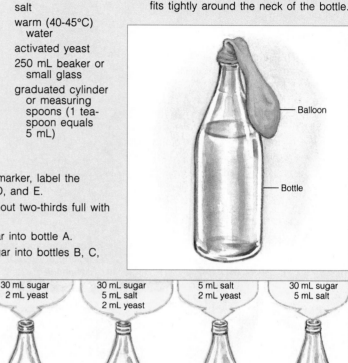

Balloon

Bottle

| 5 mL sugar 2 mL yeast | 30 mL sugar 2 mL yeast | 30 mL sugar 5 mL salt 2 mL yeast | 5 mL salt 2 mL yeast | 30 mL sugar 5 mL salt |
| A | B | C | D | E |

WHAT DID YOU LEARN?

1. The balloons on some of the bottles inflated. The yeast produced a gas that caused the balloons to inflate.

2. The balloons inflated on the bottles in which yeast was growing because the growing yeast produces a gas.

3. The gas produced by yeast forms the bubbles found in these substances.

4. Answers may vary. Students should find that yeast grows faster when more food (sugar) is present.

5. Student experiments will vary. You may need to point out to students that heat causes air to expand—a variable that they will have to take into consideration in designing their experiment.

JUST FOR FUN

In the spring, large draft horses are often tormented by small birds. The birds find the long hair on the neck and tail wonderful places to make their homes. Fortunately for the sanity of the horses and their owners, the birds do not like

8. Repeat steps 6 and 7 for bottles B, C, and D.

9. Repeat step 7 for bottle E.

10. Place the bottles in a warm spot away from drafts. Observe them carefully. Record your observations in a data table similar to the following.

What Did You Observe?

DATA TABLE

	Sugar	Salt	Yeast	Observations
A	5 mL	none	2 mL	
B	30 mL	none	2 mL	
C	30 mL	5 mL	2 mL	
D	none	5 mL	2 mL	
E	30 mL	5 mL	none	

What Did You Learn?

1. What happened to the balloons on some of the bottles? Why do you think this happened?

2. Relate what you observed to the growth of yeast.

3. Relate what you observed to how yeasts make the bubbles in bread, beer, and champagne.

4. What seem to be the best conditions for growing yeast?

5. Design an experiment for determining how temperature affects the growth of yeast.

baker's yeast. Sprinkling the stuff on the horse and rubbing it in well causes the birds to give up and seek more hospitable places to live. Which proves that yeast is yeast and nest is nest, and never the mane shall tweet.

Activity Bank

BEFORE THE ACTIVITY

1. Divide the class into groups of three to six students. This activity may also be done at home under adult supervision.
2. Gather all materials at least one day prior to the activity. You may wish to have students bring saucepans, measuring cups, wooden spoons, and shallow rectangular pans (such as 8" cake pans) from home.
3. Arrange for the use of a refrigerator.
4. If your school has a home economics classroom with kitchenettes, try to arrange for the use of these facilities.
5. Three packets of unflavored gelatin can be substituted for the agar in these recipes.

PRE-ACTIVITY DISCUSSION

Have students read the entire activity procedure. Make sure students understand the procedure by asking questions such as the following.
• **What edible materials do you need?** (Kanten, sugar, juice, food coloring.)
• **What electrical appliances do you need?** (Hot plate or stove, refrigerator.)
• **What other materials or facilities do you need?** (A sink with running water, saucepan, spoon, measuring cup, shallow pan, knife.)

TEACHING STRATEGY

You may wish to combine this activity with your favorite demonstrations and discussions of science in the kitchen. Through a careful selection of foods to make, you can emphasize cultural diversity as you teach science. For example, you might discuss oxidation in terms of why lemon juice is added to guacamole.

DISCOVERY STRATEGIES

Discuss how the activity relates to the chapter ideas by asking questions similar to the following.
• **What are some other foods in which algae or algae products are used?** (Possible answer: ice cream, chocolate milk, sushi.)
• **What are some other ways in which humans use algae?** (Possible answers: food for livestock, fertilizer, sewage treatment,

How do people use seaweed? You can experience how seaweed is used as a food by using this old family recipe to make a traditional Chinese dessert. The name of this dessert is based on the Chinese dialect called Cantonese. *Popo* means maternal grandmother. *Kanten* means agar in general and this dessert in particular. Kanten is sold in oriental grocery stores. It comes in long, thin, blocks.

Popo Thom's Kanten

2 blocks of kanten (one 0.5 oz package)

1 cup sugar

4 cups liquid (Popo Thom uses guava juice. You can use any kind of fruit juice you like.)

food coloring (optional)

Recipe Directions

1. Rinse kanten under running water.
2. Break kanten into cubes and put the cubes into a saucepan.
3. Add the liquid and sugar to the saucepan. Bring the mixture to a boil.

Kanten
1 cup sugar dissolved in 4 cups liquid

Then turn down the heat and cook, stirring, until the kanten dissolves.
4. Remove the saucepan from the heat. If you wish, stir in a few drops of food coloring to make the mixture a pretty color.
5. Pour the mixture into a shallow rectangular pan.
6. Let cool; then refrigerate until set.
7. Cut kanten into blocks and serve. Do you like the dessert, or do you think it's an acquired taste?

Another Chinese dessert, almond float, is made in a similar way. In step 3, use 2 cups of water and 2 tablespoons of almond extract as the liquid. In step 4, stir in 2 cups of evaporated milk. Almond float cubes are usually served with canned fruits in their syrup (canned lichee, or litchi nuts, are traditional; fruit cocktail is also good).

Things to Think About

1. The ingredients in a recipe often serve a particular function. For example, cornstarch thickens sauces, mustard helps the oil and vinegar in a salad dressing to mix, and egg helps to "glue" other ingredients together. What purpose does the kanten serve in this recipe? What purposes do the sugar and juice serve?
2. Share the kanten you have made with your family and/or classmates. What do they think of this dessert?
3. Working with some friends, brainstorm some other ways in which you might use kanten blocks.

---

additive in shampoo. Pressed algae is also used for decoration in dried-plant pictures, as in the Japanese art of *oshibana*.)

RECIPE DIRECTIONS

7. Answers will vary. Kanten is similar in taste and texture to gelatin desserts, and thus should not seem alien to most students.

THINGS TO THINK ABOUT

1. The kanten causes the liquid to gel and gives the dessert its distinctive tex-

ture. The sugar and juice give the dessert its flavor. The juice provides a medium in which the kanten can dissolve.
2. Answers will vary.
3. Answers will vary. Kanten could be used instead of gelatin to make jellied foods such as aspics. Soaked kanten is sliced and used in salads. Agar, which is the same thing as kanten, is used to thicken the media in which bacteria are cultured.

Activity Bank

THE INS AND OUTS OF PHOTOSYNTHESIS

Introduction

Blue-green bacteria, plantlike protists, and green plants perform a special process known as photosynthesis (foh-toh-SIHN-thuh-sihs). In photosynthesis, light energy is used to combine carbon dioxide and water to form food and oxygen. As you read on pages B111–112 in the textbook, oxygen is very important to humans and many other living things. Oxygen, which we humans take in when we breathe in, is used to break down food to release energy—energy that powers all life processes. The breakdown of food also produces carbon dioxide, which is removed from the body when we breathe out. What do organisms that can perform photosynthesis do with carbon dioxide?

In this activity you will observe the "ins and outs" of photosynthesis and the complementary process of food breakdown, or respiration.

Materials

3 125-mL flasks
3 #5 rubber stoppers
100 mL graduated cylinder
bromthymol blue solution
2 sprigs of *Elodea*, about the same size
light source
drinking straw

Procedure 🧪 📷

1. Using the graduated cylinder, measure out 100 mL of bromthymol blue solution for each of the three flasks. **CAUTION**: *Bromthymol blue is a dye and can stain your hands and clothing.* What color is the bromthymol blue solution?

2. Put the straw into one of the flasks. Gently blow bubbles into the solution until there is a change in its appearance. How does the solution change? Repeat this procedure with the other two flasks.

3. Place one sprig of *Elodea* in each of two of the flasks. Stopper all three flasks.

4. Put one flask containing *Elodea* in the dark for 24 hours. Put the other two flasks on a sunny windowsill for the same amount of time.

5. After 24 hours, examine each flask. What do you observe?

(continued)

B ■ 181

Activity Bank

THE INS AND OUTS OF PHOTOSYNTHESIS

BEFORE THE ACTIVITY

1. Divide the class into groups of three to six students.
2. Gather all materials at least one day prior to the activity.
3. Prepare enough bromthymol blue (BTB) solution for the activity (300 mL of BTB solution per group, plus a little extra, "just in case"). To do this, add enough 0.1% stock solution (0.5 g BTB dissolved in 50 mL water) to the water to give it a blue-green color. Add a drop or two of ammonium hydroxide to turn the solution a definite blue.

PRE-ACTIVITY DISCUSSION

Have students read the procedure for the activity.
• **How many flasks will you be preparing? How many of these flasks will contain** sprigs of *Elodea*? (Three; two.)
• **Why do you have to be careful when handling bromthymol blue solution?** (It is a dye that can stain things that it contacts.)

Write "Photosynthesis:" on the chalkboard.
• **What is the formula for photosynthesis in words?** (If students do not remember, instruct one student to look up the formula on page B138.) Fill in the formula after the colon.

Write "Respiration: ?" on the chalkboard. Tell students that they will be able to fill in the formula for respiration when they have completed the activity.

TEACHING STRATEGY

At the next class period, after students have obtained their results, elicit responses from the different groups to construct the formula for respiration.

DISCOVERY STRATEGIES

Discuss how the activity relates to the chapter ideas by asking questions similar to the following.
• **What would happen if you sealed an animal in an airtight jar? Why?** (The animal would die because it would eventually use up all the oxygen in the jar.)
• **What would happen if you sealed an animal in a jar with a large number of live plants?** (The animal would survive, at least for a time, because it would get oxygen from the plants and the plants would use up the carbon dioxide produced by the animal.)
• **Predict what would happen to plants sealed in an airtight jar. Explain your answer.** (Accept all logical responses; then point out that the plants would probably survive indefinitely, provided they received enough light and there was enough moisture inside the jar.)
• **Why would the plants survive?** (Plants undergo respiration as well as photosynthesis.)

INTRODUCTION

Organisms that perform photosynthesis can take in carbon dioxide and use it to produce carbohydrates in the process of photosynthesis.

PROCEDURE

1. Bromthymol blue solution is blue.
2. The solution turns yellow.

B ■ 181

5. Flask with *Elodea* in sunny spot: bromthymol blue solution has changed color from yellow to blue. Flask without *Elodea* in sunny spot: bromthymol blue solution remains yellow. Flask with *Elodea* in dark spot: bromthymol blue solution remains yellow.

ANALYSIS AND CONCLUSIONS

1. Carbon dioxide was added. Carbon dioxide is a byproduct of respiration, the process in which food is broken down to release energy. The carbon dioxide caused the solution to turn yellow.

2. The bromthymol blue solution remained yellow because the plant did not perform photosynthesis and thus did not take up carbon dioxide. Photosynthesis does not occur in the absence of light. (The light reactions that make up the first part of photosynthesis require light; the dark reactions that make up the second part require substances made in the light reactions.)

3. The bromthymol blue solution turned blue as the plant removed carbon dioxide for photosynthesis.

4. It was a control. It showed that bromthymol blue solution does not change from yellow to blue due to the presence of light.

5. Both include carbon dioxide. Some students may know that one of the products of respiration is water and that the products of respiration are the same as the reactants of photosynthesis.

6. They are the same: oxygen and food.

7. They are opposite, complementary processes.

On a Sunny Windowsill

Stopper

Elodea

In the Dark

Analysis and Conclusions

1. What substance did you add to the bromthymol blue solution when you blew bubbles into it? Where did this substance come from? What effect did this substance have on the solution?

2. What happened in the flask that was kept in the dark? Explain why this occurred.

3. What happened in the flask containing *Elodea* that was kept in a sunny spot? Why?

4. What was the purpose of the flask that did not contain *Elodea*?

5. How are the "ins," or raw materials, for photosynthesis related to the "outs," or products, of respiration?

6. How are the "outs" of photosynthesis related to the "ins" of respiration?

7. How is photosynthesis related to respiration?

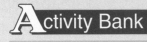

BUBBLING LEAVES

Photosynthesis uses the gas carbon dioxide and produces the gas oxygen. Tiny openings in the leaves called stomata allow carbon dioxide to get in and oxygen to get out. Where are stomata located? Are they in the same position in all leaves? Discover for yourself by doing this activity.

Procedure and Observations

1. Obtain a variety of fresh leaves and a glass container of hot water. **CAUTION:** *Be careful when handling hot substances.*

2. Pinch the stalk of a leaf. Then dip the flat part of the leaf in hot water. What do you observe? Why do you think this happens?

3. Repeat step 2 with some different kinds of leaves. How do your results differ from one another?

Analysis and Conclusions

1. What can you infer about the structure of the leaves?

2. Share your findings with your classmates. Compare your results. Are they similar? Why or why not? What are some possible sources of error in this activity?

B ■ 183

DISCOVERY STRATEGIES

Discuss how the activity relates to the chapter ideas by asking questions similar to the following.
• **What happens to air when it is warmed?** (It expands.)
• **Why did bubbles come out of the leaves?** (The hot water warmed the air inside the leaf, which expanded and exited the leaf through the stomata.
• **What would happen if you did not pinch the stalk of the leaf?** (Much of the air might have escaped through the stalk.)
• **Where would you expect to find the stomata on a lily pad? Why?** (On the top—the surface that is exposed to air.)
• **In most plants, most of the stamata are found on the lower surface of the leaves. Explain how this adaptation conserves water.** (Because water evaporates faster under warmer, drier conditions, it is more readily lost from the upper surface of the leaves, the side more exposed to wind and sun. Having few stomatas on the upper surface of the leave reduces the places from which water can escape and this acts to conserve water.)

PROCEDURE AND OBSERVATIONS

2. When the flat part of the leaf is dipped in hot water, bubbles emerge from the leaf. This happens because the air in the leaves expands due to the heat and escapes through the stomata.

3. The number of bubbles produced by the different sides of the leaves varies. In some leaves, only one side produces bubbles. In others, many more bubbles are produced on one side than on the other. In still others, the two sides of the leaf produced approximately the same number of bubbles.

ANALYSIS AND CONCLUSIONS

1. The distribution of stomata is different in various kinds of plants.

2. Accept all logical answers. Possible sources of error might include not pinching the stalk of the leaf tightly enough, using leaves of different levels of freshness or maturity, and accidentally comparing results for different kinds of plants.

Activity Bank

BUBBLING LEAVES

BEFORE THE ACTIVITY

1. Divide the class into groups of three to six students.

2. Gather all materials at least one day prior to the activity. The leaves should be as fresh as possible, so keep them in a vase with water or in a sealed plastic bag in the refrigerator. Lightly sprinkling the leaves with water before closing the bag often helps keep the leaves fresher.

PRE-ACTIVITY DISCUSSION

Have students read the complete activity procedure.
• **What do you need for this activity?** (Fresh leaves, glass container, hot water.)
• **What do you do with each leaf?** (Pinch the stalk, dip the flat part of the leaf into the water, and see what happens.)

Activity Bank

BEFORE THE ACTIVITY

1. Divide the class into groups of three to six students.

2. Gather all materials at least one day prior to the activity. Have volunteers bring in clean, empty half-gallon milk or juice cartons a day or two before the activity. Save extra cartons for other activities, such as making plant pots, birdhouses, or dioramas.

PRE-ACTIVITY DISCUSSION

Make sure that students understand the concept of stimulus and response by asking questions such as the following.

• **Why is it important for living things to be able to respond to their environment?** (Answers will vary. Possible answer: Living things need to find food, water, and shelter and avoid danger.)

• **Define the term stimulus.** (A stimulus is something in the internal or external environment to which a living thing responds.)

• **Define the term response.** (A response is an organism's reaction to a stimulus.)

DISCOVERY STRATEGIES

Discuss how the activity relates to the chapter ideas by asking questions similar to the following.

• **Why did the seedlings under the carton without a hole grow straight?** (Accept all logical answers. Possible answer: Their growth was affected only by gravity, whereas the growth of the seedlings under the carton with a hole was also affected by light.)

• **If the tip of a seedling in the carton with a hole were covered by an aluminum foil cap, the seedling would have grown straight. What can you infer from this?** (Something in the tip of the shoot controls the phototropic response in the seedlings.)

• **Experiments done by Charles Darwin and other scientists demonstrated that the tip of the shoot produces a chemical that causes stem cells to grow faster. Which side of the shoot has more of this chemical? Explain.** (The side away from the light. Because the side away from the

Living things respond to their environment. Pet dogs and cats come running when they hear the can opener or the rustle of the pet-food bag. People put up umbrellas and hurry toward shelter when it starts to rain. In this activity you will discover one way in which plants respond to their environment.

What Do I Need?

2 half-gallon milk cartons with the tops cut off, old newspapers, old artist's smock or apron, black tempera (poster) paint, paintbrush, 2 plastic cups, soil, scissors, and 10 popcorn kernels.

What Do I Do?

1. Spread some old newspapers on a work surface. Wear old clothes on which it is okay to get paint, or wear an artist's smock or apron to protect your clothes. Paint the inside of the milk cartons black. Put the cartons on a sheet of old newspaper someplace where they can dry, and then clean up your work area.

2. Fill the cups about three-fourths full with soil. Water the soil so that it is moist but not soaking wet. Plant five popcorn kernels in each cup. The kernels should be just under the surface of the soil.

3. Cut a small hole about 1 cm across in the side of one of the dry milk cartons.

4. Set the cups in a well-lighted place. Cover each cup with a milk carton. Make sure to position the carton with the hole in it so that light can enter through the hole.

5. Examine your seedlings after several days. If the soil is dry, add a little water. Replace the cartons so that they are in exactly the same position they were originally.

What Did I Find Out?

1. Describe how your seedlings appear after they have been growing for a week. How are the two groups of seedlings different?

2. Why do you think this happens?

3. Why might your results be important to someone who has houseplants?

4. Draw a picture that show your results. Using the picture, share your findings with your classmates.

light elongates at a faster rate than the side facing the light, the seedling stem bends toward the light. If you are teaching more advanced classes, you may wish to take this opportunity to introduce the topics of auxins and plant hormones in general.)

WHAT DID I FIND OUT?

1. They are tall and spindly. The ones that have been growing in the carton without a hole are growing upright, are paler in color, and may be taller than the other seedlings. The seedlings in the carton with a hole are growing crooked, almost sideways, in the direction of the hole.

2. Plants grow toward a source of light.

3. If houseplants aren't turned regularly, they will grow crooked as they bend toward windows or other sources of light.

Appendix A

The metric system of measurement is used by scientists throughout the world. It is based on units of ten. Each unit is ten times larger or ten times smaller than the next unit. The most commonly used units of the metric system are given below. After you have finished reading about the metric system, try to put it to use. How tall are you in metrics? What is your mass? What is your normal body temperature in degrees Celsius?

Commonly Used Metric Units

Length   The distance from one point to another

meter (m)   A meter is slightly longer than a yard.
1 meter = 1000 millimeters (mm)
1 meter = 100 centimeters (cm)
1000 meters = 1 kilometer (km)

Volume   The amount of space an object takes up

liter (L)   A liter is slightly more than a quart.
1 liter = 1000 milliliters (mL)

Mass   The amount of matter in an object

gram (g)   A gram has a mass equal to about one paper clip.

1000 grams = 1 kilogram (kg)

Temperature   The measure of hotness or coldness

degrees   0°C = freezing point of water
Celsius (°C)   100°C = boiling point of water

Metric–English Equivalents

2.54 centimeters (cm) = 1 inch (in.)
1 meter (m) = 39.37 inches (in.)
1 kilometer (km) = 0.62 miles (mi)
1 liter (L) = 1.06 quarts (qt)
250 milliliters (mL) = 1 cup (c)
1 kilogram (kg) = 2.2 pounds (lb)
28.3 grams (g) = 1 ounce (oz)
°C = 5/9 x (°F – 32)

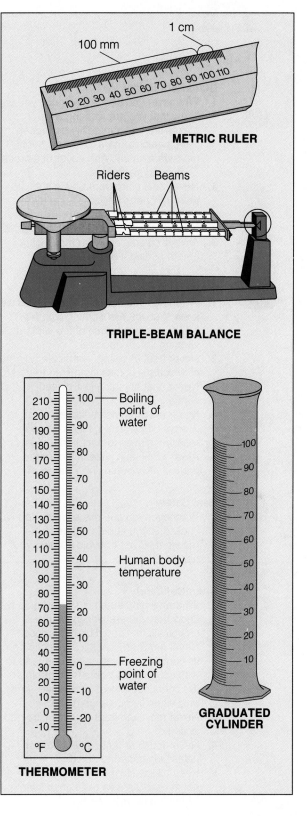

METRIC RULER

TRIPLE-BEAM BALANCE

GRADUATED CYLINDER

THERMOMETER

Glassware Safety

1. Whenever you see this symbol, you will know that you are working with glassware that can easily be broken. Take particular care to handle such glassware safely. And never use broken or chipped glassware.
2. Never heat glassware that is not thoroughly dry. Never pick up any glassware unless you are sure it is not hot. If it is hot, use heat-resistant gloves.
3. Always clean glassware thoroughly before putting it away.

Fire Safety

1. Whenever you see this symbol, you will know that you are working with fire. Never use any source of fire without wearing safety goggles.
2. Never heat anything—particularly chemicals—unless instructed to do so.
3. Never heat anything in a closed container.
4. Never reach across a flame.
5. Always use a clamp, tongs, or heat-resistant gloves to handle hot objects.
6. Always maintain a clean work area, particularly when using a flame.

Heat Safety

Whenever you see this symbol, you will know that you should put on heat-resistant gloves to avoid burning your hands.

Chemical Safety

1. Whenever you see this symbol, you will know that you are working with chemicals that could be hazardous.
2. Never smell any chemical directly from its container. Always use your hand to waft some of the odors from the top of the container toward your nose—and only when instructed to do so.
3. Never mix chemicals unless instructed to do so.
4. Never touch or taste any chemical unless instructed to do so.
5. Keep all lids closed when chemicals are not in use. Dispose of all chemicals as instructed by your teacher.

6. Immediately rinse with water any chemicals, particularly acids, that get on your skin and clothes. Then notify your teacher.

Eye and Face Safety

1. Whenever you see this symbol, you will know that you are performing an experiment in which you must take precautions to protect your eyes and face by wearing safety goggles.
2. When you are heating a test tube or bottle, always point it away from you and others. Chemicals can splash or boil out of a heated test tube.

Sharp Instrument Safety

1. Whenever you see this symbol, you will know that you are working with a sharp instrument.
2. Always use single-edged razors; double-edged razors are too dangerous.
3. Handle any sharp instrument with extreme care. Never cut any material toward you; always cut away from you.
4. Immediately notify your teacher if your skin is cut.

Electrical Safety

1. Whenever you see this symbol, you will know that you are using electricity in the laboratory.
2. Never use long extension cords to plug in any electrical device. Do not plug too many appliances into one socket or you may overload the socket and cause a fire.
3. Never touch an electrical appliance or outlet with wet hands.

Animal Safety

1. Whenever you see this symbol, you will know that you are working with live animals.
2. Do not cause pain, discomfort, or injury to an animal.
3. Follow your teacher's directions when handling animals. Wash your hands thoroughly after handling animals or their cages.

Appendix C

One of the first things a scientist learns is that working in the laboratory can be an exciting experience. But the laboratory can also be quite dangerous if proper safety rules are not followed at all times. To prepare yourself for a safe year in the laboratory, read over the following safety rules. Then read them a second time. Make sure you understand each rule. If you do not, ask your teacher to explain any rules you are unsure of.

Dress Code

1. Many materials in the laboratory can cause eye injury. To protect yourself from possible injury, wear safety goggles whenever you are working with chemicals, burners, or any substance that might get into your eyes. Never wear contact lenses in the laboratory.

2. Wear a laboratory apron or coat whenever you are working with chemicals or heated substances.

3. Tie back long hair to keep it away from any chemicals, burners and candles, or other laboratory equipment.

4. Remove or tie back any article of clothing or jewelry that can hang down and touch chemicals and flames.

General Safety Rules

5. Read all directions for an experiment several times. Follow the directions exactly as they are written. If you are in doubt about any part of the experiment, ask your teacher for assistance.

6. Never perform activities that are not authorized by your teacher. Obtain permission before "experimenting" on your own.

7. Never handle any equipment unless you have specific permission.

8. Take extreme care not to spill any material in the laboratory. If a spill occurs, immediately ask your teacher about the proper cleanup procedure. Never simply pour chemicals or other substances into the sink or trash container.

9. Never eat in the laboratory.

10. Wash your hands before and after each experiment.

First Aid

11. Immediately report all accidents, no matter how minor, to your teacher.

12. Learn what to do in case of specific accidents, such as getting acid in your eyes or on your skin. (Rinse acids from your body with lots of water.)

13. Become aware of the location of the first-aid kit. But your teacher should administer any required first aid due to injury. Or your teacher may send you to the school nurse or call a physician.

14. Know where and how to report an accident or fire. Find out the location of the fire extinguisher, phone, and fire alarm. Keep a list of important phone numbers—such as the fire department and the school nurse—near the phone. Immediately report any fires to your teacher.

Heating and Fire Safety

15. Again, never use a heat source, such as a candle or burner, without wearing safety goggles.

16. Never heat a chemical you are not instructed to heat. A chemical that is harmless when cool may be dangerous when heated.

17. Maintain a clean work area and keep all materials away from flames.

18. Never reach across a flame.

19. Make sure you know how to light a Bunsen burner. (Your teacher will demonstrate the proper procedure for lighting a burner.) If the flame leaps out of a burner toward you, immediately turn off the gas. Do not touch the burner. It may be hot. And never leave a lighted burner unattended!

20. When heating a test tube or bottle, always point it away from you and others. Chemicals can splash or boil out of a heated test tube.

21. Never heat a liquid in a closed container. The expanding gases produced may blow the container apart, injuring you or others.

22. Before picking up a container that has been heated, first hold the back of your hand near it. If you can feel the heat on the back of your hand, the container may be too hot to handle. Use a clamp or tongs when handling hot containers.

Using Chemicals Safely

23. Never mix chemicals for the "fun of it." You might produce a dangerous, possibly explosive substance.

24. Never touch, taste, or smell a chemical unless you are instructed by your teacher to do so. Many chemicals are poisonous. If you are instructed to note the fumes in an experiment, gently wave your hand over the opening of a container and direct the fumes toward your nose. Do not inhale the fumes directly from the container.

25. Use only those chemicals needed in the activity. Keep all lids closed when a chemical is not being used. Notify your teacher whenever chemicals are spilled.

26. Dispose of all chemicals as instructed by your teacher. To avoid contamination, never return chemicals to their original containers.

27. Be extra careful when working with acids or bases. Pour such chemicals over the sink, not over your workbench.

28. When diluting an acid, pour the acid into water. Never pour water into an acid.

29. Immediately rinse with water any acids that get on your skin or clothing. Then notify your teacher of any acid spill.

Using Glassware Safely

30. Never force glass tubing into a rubber stopper. A turning motion and lubricant will be helpful when inserting glass tubing into rubber stoppers or rubber tubing. Your teacher will demonstrate the proper way to insert glass tubing.

31. Never heat glassware that is not thoroughly dry. Use a wire screen to protect glassware from any flame.

32. Keep in mind that hot glassware will not appear hot. Never pick up glassware without first checking to see if it is hot. See #22.

33. If you are instructed to cut glass tubing, fire-polish the ends immediately to remove sharp edges.

34. Never use broken or chipped glassware. If glassware breaks, notify your teacher and dispose of the glassware in the proper trash container.

35. Never eat or drink from laboratory glassware. Thoroughly clean glassware before putting it away.

Using Sharp Instruments

36. Handle scalpels or razor blades with extreme care. Never cut material toward you; cut away from you.

37. Immediately notify your teacher if you cut your skin when working in the laboratory.

Animal Safety

38. No experiments that will cause pain, discomfort, or harm to mammals, birds, reptiles, fishes, and amphibians should be done in the classroom or at home.

39. Animals should be handled only if necessary. If an animal is excited or frightened, pregnant, feeding, or with its young, special handling is required.

40. Your teacher will instruct you as to how to handle each animal species that may be brought into the classroom.

41. Clean your hands thoroughly after handling animals or the cage containing animals.

End-of-Experiment Rules

42. After an experiment has been completed, clean up your work area and return all equipment to its proper place.

43. Wash your hands after every experiment.

44. Turn off all burners before leaving the laboratory. Check that the gas line leading to the burner is off as well.

Glossary

alga (AL-gah; plural: algae, AL-jee): any simple plantlike nonvascular photosynthetic autotroph

ameba (uh-MEE-bah): a bloblike protist that uses its pseudopods to move and to obtain food

angiosperm (AN-jee-oh-sperm): a plant whose seeds are contained in an ovary

annual: a plant that completes its life cycle within one growing season

antibiotic: a chemical that destroys or weakens harmful microorganisms

autotroph: an organism that obtains energy by making its own food

biennial: a plant that completes its life cycle in two years

binomial nomenclature (bigh-NOH-mee-uhl NOH-muhn-klay-cher): the naming system devised by Linnaeus in which each organism is given two names: a genus name and a species name

brown alga: an alga that contains brown acessory pigments and belongs to the phylum Phaeophyta

cilia (SIHL-ee-uh; singular: cilium): hairlike projections on the outside of cells that move with a wavelike beat; in ciliates, cilia help these organisms to move, obtain food, and sense the enviroment

ciliate (SIHL-ee-iht): an animallike protist that possesses cilia at some point in its life

cone: in plants, a reproductive structure of a gymnosperm

decomposer: an organism that breaks down dead organisms into simpler substances, thereby returning important materials to the soil and water

diatom (DIGH-ah-tahm): a plantlike protist that has a beautiful two-part glassy shell

dinoflagellate (digh-noh-FLAJ-eh-layt): a plantlike protist that typically has cell walls that look like plates of armor and possesses two flagella, one of which trails from one end like a tail and the other of which wraps around the middle of the organism like a belt

euglena (yoo-GLEE-nah): a plantlike flagellate protist that belongs to the genus *Euglena* and is characterized by a pouch that holds two flagella, a reddish eyespot, and a number of grass-green chloroplasts that are used in photosynthesis

flagellum (flah-JEHL-uhm; plural: flagella): a long whiplike structure that propels a cell through its environment

flower: a reproductive structure of an angiosperm

fruit: a ripened ovary of a plant that encloses and protects the seed or seeds

fungus (FUHN-guhs; plural: fungi, FUHN-jigh): a heterotroph, usually multicellular, that releases chemicals that digest the substance on which it is growing and then absorbs the digested food; multicellular fungi are made up of threadlike hyphae; many fungi reproduce by means of spores

genus (plural: genera): the second-smallest taxonomic group; a genus consists of a number of similar, closely related species; a number of closely related genera make up a family

green alga: an alga that gets its color from the green pigment chlorophyll and belongs to the phylum Chlorophyta

gymnosperm (JIHM-noh-sperm): a plant whose seeds are not contained in an ovary

heterotroph: an organism that cannot make its own food and thus must eat other organisms in order to obtain energy

host: a living thing that provides a home and/or food for a parasite

hypha (plural: hyphae, HIGH-fee): one of the branching, threadlike tubes that makes up the body of a multicellular fungus

leaf: a structure in vascular plants whose main function is typically photosynthesis

lichen (LIGH-kuhn): a plantlike structure that is formed by a fungus and an alga that live together

mold: a fuzzy, shapeless, fairly flat multicellular fungus

mushroom: a multicellular fungus shaped like an umbrella

nonvascular plant: a plant that lacks vascular tissue

ovule (AH-vyool): a structure in a female cone or flower that contains an eggs cell and develops into a seed

paramecium (par-uh-MEE-see-uhm; plural: paramecia): a slipper-shaped ciliate protist that belongs to the genus *Paramecium*

parasite (PAIR-ah-sight): an organism that survives by living on or in a host organism, thus harming it

perennial: a plant that lives for many years

petal: a colorful leaflike flower structure that serves to attract pollinators

phloem (FLOH-ehm): plant vascular tissue that carries food

photosynthesis (foht-oh-SIHN-thuh-sihs): the food-making process that involves chlorophyll, in which light energy is used to make food (glucose) from carbon dioxide and water

pigment: a colored chemical; plant pigments are often associated with photosynthesis

pistil (PIHS-tihl): a female reproductive organ in a flower, which consists of a stigma, style, and ovule-containing ovary

pollen (PAHL-uhn): tiny grains that can be thought of as containing sperm cells, which are produced by male cones and flowers

pollination (pahl-uh-NAY-shuhn): the process by which pollen is carried from male reproductive structures to female reproductive structures

protist: a unicellular organism that contains a nucleus

pseudopod (SOO-doh-pahd): a temporary extension of the cell membrane and cytoplasm used in feeding and/or movement

red alga: an alga that contains red accessory pigments and belongs to the phylum Rhodophyta

root: a structure in vascular plants whose main functions are typically absorption and anchorage

sarcodine (SAHR-koh-dighn): an animallike protist that possesses pseudopods

sepal (SEE-puhl): a leaflike structure that protects a developing flower

slime mold: a funguslike protist that is a microscopic amebalike cell at one stage of its life cycle and a large, moist, flat, shapeless blob at another stage; slime molds produce reproductive structures known as fruiting bodies, which contain spores

species: the smallest and most specific taxonomic group, which consists of individuals that are quite similar in appearance and behavior and that can interbreed to produce fertile offspring

spore: a cell, usually surrounded by a protective wall, that is specialized either for reproduction or for a resting stage; the reproductive cells produced by most fungi are known as spores

sporozoan (spohr-oh-ZOH-uhn): a parasitic animallike protist that has a complex life cycle involving more than one kind of host animal and that typically produces cells called spores in order to pass from one host to another

stamen (STAY-muhn): a male reproductive organ in a flower, which consists of a filament topped by a pollen-producing anther

stem: a structure in vascular plants whose main functions are typically to carry materials between the roots and leaves and to support the plant

symbiosis (sihm-bigh-OH-sihs; plural: symbioses): a relationship in which one organism lives on, near, or even inside another organism and at least one of the organisms benefits

tropism (TROH-pihz-uhm): in plants, the growth of a plant toward or away from a stimulus

vascular plant: a plant that has vascular tissue

virus: a disease-causing particle consisting of hereditary material enclosed in a protein coat that is smaller and less complex than a cell

xylem (ZIGH-luhm): plant vascular tissue that carries water and minerals from the roots up through a plant and that also helps to support a plant

yeast: a unicellular fungus

zooflagellate (zoh-oh-FLAJ-ehl-iht): an animallike protist that possesses flagella

Index

Aflatoxin, B95
African sleeping sickness, B69
Agar, B111
Algae, B106–112
  brown algae, B108–110
  classification of, B106–107
  compared to land plants, B107
  and plant evolution, B107
  green algae, B111–113
  pigments of, B107–108
  red algae, B110–111
Algins, B110
Ameba, B64–65
Angiosperms, B150–152
  flowers, B150–152
  pollination, B152
  seeds of, B144
Animallike protists, B63–72
  ciliates, B66–67
  sarcodines, B63–65
  sporozoans, B69–71
  zooflagellates, B68–69
Animals, kingdom of, B28
Annuals, B153
Antibiotics, B52, B93
Autotroph, definition, B26

Bacteria, B43–53
  and humans, B49–53
  decomposers, B46, B47
  diseases caused by, B52
  food and energy relationships,
    B47–48
  life functions of, B45–47
  nitrogen-fixing bacteria, B49
  oxygen production, B48
  structure of, B44–45
  symbiosis, B48–49
  varieties of, B43–44
Bacteriologists, B52
Bacteriophage, B39–40
Biennials, B153–154
Binomial nomenclature, B17
Brown algae, B108–110
  kelp, B109–110
  of Sargasso Sea, B108–109
  use as food, B110
Budding, yeasts, B93

Cancer, B40, B95
Canning foods, and bacteria, B50–51
Carrageenan, B111, B123
Chestnut blight, B95
Chlorophyll
  algae, B108
  and photosynthesis, B138
Cilia, B66
Ciliates, B66–67
  paramecium, B66–67
Class, B22
Classification
  binomial nomenclature, B17
  classification groups, B21–25

definition of, B12
and evolution, B19–20
five-kingdom classification system,
  B25–28
history of, B12–15
importance of, B13–14
Linnaeus' system, B16–19
naming of organisms, B17–18
taxonomy, B14–15
and technology, B20–21
Computer viruses, B42
Cones, B143
Conifers, characteristics of, B148–149
Conjugation, paramecium, B67
Cyanobacteria, B43
Cycads, characteristics of, B148

Decomposers
  bacteria, B46, B47, B48
  fungi, B87
Diatoms, characteristics of, B75–76
Dinoflagellates, B76
Disease
  caused by bacteria, B52
  caused by viruses, B40–41
  and fungi, B94–96
  and protists, B69–71
Dutch elm disease, B94–95

Embryo, of seeds, B145
Endospores, B46–47
Environmental cleanup, and bacteria,
  B51
Epidermis
  of leaves, B138–139
  of roots, B131
Ergot poisoning, B95–96
Euglenas, characteristics of, B74
Evolution
  and algae, B107
  and classification, B19–20
  and green algae, B112–113
  and plants, B115–117
  and protists, B59, B61

Family, B22
Ferns, B119–122
  characteristics of, B120–121
  uses for, B121–122
Fertilization in seed plants, B143–147
Fibrous roots, B131
Five-kingdom classification system,
  B25–28
Flagella
  bacteria, B45
  protists, B68, B73
Fleming, Sir Alexander, B93
Flowers, B143, B150–152
  definition of, B150
  parts of, B151–152
  pollination, B152

Foraminiferans, B63
Fruit, B152
Fruiting bodies
  fungi, B89–90
  slime molds, B78–79
Fuel, produced with bacteria, B51
Fungi, B27–28
  characteristics of, B27–28, B86–89
  decomposers, B87
  and disease, B94–96
  molds, B93–94
  mushrooms, B90–91
  structure of, B88–89
  symbiosis, B87, B94–98
  yeasts, B92–93
Funguslike protists, B77–79
  slime molds, B78–79

Genus, B17, B22
Germination, seeds, B147
Ginkgoes, characteristics of, B148
Glucose, and photosynthesis, B138
Gnetophytes, characteristics of, B149
Gravitropism, B155
Great Potato Famine, B77
Green algae, B111–113
  and evolution, B112
Gymnosperms, B148–149
  conifers, B148–149
  cycads, B148
  ginkgoes, B148
  gnetophytes, B149

Herbaceous plants, B134
Heterotroph, definition, B26
Hornworts, B116–117
Hosts, definition of, B38
Hyphae, B88–89

Industrial uses, bacteria, B52–53

Kelp, B109–110
Kingdom, B21, B22

Leaves, B137–140
  and photosynthesis, B137–139
  structure of, B138–139, B140
  transpiration, B140
  uses of, B140
Lichens, B97–98
Linnaeus' system, classification, B16–19
Liverworts, B116–117
Living organisms
  naming of, B17–18
  number of organisms identified,
    B14

Malaria, transmission of, B69–71

Mesophyll, leaves, B139
Mildew, B27
Molds, B90, B93–94
  and antibiotics, B93
  and food making, B93–94
Monerans
  See Bacteria
Mosquitoes
  malaria, B69–71
  protists as killers of, B72
Mosses, B116–118
Mushroom growers, B98
Mushrooms, B90–91
  poisonous mushrooms, B99
Mycorrhizae, B97

Nitrogen-fixing bacteria, B49–50

Order, B22
Ovary, plant, B144, B152
Ovules, B143, B144, B152
Oxygen production,
  algae, B112
  bacteria, B48
  in protosynthesis, B138

Paramecium, B66–67
  conjugation, B67
Parasite, definition of, B38
Pasteurization, B50
Peat, 118
Perennials, B154
Pesticides, fungal, B96, B97
Petals, B151
Petroleum, B51
Phloem, definition, B130
Photosynthesis, B137–139
  artificial forms of, B142
  equation for, B138
Phototropism, B155
Phylum, B21, B22
Phytoflagellates, B73–74
Pigments, of algae, B107–108
Pistils, B151–152
Pith, stems, B134
Plantlike protists, B73–76
  diatoms, B75–76
  dinoflagellates, B76
  euglenas, B74
Plants, B106–112, B130–155
  and evolution, B115–117
  kingdom of, B28
  land plants, characteristics of,
    B115–116
  nonvascular, B106–118
  seed plants, B130-155
  tropisms, B154–155
  vascular plants, B119–122, B130–155

Pollen, B143, B152
Pollination, B143–144
  angiosperms, B152
Protists, B26–27
  animallike protists, B63–72
  classification of, B62
  definition of, B60
  and evolution, B59, B61
  funguslike protists, B77–79
  plantlike protists, B73–76
  See also specific groups of protists
Pseudopod, B63, B64

Radiolarians, B63
Red algae, B110–111
Red tides, B76
Response, meaning of, B154
Root cap, B131
Root hairs, B131
Roots, B131–132
  fibrous roots, B131
  and fungi, B96–97
  structure of, B131
  taproots, B131
  uses of, B132

Sarcodines, B63–65
  ameba, B64–65
  foraminiferans, B63
  radiolarians, B63
Sargasso Sea, B108–109
Seed coat, B144
Seaweed, as algae, B106
Seed plants, B130–155
  angiosperms, B150–152
  fertilization in, B143–147
  gymnosperms, B148–149
  leaves, B137–140
  roots, B131–132
  stems, B133–136
Seeds, B144–145
  of angiosperms, B144
  embryo of, B145
  germination, B147
  of gymnosperms, B144
  parts of, B144–145
  seed dispersal, B146–147
Sepals, B151
Slime molds, B78–79
Sphagnum moss, 118
Species, B17, B21, B22, B23
Spores
  fungi, B89–90
  slime molds, B79
  sporozoans, B69
Sporozoans, B69–71
  and malaria, B69–71
Stamens, B151
Stems, B133–136

function of, B133
of herbaceous plants, B134
structure of, B134–135
tree rings, B134–135
uses of, B135–136
wood, B135–136
of woody plants, B134
Stigma, flower, B152
Stimulus, meaning of, B154
Stoma, leaves, B140
Style, B152
Symbiosis, definition of, B48

Taproots, B131
Taxonomists, B15
Taxonomy, B14–15
Transpiration, B140
Tree rings, B134–135
Tropisms, B154–155
  gravitropism, B155
  phototropism, B155
  positive and negative tropisms,
    B155
Truffles, B85

Vaccines, B40
Vascular cambium, stems, B134
Vascular plants, B119–120
  characteristics of, B119–120
Viruses, B36–41
  bacteriophage, B39–40
  beneficial uses of, B40–41
  and cancer, B40
  discovery of, B36–37
  and disease, B40–41
  life cycle of, B40
  as living/non-living things, B37
  nature of, B37–38
  origin of, B38
  reproduction of, B39–40
  structure of, B38–39

Wood, uses of, B135–136
Woody plants, types of, B134

Xylem
  stems, B133, B134–135
  vascular plants, B130

Yeasts, B90, B92–93
  and bread making, B92–93
  budding, B93

Zooflagellates, B68–69

Credits

Cover Background: Ken Karp
Photo Research: Natalie Goldstein
Contributing Artists: Michael Adams/ Phil Veloric, Art Representatives; Ray Smith; Warren Budd Associates Ltd.; Fran Milner; Function Thru Form; Keith Kasnot; David Biedrzycki; Gerry Schrenk; Carol Schwartz/Dilys Evans, Art Representatives
Photographs: **5** left: Dwight Kuhn Photography; right: D. Avon/Ardea London; **6** top: Lefever/Grushow/Grant Heilman Photography; center: Index Stock Photography, Inc.; bottom: Rex Joseph; **8** left: London School of Hygiene & Tropical Medicine/SPL/Photo Researchers, Inc.; right: Dr. Ann Smith/SPL/Photo Researchers, Inc.; **9** left: S. Nielsen/DRK Photo; right: D. Cavagnaro/DRK Photo; **10** and **11** Peter Scoones/Seaphoto Ltd./Planet Earth Pictures; **14** top and center: Breck P. Kent; bottom: Robert Frerck/Odyssey Productions; **15** left: Lawrence Migdale/Photo Researchers, Inc.; right: Dan Guravich/Photo Researchers, Inc.; **16** left: Stephen Dalton/ Animals Animals/Earth Scenes; right: M. P. Kahl/DRK Photo; **17** Breck P. Kent; **18** top: E.R. Degginger/Animals Animals/Earth Scenes; bottom: Stephen J. Krasemann/ DRK Photo; **20** top left: Barbara J. Wright/ Animals Animals/Earth Scenes; top center: Robert C. Simpson/Tom Stack & Associates; top right: Charles Palek/Tom Stack & Associates; bottom: Tom & Pat Leeson/DRK Photo; **21** D. Avon/Ardea London; **22** left to right, top to bottom: Phil Dotson/DPI; Jack Dermid; Larry Lipsky/ DRK Photo; T. Zywotko/DPI; Larry Roberts/ Visuals Unlimited; David M. Stone; Barbara K. Deans/DPI; T. Zywotko/DPI; Jerry Frank/DPI; DPI; J. Alex Langley/DPI; T. Zywotko/DPI; Phil Dotson/DPI; Phil Dotson/DPI; F. Erizel/Bruce Coleman, Inc.; T. Zywotko/DPI; Lois and George Cox/Bruce Coleman, Inc.; Marty Stouffer/Animals Animals/Earth Scenes; Mimi Forsyth/Monkmeyer Press; T. Zywotko/DPI; Kenneth W. Fink/Bruce Coleman, Inc.; Wil Blanche/ DPI; Phil Dotson/DPI; T. Zywotko/DPI; Jon A. Hull/Bruce Coleman, Inc.; **26** top and bottom: Peter Parks/Oxford Scientific Films/Animals Animals/Earth Scenes; center: T. E. Adams/Visuals Unlimited; **27** left: Stephen J. Krasemann/DRK Photo; top right: M & A Doolittle/Rainbow; bottom right: Michael Fogden/DRK Photo; **28** top: Michael Fogden/DRK Photo; bottom: G.I. Bernard/Oxford Scientific Films/Animals Animals/Earth Scenes; **29** top: Michael Dick/ Animals Animals/Earth Scenes; bottom: Richard Shiell/Animals Animals/Earth Scenes; **33** Michael Fogden/DRK Photo; **34** and **35** Biozentrum/Science Photo Library/Photo Researchers, Inc.; **36** top left: CNRI/Science Photo Library/Photo Researchers, Inc.; bottom left: Tektoff-RM/ CNRI/Science Photo Library/Photo Researchers, Inc.; bottom center: A. B. Dowsett/Science Photo Library/Photo Researchers, Inc.; bottom right: Dr. Oscar Bradfute/Peter Arnold, Inc.; **37** A. Jones/Visuals Unlimited; **39** Lee D. Simon/Photo Researchers, Inc.; **40** top: CNRI/Science Photo Library/Photo Researchers, Inc.; bottom: Tom Broker/Rainbow; **41** top left: Ed Reschke/Peter Arnold, Inc.; top right: Tektoff-RM/CNRI/Science Photo Library/ Photo Researchers, Inc.; bottom: CNRI/Science Photo Library/Photo Researchers, Inc.; **42** top: Sepp Seitz/Woodfin Camp & Associates; bottom left: Renee Lynn/Photo Researchers, Inc.; bottom right: Chuck O'Rear/Woodfin Camp & Associates; **43** top and bottom right: CNRI/Science Photo Library/Photo Researchers, Inc.; left: David M. Phillips/Visuals Unlimited; center: A. B. Dowsett/Science Photo Library/Photo Researchers, Inc.; **44** left: David M. Phillips/Visuals Unlimited; right: Biophoto Associates/Photo Researchers, Inc.; **45** top: Dr. Tony Brain/Science Photo Library/Photo Researchers, Inc.; bottom: A. B. Dowsett/Science Photo Library/Photo Researchers, Inc.; **46** top: Wolfgang Kaehler; bottom left: Alfred Pasieka/Science Photo Library/Photo Researchers, Inc.; bottom right: Dr. L. Caro/ Science Photo Library/Photo Researchers, Inc.; **47** left: John Cancalosi/DRK Photo; right: Alfred Pasieka/Science Photo Library/Photo Researchers, Inc.; **48** Tim Rock/Animals Animals/Earth Scenes; **49** top: Hugh Spencer/Photo Researchers, Inc.; bottom: Dr. Jeremy Burgess/Science Photo Library/Photo Researchers, Inc.; **50** top: Don & Pat Valenti/DRK Photo; bottom: B. Nation/Sygma; **51** Bonnie Sue Rauch/ Photo Researchers, Inc.; **52** top: Martin M. Rotker; bottom: Cabisco/Visuals Unlimited; **53** Dr. R. Clinton Fuller/University of Massachusetts; **58** and **59** London School of Hygiene & Tropical Medicine/Science Photo Library/Photo Researchers, Inc.; **60** left: Cabisco/Visuals Unlimited; center: A. M. Siegelman/Visuals Unlimited; right: K. G. Murti/Visuals Unlimited; **61** top: T. E. Adams/Visuals Unlimited; bottom: Cabisco/Visuals Unlimited; **62** left: A. M. Siegelman/Visuals Unlimited; right: Biophoto Associates/Photo Researchers, Inc.; **63** top: R. Oldfield/Polaroid/Visuals Unlimited; bottom: A. M. Siegelman/Visuals Unlimited; **64** left: David M. Phillips/Visuals Unlimited; right: Omikron/Science Source/Photo Researchers, Inc.; **65** top: Cabisco/Visuals Unlimited; bottom left and right: Michael Abbey/Visuals Unlimited; **66** Karl Aufderheide/Visuals Unlimited; **67** top: Michael Abbey/Photo Researchers, Inc.; bottom left: Bruce Iverson/Visuals Unlimited; bottom right: CBS/Visuals Unlimited; **68** top: David M. Phillips/Visuals Unlimited; bottom: Jerome Paulin/Visuals Unlimited; **69** top left: Michael Abbey/Visuals Unlimited; top right: Raymond A. Mendez/Animals Animals/Earth Scenes; bottom: Will & Deni McIntyre/Photo Researchers, Inc.; **71** left: © Lennart Nilsson, THE INCREDIBLE MACHINE, National Geographic Society; right: © Lennart Nilsson, National Geographic Society; **72** top: G. I. Bernard/Animals Animals/Earth Scenes; bottom: SCIENCE, "Predator-Induced Trophic Shift of a Free-Living Ciliate: Parasitism of Mosquito Larvae by Their Prey" published 05/27/88. Volume 240, beginning on page 1193. Dr. Jan O. Washburn, University of California, Berkeley. © 1988 by the AAAS.; **73** top right and bottom left and right: Dr. John C. Steinmetz; **74** top: T. E. Adams/Visuals Unlimited; bottom: David M. Phillips/ Visuals Unlimited; **75** top left and right: Cabisco/Visuals Unlimited; top center: Cecil Fox/Science Source/Photo Researchers, Inc.; bottom left: Veronika Burmeister/Visuals Unlimited; right: Nuridsany et Pérennou/Photo Researchers, Inc.; **76** top: Dr. J.A.L. Cooke/Oxford Scientific Films/Animals Animals/Earth Scenes; bottom: David M. Phillips/Visuals Unlimited; **77** left: Biophoto Associates/Science Source/Photo Researchers, Inc.; top right: Peter Parks/Oxford Scientific Films/Animals Animals/Earth Scenes; bottom right: M.I. Walker/Science Source/Photo Researchers, Inc.; **78** top: Dwight Kuhn Photography; bottom left: Cabisco/Visuals Unlimited; bottom right: CBS/Visuals Unlimited; **79** top left: V. Duran/Visuals Unlimited; top right: G. I. Bernard/Oxford Scientific Films/Animals Animals/Earth Scenes; center: Cabisco/Visuals Unlimited; bottom: CBS/Visuals Unlimited; **83** David M. Phillips/Visuals Unlimited; **84** and **85** Adam Woolfitt/Woodfin Camp & Associates; **86** top and bottom left: Michael Fogden/DRK Photo; bottom right: Charles Brewer; **87** top left: John Gerlach/DRK Photo; top right: S. Flegler/Visuals Unlimited; center right: Michael Fogden/Oxford Scientific Films/Animals Animals/Earth Scenes; bottom right: Biophoto Associates/Science Source/Photo Researchers, Inc.; **88** left: Don & Pat Valenti/DRK Photo; center: S. Nielsen/DRK Photo; right: C. Gerald Van Dyke/Visuals Unlimited; **89** top left: Michael Fogden /Oxford Scientific Films/Animals Animals/Earth Scenes; center: Michael Fogden/DRK Photo; right: P. L. Kaltreider/Visuals Unlimited; bottom left: Cabisco/Visuals Unlimited; **90** John Gerlach/DRK Photo; **91** top: Michael Fogden/Oxford Scientific Films/Animals Animals/Earth Scenes; bottom: Dick Poe/Visuals Unlimited; **92** J. Forsdyke/ Gene Cox/Science Photo Library/Photo Researchers, Inc.; **93** left: Andrew McClenaghan/Science Photo Library/Photo Researchers, Inc.; right: Dr. Jeremy Burgess/Science Photo Library/Photo Researchers, Inc.; **95** top left: William D. Griffin/Animals Animals/Earth Scenes; top right: Richard K. LaVal/Animals Animals/Earth Scenes; bottom: D. Cavagnaro/DRK Photo; **96** top: Patti Murray/ Animals Animals/Earth Scenes; bottom: Glenn Oliver/Visuals Unlimited; **97** left: Tom Bean/DRK Photo; right: Breck P. Kent; **98** Cary Wolinsky/Stock Boston, Inc.; **99** top left: Bates Littlehales/Animals Animals/Earth Scenes; top right: Stephen J. Krasemann/DRK Photo; bottom left: E. R. Degginger/Animals Animals/Earth Scenes; bottom right: Larry Ulrich/DRK Photo; **103** Charlie Palek/Animals Animals/Earth Scenes; **104** and **105** Jeffrey L. Rotman; **106** top: L. L. Sims/Visuals Unlimited; bottom left: Anne Wertheim/Animals Animals/Earth Scenes; bottom right: Robert & Linda Mitchell Photography; **107** top right: Biophoto Associates/Science Source/Photo Researchers, Inc.; center right : M. I. Walker/Science Source/Photo Researchers, Inc.; bottom left: Jeffrey L. Rotman; bottom center: Robert & Linda Mitchell Photography; bottom right: Charles Seaborn/Odyssey Productions; **109** top: Runk/Schoenberger/Grant Heilman Photography; bottom left: Heather Angel; bottom right: Robert Maier/Animals Animals/ Earth Scenes; **110** top left: Robert & Linda Mitchell Photography; top right: Charles Seaborn/Odyssey Productions; bottom: Doug Wechsler/Animals Animals/Earth Scenes; **111** top left: Jeffrey L. Rotman; top right: Jack Dermid; center right: Charles Seaborn/Odyssey Productions; bottom right: John Durham/Science Photo Library/Photo Researchers, Inc.; **112** top: Robert Frerck/Odyssey Productions; center: Charles Seaborn/Odyssey Productions; bottom: Dwight Kuhn Photography; **113** top left: M. I. Walker/Photo Researchers, Inc.; top right: Jeff Foott Productions; bottom left: Oxford Scientific Films/Animals Animals/Earth Scenes; bottom right: Charles Seaborn/Odyssey Productions; **116** top: Doug Wechsler/Animals Animals/Earth Scenes; bottom left: Robert & Linda Mitchell Photography; bottom right: Frans Lanting/Minden Pictures, Inc.; **117** Dwight Kuhn Photography; **118** top: David Muench Photography Inc.; bottom: Stephen J. Krasemann/DRK Photo; **119** left: Kjell B. Sandved; right: Jack Dermid; **120** top left: Stephen J. Krasemann/DRK Photo; top right and bottom left: Kjell B. Sandved; bottom right: Wolfgang Kaehler; **121** Robert & Linda Mitchell Photography; **122** top left: Doug Wechsler/Animals Animals/Earth Scenes; top right, bottom left and right: Kjell B. Sandved; **123** Garry Gay/Image Bank; **127** Robert & Linda Mitchell Photography; **128** and **129** John Lemker/Animals Animals/Earth Scenes; **130** top: Wolfgang Kaehler; bottom left: Robert & Linda Mitchell Photography; bottom right: Michael Fogden/DRK Photo; **132** top: P. Dayanandan/Photo Researchers, Inc.; center: J. F. Gennaro/Photo Researchers, Inc.; bottom: Robert & Linda Mitchell Photography; **133** top left: Jack Swenson/Tom Stack & Associates; top center: Wolfgang Kaehler; top right: Robert & Linda Mitchell Photography; bottom left: Kjell B. Sandved; bottom center and bottom right: Dwight Kuhn Photography; **134** top: Frans Lanting/Minden Pictures, Inc.; center: Jim Brandenburg/Minden Pictures, Inc.; bottom: T. A. Wiewandt /DRK Photo; **135** top: Robert & Linda Mitchell Photography; bottom: Ardea London; **136** top left and right: Robert & Linda Mitchell Photography; bottom left: Jeff Foott/Tom Stack & Associates; bottom center: Dwight Kuhn Photography; bottom right: D. Cavagnaro/DRK Photo; **137** left and center left: Kjell B. Sandved; center right: Wolfgang Kaehler; top right and bottom right: Robert & Linda Mitchell Photography; **138** DPI; **140** left and right: Dr. Jeremy Burgess/Science Photo Library/Photo Researchers, Inc.; **141** top left: Breck P. Kent; top center and top right: Robert & Linda Mitchell Photography; bottom left: Jeff Foott/Tom Stack & Associates; bottom center: Dwight Kuhn Photography; bottom right: Kjell B. Sandved; **142** left: Tom Bean/DRK Photo; right: Dwight R. Kuhn/DRK Photo; **143** left: Robert & Linda Mitchell Photography; top right and bottom right: Wolfgang Kaehler; **144** top left: Dr. Jeremy Burgess/ Science Photo Library/Photo Researchers, Inc.; right top, center bottom, and bottom right: Robert& Linda Mitchell Photography; center top: Patti Murray/Animals Animals/Earth Scenes; bottom left: CNRI/ Science Photo Library/Photo Researchers, Inc.; **145** left, center, and right: Kjell B. Sandved; **146** left and bottom right: Robert & Linda Mitchell Photography; top right: Jeff Foott/Tom Stack & Associates; center: Adrienne T. Gibson/Animals Animals/ Earth Scenes; **147** top left: Mickey Gibson/Animals Animals/Earth Scenes; top right: Robert & Linda Mitchell Photography; bottom: Coco McCoy/Rainbow; **148** top: Kjell B. Sandved; bottom: E. R. Degginger/Animals Animals/Earth Scenes; **149** top left: William E. Ferguson; top center and top right: Wolfgang Kaehler; bottom: Peter Pickford/DRK Photo; **150** top left: D. Cavagnaro/DRK Photo; top right and bottom left: David Muench Photography Inc.; **151** top left: Michael Fogden/ DRK Photo; top right: J.A.L. Cooke/Oxford Scientific Films/Animals Animals/Earth Scenes; center right: G. I. Bernard/Oxford Scientific Films/Animals Animals/Earth Scenes; bottom left: Alastair Shay/Oxford Scientific Films/Animals Animals/Earth Scenes; bottom right: Stanley Breeden/ DRK Photo; **154** top left: William E. Ferguson; right: John Gerlach/DRK Photo; bottom left: Don & Pat Valenti/DRK Photo; **155** Runk/Schoenberger/Grant Heilman Photography; **159** Michael J. Doolittle/ Rainbow; **160** Michael Elias; **161** top and center: Dudley Foster/Woods Hole Oceanographic Institution; bottom: Colleen Cavanaugh; **162** David Scharf/Peter Arnold, Inc.; **163** Annie Griffiths/DRK Photo; **164** Michael Habicht/Animals Animals/Earth Scenes; **166** Terrence Moore/Woodfin Camp & Associates; **167** Runk/Schoenberger/Grant Heilman Photography; **168** Tom & Pat Leeson/DRK Photo; **189** Kjell B. Sandved; **191** D. Cavagnaro/DRK Photo